ADVANCES IN EXPERIMENTAL MEDICINE AND BIOLOGY

Volume 719

For further volumes:
http://www.springer.com/series/5584

Nigel Curtis · Adam Finn · Andrew J. Pollard

Editors

Hot Topics in Infection
and Immunity in Children VIII

 Springer

Editors
Nigel Curtis
Professor of Paediatric Infectious Diseases
University of Melbourne
Murdoch Children's Research Institute
Royal Children's Hospital
Parkville, Australia
nigel.curtis@rch.org.au

Andrew J. Pollard
Professor of Paediatric Infection & Immunity
Director, Oxford Vaccine Group
University of Oxford
Oxford, United Kingdom
andrew.pollard@paediatrics.ox.ac.uk

Adam Finn
David Baum Professor of Paediatrics
University of Bristol
Bristol, United Kingdom
adam.finn@bristol.ac.uk

ISBN 978-1-4614-0203-9   e-ISBN 978-1-4614-0204-6
DOI 10.1007/978-1-4614-0204-6
Springer New York Dordrecht Heidelberg London

Library of Congress Control Number: 2011938790

Printed on acid-free paper

Springer is part of Springer Science+Business Media (www.springer.com)

Preface

Each of the chapters in this book is based on a lecture given at the eighth 'Infection and Immunity in Children' (IIC) course held at the end of June 2010 at Keble College, Oxford. Thus, it is the eighth book in a series, which collectively provide succinct and readable updates on just about every aspect of the discipline of Paediatric Infectious Diseases.

The ninth course in 2011 has another exciting programme delivered by renowned top-class speakers, and a further edition of this book will duly follow.

The clinical discipline of Paediatric Infectious Diseases continues to grow and flourish in Europe. The University of Oxford Diploma Course in Paediatric Infectious Diseases, started in 2008, is now well established with a large number of trainees enrolled from all parts of Europe. The Oxford IIC course, as well as other European Society for Paediatric Infectious Diseases (ESPID)-sponsored educational activities, is an integral part of this course.

We hope this book will provide a further useful contribution to the materials available to trainees and practitioners in this important and rapidly developing field.

Melbourne, Australia   Nigel Curtis
Bristol, UK   Adam Finn
Oxford, UK   Andrew J. Pollard

Acknowledgments

We thank all the contributors who have written chapters for this book, which is based on lectures given at the 2010 Infection and Immunity in Children (IIC) course. We are grateful to the staff of Keble College, Oxford, UK where the course was held.

Sue Sheaf has administered and run the course for several years now. As course organisers we are extremely grateful for the work and effort that Sue puts into making the course such a success. We are constantly impressed by her ability to maintain a quiet, calm and efficient manner under pressure and her can-do approach. We are indebted to Sue and enormously appreciative of her contribution. We also thank Sue on behalf of all the speakers and delegates who have benefited from her behind the scenes administrative, organisational and diplomatic skills

Pamela Morison administered the production of this book. This involved painstaking checking, correcting and formatting of the chapters as well as liaising with authors and the publishers. Pam has mastered the art of persuading authors (and editors!) to meet deadlines, read formatting instructions and answer emails. Using a combination of gentle encouragement, persistent coaxing and cajoling, and high diplomacy, Pam has dealt admirably with the challenges this volume has presented. We thank Pam for her patient and cheerful approach to this difficult task, and we gratefully share with her the credit for this book's production.

We thank the European Society for Paediatric Infectious Diseases (ESPID) for consistent support and financial assistance for this and previous courses and for providing bursaries, which have paid the costs of many young ESPID members' attendance. We also acknowledge the recognition given to the course by the Royal College of Paediatrics and Child Health.

Finally, we are grateful to several pharmaceutical industry sponsors who generously offered unrestricted educational grants towards the budget for the meeting.

Contents

Contributors

Philip Alcabes Hunter College, City University of New York, New York, USA
palcabes@hunter.cuny.edu

Mona Al-Dabbagh Division of Infectious and Immunological Diseases,
Department of Pediatrics, BC Children's Hospital, Vancouver, Canada
maldabbagh@cw.bc.ca

Wendy Barclay Barclay Influenza Group, Imperial College London, London, UK
w.barclay@imperial.ac.uk

Maria Bitsori Department of Paediatrics, University of Crete, Heraklion, Greece
bitmar@hol.gr

Chantal P. Bleeker-Rovers Department of Internal Medicine and Radboud Expertise
Centre for Q fever, Nijmegen Institute for Infection, Inflammation and Immunity;
Radboud University Nijmegen Medical Centre, Nijmegen, The Netherlands
C.Bleeker-Rovers@aig-umcn.nl

Julia E. Clark Great North Children's Hospital, Newcastle, UK
Julia.Clark@nuth.nhs.uk

Nigel Curtis Department of Paediatrics, The University of Melbourne; Infectious Diseases Unit,
Department of General Medicine; Murdoch Children's Research Institute; Royal Children's
Hospital Melbourne, Parkville, Australia
nigel.curtis@rch.org.au

Corine E. Delsing Department of Internal Medicine and Radboud Expertise Centre for Q fever,
Nijmegen Institute for Infection, Inflammation and Immunity, Nijmegen, The Netherlands
Radboud University Nijmegen Medical Centre, Nijmegen, The Netherlands
C.Delsing@AIG.umcn.nl

Simon Dobson Division of Infectious and Immunological Diseases, Department of Pediatrics,
BC Children's Hospital, Vancouver, Canada
sdobson@cw.bc.ca

Ruth Elderfield Barclay Influenza Group, Imperial College London, London, UK
ruth.elderfield07@imperial.ac.uk

Adam Finn Paediatrics, University of Bristol, Bristol Royal Hospital for Children, Bristol, UK
adam.finn@bristol.ac.uk

Ron Fouchier Department of Virology, Erasmus Medical Centre, Rotterdam, The Netherlands

Christina Gagliardo Department of Pediatrics, Division of Infectious Diseases, Morgan Stanley Children's Hospitalof New York-Presbyterian, Columbia University, New York, USA
cg2406@columbia.edu

Emmanouil Galanakis Department of Paediatrics, University of Crete, Heraklion, Greece
emmgalan@med.uoc.gr

Philip J.R. Goulder Department of Paediatrics, University of Oxford, Oxford, UK
philip.goulder@paediatrics.ox.ac.uk

Paul T. Heath Child Health and Vaccine Institute, St Georges, University of London, London, UK
pheath@sgul.ac.uk

Thijs Kuiken Department of Virology, Erasmus Medical Centre, Rotterdam, The Netherlands
t.kuiken@erasmusmc.nl

Ben J. Marais Department of Paediatrics and Child Health, Stellenbosch University, Tygerberg, South Africa
bjmarais@sun.ac.za

Clarissa Oeser Child Health and Vaccine Institute, St Georges, University of London, London, UK

Ifeanyichukwu O. Okike Child Health and Vaccine Institute, St Georges, University of London, London, UK

Albert Osterhaus Department of Virology, Erasmus Medical Centre, Rotterdam, The Netherlands

Markus Pääkkönen Turku University Hospital, Turku, Finland
markus.paakkonen@helsinki.fi

Heikki Peltola Children's Hospital, Helsinki University Central Hospital, University of Helsinki, Helsinki, Finland
heikki.peltola@hus.fi

Andrew J. Pollard Department of Paediatrics, Oxford Vaccine Group, University of Oxford, Oxford, UK
andrew.pollard@paediatrics.ox.ac.uk

Andrew J. Prendergast Centre for Paediatrics, Queen Mary, University of London, London, UK
a.prendergast@qmul.ac.uk

Guus Rimmelzwaan Department of Virology, Erasmus Medical Centre, Rotterdam, The Netherlands

Manish Sadarangani Department of Paediatrics, Oxford Vaccine Group, University of Oxford, Oxford, UK
manish.sadarangani@paediatrics.ox.ac.uk

Lisa Saiman Department of Pediatrics, Columbia University, Department of Infection Prevention & Control, NewYork-Presbyterian Hospital, New York, USA
ls5@columbia.edu

Marc Tebruegge Department of Paediatrics, The University of Melbourne; Infectious Diseases Unit, Department of General Medicine; Murdoch Children's Research Institute; Royal Children's Hospital Melbourne, Parkville, Australia
marc.tebruegge@rch.org.au

Judith van den Brand Department of Virology, Erasmus Medical Centre, Rotterdam, The Netherlands

Debby van Riel Department of Virology, Erasmus Medical Centre, Rotterdam, The Netherlands

Ken B. Waites Department of Pathology, University of Alabama at Birmingham, Alabama, USA
waiteskb@uab.edu

Adilia Warris Department of Pediatric Infectious Diseases, Nijmegen Institute for Infection, Inflammation and Immunity, Radboud University Nijmegen Medical Centre, Nijmegen, The Netherlands
A.Warris@cukz.umcn.nl

Inbal Weiss-Salz Department of Health Services Research, Ministry of Health, Jerusalem, Israel
dr.isalz@gmail.com

Pablo Yagupsky Clinical Microbiology Laboratory, Soroka University Medical Center, Ben-Gurion University of the Negev, Beer-Sheva, Israel
yagupsky@bgu.ac.il

Course Organisers Prof Adam Finn, Prof Nigel Curtis, Prof Ronald de Groot, Prof Andrew Pollard.

Our Time of Pestilence: Purchasing Immunity and Ignoring the Misery of Others

Philip Alcabes

1 Prisoners of Life

In 1593, Thomas Nashe's poem "In Time of Pestilence" admonished the wealthy not to trust in their riches [1]:

Gold cannot buy you health;
Physic himself must fade;
All things to end are made;
The plague full swift goes by;
I am sick, I must die.

The Royal College of Physicians had already been in place for 75 years at that point. Physicians were making strides in identifying and differentiating plague from poxes and other maladies. But it was true that doctors—"physics," Nashe called them—were of little help, even for the wealthy.

How different today. The physic no longer fades. Residents of so-called developed countries believe that, as much as each of us must die, we must not die too soon. Gold *can* buy health, at least here in the wealthy world.

Median life expectancy is long in the affluent world today: over 82 years in Japan and Singapore, 81 in France and Canada, 80 in New Zealand and Spain (79 in the European Union generally) [2]. Even amid the capitalist muddle of American healthcare, half of U.S. residents live past age 78. Twenty percent or more of the babies born in the developed world in 2010 will still be alive at the start of the twenty-first century. We have used our wealth well, it could be said, using it for wellness.

But we pay a moral price. We remain "prisoners of life," as the eastern European writer Joseph Roth once put it: the implacable unpredictability of the universe is everyone's lot. The affluent are prisoners no less so than are the poor.

We who can buy our health and longevity easily imagine we can escape all harms. Striving to preserve the increasingly protracted future to which the public feels entitled, health authorities in affluent countries are expected to foresee the coming plague, the pandemic in waiting. The U.S. government awarded a contract to a private firm in 2006 for USD 363 million for development of multi-type botulinum antitoxin, for instance (there are only about 100 botulism cases per year in the

P. Alcabes (✉)
Hunter College, City University of New York, New York, USA
e-mail: palcabes@hunter.cuny.edu

N. Curtis et al. (eds.), *Hot Topics in Infection and Immunity in Children VIII*,
Advances in Experimental Medicine and Biology 719, DOI 10.1007/978-1-4614-0204-6_1,
© Springer Science+Business Media, LLC 2011

U.S., almost all of them caused by just two of the seven possible botulinum neurotoxin types [3], but the need for antitoxin against all seven types was rationalized on the basis of future risk of bioterror events). The botulism contract of 2006 was dwarfed by the USD 5.6 billion spent that year for "pandemic influenza preparedness"—this was at the time of fears about H5N1 avian flu, which had killed about 200 people worldwide although none in the U.S. [4]. One American health official estimated that between 7% and 8% of U.S. health expenditures went for emergency preparedness that year [5]—even though the vast majority of fatalities are caused by non-emergent conditions like smoking, common infections, and so forth. And the expenditure has increased since then.

Emergency preparedness isn't the only preoccupation. The affluent also seek to stem the forecast tide of dire consequences of the epidemics of modernity. Obesity keeps increasing, threatening to "reduce quality of life and increase the risk for many serious chronic diseases and premature death," according to the U.S. Centers for Disease Control and Prevention [6]. The World Health Organization, lumping overweight with obesity and pointing out that two-thirds of adult men in the U.K. and other parts of northern Europe fall into this category, emphasizes that obesity is a malaise of modernity [7]:

> Obesity and overweight pose a major risk for serious diet-related chronic diseases, including type 2 diabetes, cardiovascular disease, hypertension and stroke, and certain forms of cancer... The rising epidemic reflects the profound changes in society and in behavioural patterns of communities over recent decades.

Note the recurrent appeal to *risk*. Preventing the ever-more-distant and ever-less-predictable future calamity means that risk becomes the grammar for our conversation about health. Meanwhile, our sense that freedom from epidemic threats is our due deafens us to others' misery.

2 Purchasing Health

In Thomas Nashe's day, English life expectancy was between 35 and 39 years [8]. Rates of infant and childhood mortality were well over 10%, and a man who reached age 30 had only even odds of living to 60. Many died of plague in plague years, so affecting to Nashe. But they died of consumption, spotted fevers, poxes great and small, purples, apoplexy, consumption, or a host of other conditions—some no longer part of our lexicon—in non-plague years. Horribly often, women died in childbirth.

Even in the early 1800s, even in the world's wealthiest places, death in childhood or by childbearing was common. One out of every four infants born in New York City died before its first birthday then, and only half of those Americans who survived childhood and adolescence lived past the age of 50 in 1810 [9].

Changes in social structure and expanded choices for women and laborers made life in the wealthy world longer and healthier by the twentieth century. Sanitarianism, primarily in the form of urban sewerage systems and clean-water supplies, had put paid to cholera outbreaks. Housing reform, along with improved nutrition, had begun to limit the spread of contagion. The workplace safety movement was reducing injury rates. Perhaps most important of all, affording women better control over their reproductive cycles was leading to reduced family size, lowering both infant and maternal mortality rates, and providing fewer opportunities for infectious foci to extend into severe outbreaks. Death began to seem escapable to the middle classes.

It was social reform that generated the healthful transformation of modern life, but consumerism was given credit for it. From roughly 1900 onward, buying the ingredients of the more salubrious life and the delayed death seemed increasingly possible. Products were available for this: disinfectants and deodorants, foods that had been inspected for purity, sanitary facilities, pasteurized milk [10]. In 1905, the director of New York City's Bureau of Laboratories, Herman Biggs, asserted that health could be purchased. "Within natural limitations, a community can determine its own death rate," Biggs said [11].

Today, we are followers of Biggs. Our credo is that we can purchase our immunity. We can buy longevity, and we do. Advanced life expectancy is the hallmark of the wealthy country today.

Everywhere else, the poor die in droves. They die of the chronic effects of malaria, schistosomiasis, onchocerciasis, filariasis, trypanosomiasis, or AIDS. They die during war or of hunger. They die in disasters. In the poorest parts of the world, half the population is dead by the age of 40—just as in England in Nashe's day.

Today's health conversation in the affluent world is wrapped up self-indulgently in the protracted future. The talk is of soda taxes, secondhand smoke, and obesity, of vaccines and autism, of diet and databases to track diabetes control, of *preparedness*. While everywhere else, people die badly. Focused on threats to the longevity to which we are now entitled, we manage to avoid, as Susan Sontag put it, "reflect[ing] on how our privileges … may … be linked to their suffering, as the wealth of some may imply the destitution of others" [12]. By speaking of the contemporary world only in the language of risk to our health, we allow ourselves to live with, and generally ignore, this fact: We of the affluent nations are party to depriving the rest of the world, the dollar-a-day world, of health and longevity.

3 The Deceptive Language of Risk

When we talk about epidemics, we are talking the language of risk, not of health. Specifically, not of *humane* health—the health of humankind. Those who do not have the gold for the risk-free life do *not* die in epidemics: the term "epidemic" is not accorded to the agents that kill the poor *en masse*. As the textbooks on public health or tropical medicine adumbrate, the diseases that scourge the have-nots (malaria, et cetera) are considered to be *endemic* problems in the so-called developing world. "Epidemic" means "meaningful to us"; "endemic" means "sorry, not our problem."

For us in the affluent world, and especially our kids, today's epidemics are of obesity, autism, attention-deficit/hyperactivity disorder, eating disorders, and binge drinking. We have, famously and controversially, experienced a pandemic of H1N1 flu. These are not cut of the same cloth as the pestilence of Nashe's day. No, we call these "epidemics" because they capture meaning and reflect it back—to a society always eager to see a glimpse of what, we hope, is our true self.

3.1 The Obesity Epidemic

Obesity seems meaningful. To some people, it bespeaks hypertrophy, overindulgence, and a loss of a sense of proportion—the defects of excess allegedly inevitable in a consumer society. It seems to others to point up a putative toxicity to modern life—a noxious, or at least unhealthy, "food environment" that is supposedly of a piece with oil spills and coal-fired smogs in the natural environment.

To still others obesity has a moral tone: it represents bad parenting. To attend to epidemic obesity is to utter the contemporary version of the timeless complaint that this generation's moms and dads just don't enforce moral codes the way earlier ones did. It's as if fat children were evidence of individual parents' moral turpitude. A great proportion of children are now born to unmarried mothers: 44% of all births in the U.K. in 2006 [13], about half of all births in France, 55% in Sweden [14], for instance. It's as if that were evidence of a deep moral failure in our civilization, of which rampant obesity were the inescapable resultant and catastrophe the impending final outcome.

The terms the health profession uses to speak of the epidemic of childhood obesity implicate our own childrearing. The language of risk raises a warning finger at how we moderns create, conduct, and end our marriages. It laments the decline of home cooking, the rise of restaurant meals, snack meals, solitary meals, or on-the-go meals. Anxieties about the culture are shaped into epidemic fears.

For instance, a Stanford University researcher told the *Washington Post* a couple of years ago that "we have taught our children how to kill themselves" [15]. In 2002, Dr. Howard Stoate, then chair of the All-Parliamentary Group on Primary Care and Public Health, referred to childhood obesity as a "time bomb" [16]. The American Public Health Association calls obesity "the biggest problem facing children today" [17]—as if American kids' fat future were more dire than the mere misfortune of children in the world's poor nations. Those children in the teeming slums of Lagos or Lahore might be filthy, hungry, disease-ridden, orphaned, or all of those—but they aren't fat.

It is true that obesity *can* lead to medical problems [18–20], and is said to be responsible for both early death [21–25] and, not incidentally, low self-esteem and poor school performance [26].

But the chances of dying from the effects of obesity are slim enough that the great majority of people who are considered obese by current standards—BMI above 30, that is—will suffer no shortening of life because of it.

In the U.S., famous now for fatness, 112,000 deaths in the year 2000 could be attributed to obesity (BMI ≥ 30) [25], amounting to roughly 5% of all U.S. deaths. A large-population epidemiologic study cannot rule out contributing causes of death of all sorts, so the true number of obesity-provoked deaths might be smaller. But accept the 5% figure for the sake of argument. According to the CDC, obesity prevalence among U.S. adults is far higher now even than in the 1990s [6], stood at 26.6% overall in 2007 [27], and is now above 20% in every U.S. state except Colorado [28]. Some prevalence-incidence bias must be considered, since deaths in year 2000 were incident events among a virtual cohort of earlier years' obese. Therefore, compare the proportionate mortality of 5% to earlier estimates of obesity prevalence, roughly 20%. Even with this correction, three quarters of obese people died of disorders unrelated to their fatness. With increasing obesity prevalence, the proportion of obese people whose death is not attributable to obesity is certainly even greater now. Almost all excess deaths attributed to obesity in the U.S. occur among people in their 60s (after age 69 obesity has no impact on mortality rates) [25] and American obesity mortality is accounted for largely by incomplete management of diabetes and/or hypertension [29, 30], two of the main adverse accompaniments of high BMI. Some, and possibly a great many, of the deaths blamed on excess body mass might equally plausibly be attributed to America's porous health-care net.

With obesity, in other words, it is risk itself that is epidemic. Some health professionals have begun to talk about obesity as a kind of contagion. In 2005 the CDC sent one of its epidemic intelligence teams to study an "outbreak" of obesity in West Virginia, as if it were cryptosporidiosis or dengue fever [31]. A 2007 article in the *New England Journal of Medicine* reported research findings showing that obesity can "spread through social networks" [32]. Fat people, in other words, make their friends fat. As Freud observed, when spirits and demons are believed to animate all living things, "these souls which live in human beings can leave their habitations and migrate into other human beings" [33]. Freud was referring to the beliefs of primitive peoples; evidently, contemporary health researchers' thinking is not as far removed from that of the ancients as we are prone to think.

Obesity would not be epidemic were it not for the availability of things you can purchase to fight it. The responsible citizen is supposed to buy the low-fat foods, the fitness-center membership, and the diet books, send her kids to fat camp, take tennis lessons. The anti-obesity crusade is a sales campaign for a multi-billion dollar industry [34]. Both mirroring our anxieties and affording a rationale for increasing corporate profits is a combination that defines an epidemic in the wealthy world today.

3.2 *Influenza and Pandemic Risk*

Unlike obesity—or autism, ADHD, eating disorders, or any of the other popular epidemics of the affluent—influenza really does spread from person to person and can be directly responsible for substantial mortality. But is that what makes a flu pandemic worthy of so vigorous a response as in

2009, and why governments of rich countries spent so much money on it? No. The reason for that was the same combination as for other so-called epidemic threats: the mirroring of contemporary anxieties in tandem with expanding markets for products.

The U.S. Centers for Disease Control and Prevention lumps all sorts of respiratory infections together, along with pneumonitides and other influenza sequelae, so as to claim that 36,000 Americans die from "flu" each year—probably several times higher than the true number of deaths directly caused by influenza [35, 36]. Would anyone claim that this overemphasis on flu as a cause of death has *nothing* to do with the availability of vaccines and antivirals against flu—in contrast to the paucity of products for responding to other respiratory viruses, respiratory syncytial virus, and so forth?

Officials have issued dire forecasts about the possibilities for widespread human mortality in a flu pandemic—even though flu outbreaks with pandemic strains, apart from the outlier of 1918, have all been relatively mild. Flu seems to bring out the Cassandra in public health professionals, whereas the far higher tolls taken by malaria, TB, diarrhea, or AIDS don't. In part, that contrast has something to do with the sense that flu seems like a problem of the developed world—one can catch flu in the office, riding the commuter train, or at the shopping mall, whereas the diseases that are the misfortunes of villagers in Bangladesh, Burundi, or Bolivia, however much more baleful their toll, seem to speak not of modernity but of dirty water, mosquitoes, and shanty towns.

But much of the contrast between the flu response on the one side and the relative nonresponse to diarrhea and TB on the other has to do with the urgency with which officials press for product-heavy responses. Officials overstate problems for which the corporate world has solutions, and understate the ones that don't expand markets.

Thus health authorities responded with alacrity to their own allegations that there would be an onslaught of flu in 2009—while giving short shrift to worse threats. In the U.S., federal funds to combat methicillin-resistant *S. aureus* (MRSA) and other hospital-acquired infections, which kill roughly 100,000 Americans each year [37], amounted to USD 17 million in 2009 (the amount was increased to USD 34 million in 2010) [38]. By contrast, the U.S. government allocated USD 6.1 *billion* from 2006 to 2009—prior to the outbreak of H1N1 flu, that is—for influenza pandemic "preparedness"; an additional USD 4.86 billion was allocated during the 2009 H1N1 outbreak [39]. Even though the 2009 H1N1 flu, even in the worst-case scenario, would have led to fewer deaths than the normal, year-in-and-year-out, hospital-acquired infection mortality. In fact, even by the CDC's inflated flu calculations, H1N1 flu killed half as many Americans as did MRSA alone.

Among the funds moved in the context of the H1N1 flu outbreak, the U.S. government transferred USD 2 billion from public coffers directly to private vaccine manufacturers, in recompense for 250 doses of vaccine, sufficient for almost all Americans [40]. It is a reminder of the intimacy with which the "not our problem" sensibility colludes with bolstering markets, to discover that the funding donated by U.S. agencies to the millions of citizens of Haiti in response to the January 2010 earthquake, about USD 800 million [41], amounted to half the sum that America donated to seven pharmaceutical companies for flu vaccine.

Perhaps everyone expects the corporate-friendly U.S. to seize on rumors of epidemic threat as an excuse to shift public monies into private hands. But it is not an American phenomenon alone. France appropriated 869 million euros to buy vaccine, Germany transferred 500 million euros for the same purpose, Canada paid 400 million Canadian dollars, and the U.K. spent hundreds of millions of pounds (exactly how much has been strictly secret) [42–44]. Collectively, taxpayers in wealthy countries subsidized a major part of a global vaccine market now estimated at well over USD 20 billion [45]. Ratepayers' tax monies become pharmaceutical company profits.

I have no argument with private companies manufacturing pharmaceutical products. Nor do I allege that Pharma twists official arms.

My point is that no arm twisting is needed. The officials are already on board, have already forsworn any skepticism about the role of for-profit corporations in public health efforts to contain alleged epidemic threats. As World Health Organization Director-General Margaret Chan said, in response to recent accusations that pharmaceutical companies had too much influence in WHO policy making around flu, "At no time, not for one second, did commercial interests enter my decision-making" [46]. Of course not. There was no need for "commercial interests" to intercede in decision making about flu. There never is. Officials look out for corporate interests without being asked.

Rather than split hairs about whether WHO's influenza advisors revealed so-called conflicts of interest, as the Council of Europe has asked, members of the medical and public health professions should ask a more challenging question: What did national health officials in powerful countries like the U.S., Canada, the U.K., Germany, and others decide *not* to pay attention to. Professionals should ask why, in the U.S., national health agencies collect data on firearm violence, an epidemic that verifiably kills as many Americans as AIDS and more than flu [47], but they finance not a single gun-violence-control initiative. Professionals should ask about hospital-acquired infections, mentioned earlier, and about roadway accidents, which kill more people worldwide than flu does [48, 49].

With respect to flu, health professionals must be more exacting of officials. Why, during the 2009 flu outbreak, did no official or official advisor say, "in the worst-case scenario, H1N1 flu is going to hospitalize or kill far fewer people in Europe and the U.S. in the coming year than malaria does *in one month* among African children, fewer than diarrhea does, fewer than AIDS." Expert science advisors, the people who, it is commonly said, cannot be expected *not* to have ties to pharmaceutical companies should be confronted, asked why none said to his or her country's health authority, "whatever you do, *don't* recommend shifting millions of euros into private hands for the purchase of an imperfect vaccine that most people don't want to get anyway." Or why so few said, "whatever you do, don't recommend spending lots of public money to increase the stockpiles of oseltamivir." Few of the masters of science managed to tell the truth. They were swept up in managing the imagined epidemic.

Without disrespect to the Parliamentary Assembly of the Council of Europe or the *BMJ* [50], I assert that the problem is *not* merely that pharmaceutical companies or vaccine manufacturers influenced WHO officials' decision to raise a pandemic alert about flu. Nor is the problem that the words "epidemic" and "pandemic" have lost some imagined denotative meaning from the good old days of plague and poxes.

The problem is that people in rich countries demand long life and expect public officials to spend money to ensure it. The health sector is complicit. Health professionals define epidemic threats not on the basis of real harm, let alone real suffering, but on the basis of what will serve as a rationale for the sale and purchase of products that allay people's anxieties about the culture we live in. When Pakistani or Congolese or Peruvian kids die of diarrhea it's a *shame*, but when American or European kids get fat it's an epidemic.

4 Beware the Forecast Epidemic

The warding off of infection, the resistance to environmental toxins, the lifesaving medical interventions, the consequent opportunity to live to the limit of our capabilities—those are our society's achievements. But long life is not an entitlement. We of the affluent world do not *merit* long life more than do the poor.

Medical and other health professionals must beware the power of epidemics, particularly the epidemic that is forecast, envisioned, or merely imagined, to promote the concentration of wealth and power in the hands of the privileged. We should be loath to embrace the assertion of Herman Biggs. It is not an unalterable fact that the wealthy will and should buy their way into health. It is not an unassailable law of nature that the poor will die young.

Just over a half-century ago, in the Rede Lecture at Cambridge University, C.P. Snow adumbrated that science held the promise to fix the world's problems [51]. I do not see two distinct cultures, nor an abyss gaping between them, as Snow perceived. I question Snow's assertion that, even in the 1950s, nonscientists, particularly the educated humanists from among whom the governing classes sprang, were always too ignorant of science to fulfill science's promise. But I do think Snow was right in outlining a mission that scientists should take up, most certainly medical scientists, who are surely the humanists of the scientific sphere. Scientific knowledge gives its possessors an opportunity to make the world a bit less odious for our fellow prisoners of life. But if scientists are beguiled by the prospect of a long life free of risk, science and society will miss this chance.

Private, corporate interests have discovered how to turn the scientific project of improving everyone's life chances into corporate profits. Policy makers have intuited how to turn science's project into legitimizing the continuance of their political power. The question for the health profession is, How will we pursue the scientific project not for profit or legitimation, but to make the world a bit less unjust, to make more people less miserable?

Acknowledgments The author thanks Dr. Tom Jefferson for expert opinion and conversations about influenza immunization. Material in this chapter was originally presented at the Infection and Immunity in Children conference, Oxford, UK, 28 June 2010.

References

1. Nashe T. In time of pestilence, 1593. The quotation is from the second stanza.
2. US Central Intelligence Agency. Country comparisons: Life expectancy at birth. World Factbook 2009. Washington, DC, Central Intelligence Agency. at https://www.cia.gov/library/publications/the-world-factbook/index.html accessed 5 August 2010.
3. Shapiro RL, Hatheway C, Swerdlow DL. Botulism in the United States: A clinical and epidemiologic review. Ann Intern Med 1998;129(3): 221–228.
4. Vanderwagen WC. Testimony to U.S. Congress on safeguarding our nation: U.S. Dept. of Health and Human Services emergency preparedness efforts, 22 July 2010. Available at http://www.hhs.gov/asl/testify/2008/07/t20080722a.html accessed 5 August 2010.
5. Lipsman J. Disaster preparedness: Ending the exceptionalism, Medscape 3 Oct 2006, at http://www.medscape.com/viewarticle/544741 accessed 5 August 2010.
6. US Centers for Disease Control and Prevention. US obesity trends by state, 1985–2009, July 2010, at http://www.cdc.gov/obesity/data/trends.html#State accessed 12 August 2010.
7. World Health Organization. Information sheet on obesity and overweight, 2003, at http://www.who.int/entity/dietphysicalactivity/media/en/gsfs_obesity.pdf accessed 7 Aug 2010.
8. Wrigley EA, Schofield RS. The population history of England 1541–1871: A reconstruction. Cambridge, MA, Harvard University Press; 1981, pp. 234, 250.
9. Drolet GJ, Lowell AM. A half-century's progress against tuberculosis in New York City, 1900–1950. New York: New York Tuberculosis and Health Association; 1952, p. 7.
10. Tomes N. The gospel of germs: Men, women, and the microbe in American life. Cambridge MA: Harvard; 1998, esp. 48–87 and 157–182.
11. Winslow C-EA. (1929) The life of Hermann M. Biggs, MD, DSc, LLD, physician and statesman of the public health. Philadelphia PA: Lea & Febiger; 1929, p. 120.
12. Sontag S. Regarding the pain of others. New York: Farrar, Straus & Giroux; 2003, 102–103.
13. UK National Statistics. Annual update: Births in England and Wales, 2006 Population Trends 2007;130:2 at http://www.statistics.gov.uk/downloads/theme_population/Births_update_web_supplement.pdf accessed 10 Aug 2010.
14. US Centers for Disease Control and Prevention. Press release: Increase in unmarried childbearing also seen in other countries, 13 May 2009, at http://www.cdc.gov/media/pressrel/2009/r090513.htm accessed 10 Aug 2010.
15. Hoffman E, quoted in Schulte B. The search for solutions. Washington Post 22 May 2008, at http://www.washingtonpost.com/wp-dyn/content/article/2008/05/09/AR2008050900666.html?sid=ST2008050900732, accessed 10 August 2010.
16. Stoate H. Quoted in (no author) Obesity threatens children's lifespan. BBC News World Edition May 2002, available at http://news.bbc.co.uk/2/low/health/2606323.stm, accessed 10 July 2010.

17. American Public Health Assocation. Obesity and overweight children: The hidden epidemic. May 2008, at http://www.apha.org/programs/resources/obesity/defaulttest.htm accessed 12 August 2010.
18. Grundy SM, Brewer HB, Cleeman JI, Smith SC Jr., Lenfant C. Definition of metabolic syndrome: Report of the National Heart, Lung, and Blood Institute/American Heart Association conference on scientific issues related to definition. Circulation 2004;109:433–438.
19. Ford ES, Giles WH, Dietz WH. Prevalence of the metabolic syndrome among US adults: Findings from the third national health and nutrition examination survey. JAMA 2002;287:356–359.
20. Trevisan M, Liu J, Bahsas FB, Menotti A. Syndrome X and mortality: A population based study. Am J Epidemiol 1998;148:958–966.
21. Manson JE, Willett WC, Stampfer MJ, Colditz GA, Hunter DJ, Hankinson SE, Hennekens CH, Speizer FE. Body weight and mortality among women. N Engl J Med 1995;333(11):677–85.
22. Must A, Spadano J, Coakley EH, Field AE, Colditz G, Dietz WH. The disease burden associated with over-weight and obesity. JAMA 1999;282:1523–29.
23. Allison DB, Fontaine KR, Manson JE, Stevens J, Vanitallie TB. Annual deaths attributable to obesity in the United States. JAMA 1999; 282:1530–38.
24. Mokdad AH, Marks JS, Stroup DF, Gerberding JL. Actual causes of death in the United States, 2000. JAMA 2002;291:1238–45 (correction published in JAMA 2005;293:298).
25. Flegal KM, Graubard BI, Williamson DF, Gail MH. Excess deaths associated with underweight, overweight, and obesity. JAMA 2005;293(15):1861–67.
26. Dietz WH. Health consequences of obesity in youth: childhood predictors of adult disease. Pediatrics 1998;101:518–525.
27. US Centers for Disease Control and Prevention. Early release of selected estimates based on data from the January-June national health interview survey, 20 December 2007, at http://www.cdc.gov/nchs/data/nhis/earlyrelease/200712_06.pdf accessed 12 August 2010.
28. US Centers for Disease Control and Prevention. Vital signs: State-specific obesity prevalence among adults—United States, 2009. MMWR 2010; 59 (early release):1–5.
29. Slynkova K, Mannino DM, Martin GS, Morehead RS, Doherty DE. The role of body mass index and diabetes in the development of acute organ failure and subsequent mortality in an observational cohort. Critical Care 2006;10(R137):1–9.
30. Livingston EH, Ko CY. Effects of obesity and hypertension on obesity-related mortality. Surgery 2005;137:16–25.
31. Gerberding J. CDC Director's press conference, 2 June 2005, reported in Kolata G, CDC investigates outbreak of obesity. NY Times, 3 June 2005, A18.
32. Christakis NA, Fowler JH. The spread of obesity in a large social network over 32 years. N Engl J Med 2007;357(4):370–79.
33. Freud S. Totem and taboo, tr. James Strachey. NY: Norton; 1950, p. 76.
34. Campos P. The obesity myth: Why America's obsession with weight is hazardous to your health. NY: Gotham; 2004.
35. Doshi P. Are US flu death figures more PR than science? BMJ 2005;331:1412.
36. Doshi P. Trends in recorded influenza mortality: United States, 1900–2004. Am J Public Health 2008;98:939–945.
37. Klevens RM, Edwards JR, Richards CL Jr, Horan TC, Gaynes RP, Pollock DA, Pardo DM. Estimating health care-associated infections and deaths in U.S. hospitals, 2002. Public Health Rep 2007;122(2):160–166.
38. US Agency for Healthcare Research and Quality. Budget estimates for appropriations committees FY 2011, online performance appendix. February 2010, at http://www.ahrq.gov/about/cj2011/cj11opa7.htm accessed 13 August 2010.
39. Lister SA, Redhead CS. The 2009 influenza pandemic: An overview. Washington, DC: Congressional Research Service Report; 2009.
40. Stein R. First swine flu vaccine arriving in cities. *Washington Post.* October 6 2009, A1.
41. USAID, US Department of State. Remarks of Paul Weisenfeld. U.S. relief efforts in Haiti six months after earth-quake (press conference transcript) 19 July 2010, at http://www.america.gov/st/texttrans-english/2010/July/20100720141102su0.1158869.html accessed 13 August 2010.
42. Deutsche-Welle. France joins neighbors in sell-off of swine flu vaccine, 4 January 2010, available in English at http://www.dw-world.de/dw/article/0,,5079423,00.html accessed 13 August 2010.
43. Der Spiegel (no author). The swine flu business: Should Germany gamble millions on more vaccine? 9 September 2009, available in English at http://www.spiegel.de/international/world/0,1518,647666,00.html accessed 13 August 2010.
44. CBC News. Canada to order 50.4 million vaccine doses. 6 August 2009, at http://www.cbc.ca/health/story/2009/08/06/swine-flu-vaccine.html accessed 13 August 2010.

45. RNCOS. Global vaccine market forecast to 2012, 1 March 2010, at http://www.marketresearch.com/product/display. asp?productid=2621517&xs=r&SID=43635241-485810916-510472397&curr=USD accessed 13 August 2010.

46. Chan M. WHO Director-General's letter to BMJ editors. 8 June 2010, at http://www.who.int/mediacentre/news/ statements/2010/letter_bmj_20100608/en/index.html accessed 13 June 2010.

47. US Bureau of Justice Statistics. Gun violence, 2010, at http://bjs.ojp.usdoj.gov/content/glance/tables/guncrime- tab.cfm accessed 14 June 2010.

48. Menken M, Munsat TL, Toole JF. The global burden of disease study. Arch Neurol 2000;57:418–420.

49. World Health Organization. Road safety is no accident, 7 April 2004, at http://www.paho.org/English/dd/pin/ whd04_info.htm accessed 13 August 2010.

50. Cohen D, Carter P. WHO and the pandemic flu "conspiracies" BMJ 2010; 340:c2912 doi doi: 10.1136/bmj. c2912.

51. Snow CP. The two cultures. The Rede lecture, Senate House, Cambridge University, 7 May 1959. Text available in Snow CP. The Two Cultures and the Scientific Revolution. Cambridge: Cambridge University Press; 1960.

Neonatal Meningitis: Can We Do Better?

Paul T. Heath, Ifeanyichukwu O. Okike, and Clarissa Oeser

1 Incidence

There are few data available from large, prospective, population-based neonatal surveillance studies.

One of the earliest regional studies in the UK showed an incidence of neonatal bacterial meningitis of 0.5/1,000 live births over the years 1947–1960. Nearly 10 years later (1969–1973), a retrospective study of acute bacterial meningitis in the North West Metropolitan region reported a lower incidence of meningitis in neonates of 0.26/1,000 live births [1].

The first prospective, national neonatal surveillance study in the UK was performed in England and Wales over the period 1985–1987 [2]. The incidence of proven bacterial meningitis in neonates over these years was 0.22/1,000 live births, consistent with an incidence of 0.25/1,000 live births reported in a regional UK study around the same time. (1988–1991) [3].

Low birth weight and prematurity were found to be associated with a tenfold higher incidence (2.5/1,000 live births).

This study was repeated 10 years later to determine whether changes in healthcare over this period had improved the outcome [4]. The overall incidence of proven bacterial meningitis however did not differ between the studies. Over the years 1996–1997 there were 0.21 cases/1,000 live births. The incidence of meningitis in neonates with low birth weight was 1.7/10,000 live births, and low birth weight and gestational age <33 weeks remained a significant risk factor.

The incidence of neonatal meningitis in other developed countries is comparable to that in the UK. Surveillance undertaken by the National Institute of Health (NIH) in the USA between 1959–1966 reported an incidence of 0.46/1,000 live births [5] while a later study conducted in California (1962–1987) showed a lower incidence of 0.3/1,000 live births. These data are consistent with European studies, for example a regional retrospective study conducted in Sweden between 1987–1996 revealed an incidence of 0.3/1,000 live births [6].

Reports from less developed countries are variable and tend to reveal higher incidences. Community based studies in Pakistan showed an incidence of 0.81/1,000 live births, whilst in some areas in India the incidence of neonatal meningitis was reported as 4.9/1,000 live births. In Brazil, a population based study reported 4.2 cases per 1,000 child years, whilst the incidence in a rural community in Guatemala was as high as 6.1/10,000 live births [7].

P.T. Heath (✉) • I.O. Okike • C. Oeser
Child Health and Vaccine Institute, St Georges, University of London, London, UK
e-mail: pheath@sgul.ac.uk

N. Curtis et al. (eds.), *Hot Topics in Infection and Immunity in Children VIII*,
Advances in Experimental Medicine and Biology 719, DOI 10.1007/978-1-4614-0204-6_2,
© Springer Science+Business Media, LLC 2011

2 Aetiology

There has been a change in the range of causative organisms over time in industrialized countries. Group B Streptococcus (GBS) emerged as a cause of neonatal infection in the early 1970s and from 1981 it displaced *Escherichia coli* as the leading cause [8–10]. Since then, most cases in Europe have been caused by GBS and *E.coli*, which together now account for at least two thirds of all deaths from neonatal meningitis. Other pathogens include *Streptococcus pneumoniae* and *Listeria monocytogenes* [4, 11] while *Enterobacter* spp, *Citrobacter* spp, *Pseudomonas* spp and *Serratia* spp. are relatively less common. Commensal organisms such as coagulase negative Staphylococci are more commonly seen in very premature neonates who require prolonged hospitalization, central venous catheters and ventilatory support.

Population based surveillance studies on GBS have been conducted in a number of European countries including Finland [12], Germany [13], Portugal [14] and the UK [15] and reveal similar incidence figures. The proportion of GBS cases presenting with meningitis ranges from 17–30% across these studies.

3 Mortality and Morbidity

A significant decline in the overall mortality of neonatal meningitis was noted between the two national UK surveillance studies (1985–7 vs. 1996–97). While GBS in the earlier study was associated with a mortality rate of 22% and *E. coli* of 25%, 10 years later, these figures were reduced to 12% and 15% respectively. Overall, there was a decline in mortality for all neonatal meningitis from 25% to 10%.

Consistent with these results, a more recent national surveillance study, directed at all GBS, showed a case fatality rate of 12.4% for meningitis specifically [15]. Mortality rates in other developed countries have been reported in the range of 20–25% [16, 17].

Following both surveillance studies cases were followed up to 5 years of age to determine the prevalence of serious sequelae [18, 19]. This revealed that the proportion of children with severe or moderate disability in the two studies had not changed significantly (Table 1). Disabilities found included hydrocephalus, developmental delay, cerebral palsy, seizures requiring anticonvulsant therapy, decreased visual acuity and, most commonly, sensorineural deafness. In both studies only half of all children who suffered from meningitis in the neonatal period had no disability at the age of five. However, significantly fewer children had cerebral palsy (9% vs. 16.4%, $p < 0.05$) or seizure disorders (2.4% vs. 12%, $p < 0.005$) than in the earlier study [19].

The later study showed that isolation of bacteria from the CSF was the best single predictor of serious long-term disability. Children with a positive CSF culture accounted for the majority of cases of severe (8/9 cases) and moderate (23/30 cases) disability, significantly more than those with negative CSF cultures or where CSF was not collected ($p < 0.002$).

Table 1 Neonatal meningitis: disability at 5 years	1985–1987	1996–1997
	n = 274 (%)	n = 166 (%)
Severe	7	5
Moderate	18	18
Mild	24	26
None	50	51

In Australia and the USA, similar levels of long–term morbidity have been reported. Long term sequelae occurred in at least 23% of survivors in Australia whilst in the US 38% had mild and 24% moderate to severe disability at follow up [10, 17].

Rates of neonatal infection increase with decreasing birth weight and gestational age as do the rates of neurodevelopmental impairment. A study of 6,314 extremely low birth weight infants (ELBW <1,000 g) sought to determine if neonatal infections were specifically associated with adverse outcomes in later childhood. This demonstrated that ELBW infants who had culture confirmed neonatal infections (including meningitis) had higher rates of adverse neurodevelopmental outcomes including impaired mental and psychomotor development, cerebral palsy, vision and hearing impairment compared to ELBW infants who had not had an infection during the neonatal period [20].

In summary, the incidence of neonatal meningitis has declined over the last four decades and the overall mortality from neonatal meningitis has also declined. There has however, been little change in long-term morbidity. It is this fact that provides the stimulus for seeking new strategies for prevention of infection and for improving the management of this condition.

4 Presentation

The signs and symptoms that are observed in babies with meningitis may be subtle and non specific, especially so in premature babies. These same signs and symptoms are also seen in sepsis. From a large series of 255 babies with meningitis, fever/hypothermia, lethargy, vomiting and respiratory distress were all seen in more than 50% of cases, convulsions in 40%, irritability in 32% and bulging/full fontanel in 28% [21].

What is missing from this list of symptoms and signs are their timing of onset. From the work done on meningococcal disease in children [22] for example, it is apparent that the timing of the onset of clinical features can be crucial for early recognition, prompt management and potentially, better outcome. The well known "classical" symptoms often appear late in the course of the disease and their presence therefore predicts a worse outcome. An example of this is low level of consciousness at hospital admission which is known to be a predictor of poor outcome [23]. Such features do not therefore allow the opportunity for early intervention.

5 Diagnosis

Given that the clinical signs are rather nonspecific and similar to those seen in sepsis a diagnostic test (s) is required. Examination of cerebrospinal fluid via lumbar puncture (LP) is currently the gold standard test for making a definitive diagnosis of bacterial meningitis. The CSF that is obtained from the procedure is cultured and a positive growth not only confirms diagnosis but also identifies the responsible bacteria and therefore directs definitive antibiotic treatment. The CSF also provides a clue in the first instance in terms of cell count and cell type, Gram stain and glucose and protein concentrations. Newer tests such as PCR are also increasingly important in defining the relevant pathogen, especially when antibiotics are given prior to LP.

How often do clinicians perform an LP during the evaluation of sepsis in neonates? In a large series from an Australasian neonatal study group involving nearly 4,000 babies, LP was performed in 51% of babies who were evaluated for sepsis. Of those evaluated, a final diagnosis of meningitis was made in 8% [24]. In a recent UK review of how Paediatricians managed neonates with confirmed GBS infection, 109/138 (79%) had a lumbar puncture [25]. More contemporary data are required but it is possible that babies with meningitis are being missed because lumbar punctures are not being performed in all cases.

In trying to rationalise the need for an LP as part of a sepsis screen there are several approaches that clinicians have used.

5.1 Should an LP Only Be Performed on Symptomatic Babies?

In a study by Merenstein et al. [26] involving 789 symptomatic babies (with or without maternal risk factors for sepsis) 13 infants (1.6%) were found to have bacterial meningitis. Johnson et al. [27] in a retrospective review in 1997 found 11/1,712 (0.7%) term babies that were evaluated for symptoms of sepsis (respiratory distress, poor perfusion, temperature instability, bloody stools, lethargy and recurrent hypoglycaemia) to have culture proven bacterial meningitis. In the same study none of the 3,423 asymptomatic babies (who had maternal risk factors only i.e. maternal colonization with GBS, maternal fever, prolonged rupture of fetal membranes at more than 18 h, foul-smelling amniotic fluid, unexplained fetal tachycardia and elevated maternal WBC count) had meningitis. In another study by Fielkow et al. of 284 asymptomatic babies with obstetric risk factors only (rupture of membrane ≥12 h before delivery, clinical amnionitis characterized by maternal fever >38.0°C, persistent fetal tachycardia >160/min, amniotic or gastric fluid Gram stains showing leukocytes or bacteria, positive cultures from amniotic fluid and prematurity) none were found to have meningitis or significant positive blood culture [28]. Conversely, in a small study, Wiswell et al. [29] found seven of 43 asymptomatic babies with maternal risk factors to have meningitis. Overall however, the published data would suggest that the yield of an LP from babies who are asymptomatic is likely to be very low.

5.2 Should an LP Be Performed Only When Blood Cultures are Positive?

It is recognised that up to 25% of babies with bacteraemia are likely to have meningitis as well. However, a range of studies have shown that up to 50% of babies with bacterial meningitis may have negative blood cultures (Visser et al. 6/39 [30], Wiswell et al. 12/43 [29], Garges et al. 35/92 [31], Ansong 9/46 [32] and Vergnano et al. 9/27 [33]). Furthermore, in one study a number of cases were described in which both blood and CSF cultures were positive, but with discordant organisms [31]. A strategy of performing an LP only when blood cultures are positive may therefore lead to a delay in the diagnosis of bacterial meningitis as well as the possibility of targeting the wrong organism.

5.3 Making the Diagnosis

Clinically one cannot easily make a specific diagnosis of bacterial meningitis in neonates. Thus in babies with any clinical features suggestive of infection (and certainly in babies with positive blood cultures) an LP and evaluation of CSF is vital. Interpretation of CSF findings may however be problematic. One issue is what constitutes "normal" CSF cytology and biochemistry. Traditionally, paediatricians use reference values for CSF white cell counts (WBC) and biochemistry as reported in standard textbooks but these may have been based on older studies with methodological limitations such as small sample size, inclusion of traumatic LPs and failure to exclude other conditions that may cause CSF pleocytosis (summarized in Table 2). Garges et al. [31] in an attempt to develop an algorithm for predicting neonatal meningitis performed an analysis of the CSF WBC values of 9,111 LPs of whom 95 babies had culture-proven meningitis. The mean estimated gestational age at birth was 38 weeks (range 34–44 weeks) and the majority of LPs 7,907/9,111 (86.7%) were

Table 2 Summary of studies to establish "normal" CSF white cell counts (WBC)

Study	Number of patients, age	CSF WBC (mm^3) value	Included patients	Excluded patients	Limitations of study
Sarf et al. [64]	117 Term, <10d	Mean 8.5	High risk neonates	Those with positive b/c, urine, CSF cultures	Traumatic LP was defined as gross blood
Portnoy and Olson [65]	64, <42d	Mean 3.73	High risk babies	Those with positive CSF bacterial and viral cultures	LPs of babies with Sepsis, UTI, Seizures and traumatic LP were included
Bonadio et al. [66]	35, 0–28d; 40, 29–56d	Med 8.5, 90th C=22; Med 4.5, 90th C=15	Babies evaluated for fever	Those with positive b/c, urine, CSF cultures	Small numbers
Ahmed et al. [67]	108, <30d	Med 4, 90th C=11	Babies evaluated for fever	PCR test positive. Babies with evidence of CNS infection	Only includes babies less than 30 days old
Garges [31]	8,912, 34–44 wks gestational age	Med 6, IQR 2–15	Babies with neg CSF culture out of 9,111	CSF from babies with VP shunts	Only first LP was included. Some babies had antibiotics pre LP.
Kestenbaum et al. [34]	142, <28d; 238, 29–56d	Med 3, 95th C=19; Med 3, 95th C=9	Babies evaluated for fever	Conditions that "might lead to CSF pleocytosis"	Not all patients had PCR testing for enteroviruses. CSF viral culture was not performed

90th C 90th Centile, *IQR* Interquartile range, *95th C* 95th Centile, *wks* weeks, *d* days, *gest* gestational, *VP* ventriculoperitoneal, *b/c* blood culture, *neg* negative [31, 34, 64–67] performed within the first week of life. This study concluded that CSF WBC count of >21 cells/mm^3 had a sensitivity at 79% and specificity at 81%.

In a recent attempt to address the limitations of the studies summarized in Table 2 and provide age specific reference values for CSF WBC, Kestenbaum et al. [34] analysed the CSF WBC values of 1,064 babies less than 56 days who had a LP in the Emergency department as part of their evaluation for fever. Enterovirus infections were specifically excluded using a PCR. They concluded that the 95th centile CSF WBC value for neonates (0–28 day old) without bacterial meningitis was 19 cells/mm^3 and for 29–56 day old infants it was 9 cells/mm^3 [34].

5.4 Other Diagnostic Issues

It is important to bear in mind that a normal initial CSF white cell count, glucose and protein does not exclude bacterial meningitis. In a large series of 9,111 neonates who had an LP performed, 95 (1%) were found to have culture proven bacterial meningitis of whom 12/95 (13%) had normal CSF parameters [31]. It is also possible that a neonate with an initially negative CSF culture might develop evidence of meningitis in the context of ongoing bacteraemia. In a case series of six infants with Gram negative bacteraemia and initially clear CSF findings, repeat LPs done after intervals of 18–84 h had developed pleocytosis suggestive of meningitis [35]. Thus a high index of suspicion of meningitis is necessary even when an early LP does not support this diagnosis.

Another question of clinical relevance is whether pretreatment with antibiotics will prevent a diagnosis of bacterial meningitis being made. This is pertinent whenever it is not possible to obtain a CSF or when the LP is delayed for other reasons. In a study involving 245 children (including neonates), those who received antibiotics 12–72 h before the LP was performed had significantly increased glucose and decreased protein as compared with those who did not receive them or received them <4 h before delivery. However, there was no influence of antibiotics on the CSF WBC so that pretreatment with antibiotics did not prevent a diagnosis of bacterial meningitis being made [36]. Pre-treatment will however, have an impact on CSF culture results and may therefore impair a specific etiological diagnosis being made. Blood cultures are usually obtained prior to antibiotics and will be positive in a significant proportion of cases of neonatal meningitis. However, this is a situation in which non culture methods of diagnosis may offer great potential.

PCR is the most widely used non-culture method of pathogen detection although its routine use in the context of neonatal infection is currently limited. Using real-time multiplex PCR in a series of 168 CSF samples (including 21 babies less than 3 months of age), Chiba et al. showed that of those who had pre LP antibiotics, only 29% had a positive CSF culture, but using the PCR an organism was identified in 58%. Amongst those who had an LP before antibiotics, routine culture identified an organism in 70% of the cases whilst PCR identified an organism in 89%. The use of this method also allowed a more rapid detection of causative organisms (total time in this series 1.5 h) as well as an opportunity to detect antibiotic resistance genes [37].

5.5 Does a Delay in the Laboratory Evaluation of the CSF Affect the Results?

Once CSF is obtained from a lumbar puncture, antibiotics should be started promptly and the sample processed as soon as possible by the laboratory. Evidence suggests that a delay in laboratory analysis may affect the CSF results and even prevent an early diagnosis being made. Investigators compared CSF analysed immediately after it was obtained with the same samples of CSF analysed after a lag time of 2 h and 4 h. They showed that in 19 cases where the baseline WCC was >30 cells/mm^3, an early diagnosis of meningitis would have been missed in 53% if the samples had been analysed at 2 h and in 79% if analysed at 4 h because of the decline in measured white cell count [38].

5.6 What Influences the Success of Obtaining an LP?

The subject's position might influence the success of a lumbar puncture. Traditionally LPs in babies and children are performed in the lateral position with hip and neck flexion. A recent study was undertaken using bedside ultrasonographic measurement of the interspinous space in various positions suitable for LP. The position associated with the largest measurement was the sitting position with hip flexion followed by the sitting position without hip flexion, the lateral position with hip flexion and without neck flexion, the lateral position with both hip and neck flexion and finally, the lateral position without hip or neck flexion. An important finding was that neck flexion did not increase the size of the interspinous space. Because neck flexion is associated with desaturation in the neonatal setting this provides further evidence that it should be avoided [39].

6 Management

6.1 Empiric Antibiotics

Once bacterial meningitis is suspected, there should be no delay in starting appropriate empiric antibiotics. This requires knowledge of the likely pathogens and their antibiotic susceptibilities. In most developed countries, group B streptococcus and *E. coli* are by far the most common pathogens isolated in bacterial meningitis in babies, followed by other Gram negative bacilli, *Streptococcus pneumoniae* and Listeria. The other factor to be considered is the ability of the antibiotics to penetrate the CSF. The choice of empiric treatment may also be influenced by the location of the neonate at the time of their presentation. For babies <3 months of age in the community an empiric antibiotic combination of amoxicillin and cefotaxime will cover the likely pathogens and afford good CSF penetration. For babies who are on a neonatal unit, a range of other factors may influence the likely spectrum of causative pathogens, in particular their risk of unusual or multiresistant bacteria. These include prior exposure to broad spectrum antibiotics, comorbidities such as surgery and chronic lung disease (use of dexamethasone treatment), the presence of central venous lines and TPN and their risk of acquiring infections through nosocomial transmission. In general, empiric cover with cefotaxime and amoxycillin will be adequate but such babies may be at risk of multiresistant Gram negative bacteria so the addition of an aminoglycoside may be prudent [40].

Additionally, although rare, meningitis due to Coagulase negative staphylococci may be encountered in very low birth weight babies, especially those with central lines in situ and repeated or persistent bacteraemia. Thus the addition of vancomycin may be required until the causative bacteria are revealed. As always clinicians should be aware of any epidemiological issues relevant to their own unit that must dictate the choice of antibiotics.

A distinction between babies admitted from home and those still in hospital at the time of meningitis may of course become blurred. It is becoming evident that neonates who have had periods of time on the neonatal unit may become colonised with resistant bacteria and then remain colonised after discharge [41]. They may then conceivably present from the community with meningitis due to these "hospital-acquired" bacteria. Ongoing surveillance is therefore essential. A recent and worrying report from the Asia-Pacific Neonatal Infections Study evaluated the Gram negative bacteria causing late onset-sepsis in several South-East Asian neonatal units. This revealed that one third of the isolates were resistant to both third generation cephalosporins (cefotaxime and ceftazidime) and gentamicin and 50% were resistant to at least one or the other [42].

6.2 Empiric Cover for Listeria Infection

Listeria requires a special mention because of the need to use specific antibiotics. Infection is rare although it contributes significantly to bacterial meningitis in babies, implicated in 5–7% of cases. Evidence suggests that most cases occur in babies less than 7 days of age and in premature babies. Usually there is a history of maternal illness (75%).

The optimal antibiotic therapy for this pathogen requires a penicillin, hence the inclusion of amoxicillin/ampicillin or penicillin in empiric antibiotic guidelines. Amoxicillin/ampicillin is often preferred to penicillin, based mainly on tradition and a slightly lower minimum inhibitory concentration, but there appears to be no clinical evidence of superiority and empiric therapy with penicillin is likely to be satisfactory. Combination with an aminoglycoside is used for its synergistic effect, demonstrated only in animal studies [43].

7 What are the Predictors for a Poor Outcome?

There are certain factors that may predict a worse outcome from meningitis. A retrospective study of 101 cases of neonatal bacterial meningitis admitted between 1979 and 1998 identified early predictors of adverse outcome at 1 year of age (death or moderate/severe disability). At 12 h after admission these were seizures, coma, use of inotropes, leucopaenia $\leq 5,000 \times 10^9$ and at 96 h after admission, seizure duration >72 h, coma, use of inotropes, leucopaenia $\leq 5,000 \times 10^9$. These factors did not change when stratified according to causative pathogen [44].

EEG may be a useful tool also. A retrospective case review has demonstrated that infants who had a normal or mildly abnormal EEG had normal outcomes (at a mean of 34 months), whereas those with notably abnormal EEGs died or had severe neurological sequelae [44]. Knowledge of these potential risk factors can then help in stratifying cases on admission but also more aggressive or specific management of those with these risk factors might potentially improve outcome. They are also of value when counseling parents.

8 Improving the Outcome of Neonatal Meningitis

As discussed earlier, it is reasonable to believe that the earlier appropriate antibiotics are commenced, the better will be the outcome. A survey of UK neonatologists in 2008 assessed their choice of empiric antibiotic cover for suspected bacterial meningitis. Overall, only 45% include a cephalosporin at all and in 19% no penicillin was used in the empiric combination, thus providing no cover for listeria meningitis [45].

Other assessments of current practice suggest that an LP is not being performed in all cases where meningitis should be excluded and indicate considerable uncertainty in antibiotic choice and antibiotic duration in established meningitis [25].

This uncertainty reflects the lack of high quality data in this field. Duration of therapy is based more on tradition than on evidence and in particular the concept that antibiotics should be continued for at least 2 weeks after CSF sterilisation has been achieved. Because rapid sterilisation is expected in GBS meningitis a total duration of at least 14 days of therapy is therefore recommended.

CSF sterilisation may be slower in Gram negative meningitis (up to 7 days historically) which is reflected in a recommendation for a total duration of at least 21 days of therapy [40].

Adult studies have shown that circulatory support in shock i.e. fluid resuscitation to restore intravascular volume, stabilize blood pressure and maintain adequate oxygenation improves outcome. Strict and early goal-directed fluid resuscitation, vasopressor therapy and transfusion of adults with severe sepsis significantly decreases mortality [46]. Similarly, studies in older children have shown that early and aggressive fluid resuscitation improves outcome [47]. Indeed, delayed reversal of shock is associated with worse outcome; every hour of failure to reverse shock results in doubling of risk of death [48]. There are no standardised guidelines for fluid management in neonates and few high-quality studies that have assessed initial fluid therapy in neonates with suspected or confirmed bacterial sepsis/meningitis. Anecdotally, neonatologists may be more cautious when resuscitating neonates with features of shock citing concerns of cerebral vascular fragility and fluid overload. More research is required in this area.

8.1 What is the Role of New or Different Antibiotics in the Management of Neonatal Bacterial Meningitis?

We know that in spite of the use of third generation cephalosporins in the last two decades both mortality and morbidity still remains unacceptably high. New or different antibiotics may have a role in improving the outcome because of better coverage of likely pathogens e.g. meropenem or because of different modes of action. One of the major factors leading to poor outcome is the intensity of the host inflammatory response. Attempts to reduce this include the use of dexamethasone (reviewed below). Some antibiotics may have a mechanism of action that avoids or minimises the release of pro-inflammatory bacterial components and thereby reducing the inflammatory response. This has been demonstrated in an animal model of pneumococcal meningitis for example, where therapy with rifampicin or daptomycin shows improved survival compared with cefotaxime [49, 50].

Other studies have also addressed the possibility that antibiotics may not penetrate adequately and considered intrathecal and intraventricular administration. The first of these demonstrated no difference in mortality (32%) or morbidity (36%) between intrathecal and intravenous administration of gentamicin in Gram negative meningitis [51]. This was followed by another study by McCracken and Mize in 1980 [52] evaluating the intraventricular administration of gentamicin. This study was terminated early because of excess mortality (43% vs. 13%) in the intraventricular arm. Hence there is no place for the routine use of intrathecal or intraventricular gentamicin.

8.2 What is the Role of Other Adjunctive Therapy in the Management of Bacterial Meningitis?

8.2.1 What is the Role of Immunoglobulin in Neonatal Bacterial Sepsis/Meningitis?

A systematic review by Ohlsson in 2004 [53] on the use of immunoglobulin in sepsis included seven RCT's and 262 neonates with proven infection and showed a reduction in mortality (RR 0. 55 (0.31–0.98), NNT 11 (5.6–100)). The conclusion was that there is insufficient evidence to support routine IVIG for treatment; therefore, further research is needed. Its place in therapy will be better defined when the results of the recently concluded International Neonatal Immunotherapy Study trial (http://www.npeu.ox.ac.uk/inis) [54] are released. As bacteraemia is a pre-requisite for bacterial meningitis, any intervention that can improve the outcome from bacteraemia/septicaemia may also translate into a better outcome for meningitis.

8.2.2 G- or GM: CSF

The rationale for the use of G or GM–CSF is to increase the number and improve the function of neutrophils. Neutropenia is a common feature of neonatal sepsis. In one randomized placebo controlled trial in very low birth weight with clinical sepsis and neutropenia, G–CSF was shown to be safe and reduce mortality: 1/13 versus 7/15 at 12 months [55]. However, in a larger trial GM–CSF was used as prophylaxis in infants <32 weeks and small for gestational age [56]. Although, the number of neutrophils were significantly increased in recipients there was no overall effect in preventing sepsis. This result is discouraging and suggests that simply preventing the neutropenia associated with sepsis is not enough. The deficits in neonatal immunity are multiple and it is perhaps naïve to expect that any single intervention will provide the solution. Combinations (e.g. IVIG and GCSF or neutrophil infusions etc.) may be more logical but will be clearly more difficult to assess in clinical trials.

8.2.3 The Use of Corticosteroids in Meningitis

Dexamethasone treatment in childhood meningitis has become standard of care and is therefore an obvious consideration for the management of neonatal meningitis. Fifty years ago, a review of 47 neonatal meningitis cases treated with corticosteroid (CS) and chloramphenicol/sulphonamide/streptomycin showed a reduction in mortality: 41% vs. 75% (p=0.05). [57]. More recently in Jordan, Daoud et al. [58], assessed the use dexamethasone (DXM) in a double blind, randomised, placebo controlled trial. DXM or placebo was administered prior to the first dose of antibiotics (cefotaxime and ampicillin). The study showed a mortality of 22% in the DXM group vs. 28% in the placebo group and no difference in morbidity (30% in DXM group vs. 39% in placebo group). It was therefore concluded that adjunctive dexamethasone therapy does not have a role in neonatal meningitis. However, it is not certain that the results of this study, where GBS was a rare pathogen (3 cases) can be widely extrapolated to other settings.

8.2.4 What About Oral Glycerol?

A recent paediatric bacterial meningitis study by Peltola in Latin America indicates better outcome with oral glycerol compared to IV dexamethasone [59]. This appears to be a safe and cheap therapy and its role in neonatal meningitis now needs to be explored.

9 Prevention

Intrapartum antibiotic prophylaxis (IAP) is efficacious against early onset GBS disease but has no impact on late onset disease, when most GBS meningitis occurs. A GBS vaccine has great potential in this regard and clinical trials of candidate vaccines, including those in pregnant women, are extremely encouraging [60, 61].

In the case of pneumococcal meningitis, the introduction of the 7-valent pneumococcal conjugate vaccine (PCV7) into national infant immunisation programmes is likely to offer some protection against invasive pneumococcal disease (IPD) in neonates and young infants through herd immunity [62]. In England and Wales, routine immunisation with PCV7 has resulted in a 60% decline in IPD incidence caused by PCV7 serotypes in infants aged <3 months. The overall impact of IPD in this age group however, did not reach statistical significance because the dominant serotypes in young infants are less well covered in PCV7 as compared with those causing IPD in older children [63]. The 10 and 13 valent conjugate vaccines however, may close this gap and afford better protection for those infants too young to be vaccinated.

Prevention of neonatal listeria can be achieved by avoidance of foods potentially contaminated with listeria. These include ready-to-eat-meat, paté, raw milk, coleslaw and soft cheese. Additionally, aggressive assessment and management of mothers with potential features of listeria infection (fever and flu-like illness, gastrointestinal symptoms during labour) may also allow prevention of neonatal infections although this is not proven.

In a neonatal unit any interventions that can reduce nosocomial bacteremia also have the potential to reduce bacterial meningitis. A number of strategies, either individually or wrapped together in a care bundle, have been shown to be efficacious [46].

10 Conclusions

10.1 Improving the Outcome of Neonatal Meningitis...Can We Do Better?

Current evidence shows that the mortality and morbidity from neonatal bacterial meningitis remains unacceptably high despite advances in antibiotics and intensive care.

There may be a number of opportunities for improving the outcome and further research is required. Specific questions include: what is the impact of earlier recognition and earlier diagnosis and of earlier use of appropriate empiric antibiotics? Is there a role for new antibiotics, for earlier and more aggressive supportive care and for new adjunctive therapies? The ultimate goal however, is prevention and more work is also required here, particularly on vaccines against Group B Streptococcus.

References

1. Goldacre MJ. Acute bacterial meningitis in childhood. Lancet. 1976;1:701.
2. de Louvois J, Blackbourn J, Hurley R, Harvey D. Infantile meningitis in England and Wales: a two year study. Arch Dis Child. 1991;66:603–7.
3. Hristeva L, Booy R, Bowler I, Wilkinson AR. Prospective surveillance of neonatal meningitis. Arch Dis Child. 1993;69:14–8.
4. Holt DE, Halket S, de Louvois J, Harvey D. Neonatal meningitis in England and Wales: 10 years on. Arch Dis Child Fetal Neonatal Ed. 2001;84:F85–9.
5. Goldacre MJ. Neonatal meningitis. Postgrad Med J. 1977;53:607–9.
6. Persson E, Trollfors B, Brandberg LL, Tessin I. Septicaemia and meningitis in neonates and during early infancy in the Goteborg area of Sweden. Acta Paediatr. 2002;91:1087–92.
7. Thaver D, Zaidi AK. Burden of neonatal infections in developing countries: a review of evidence from community-based studies. Pediatr Infect Dis J. 2009;28:S3–9.
8. Synnott MB, Morse DL, Hall SM. Neonatal meningitis in England and Wales: a review of routine national data. Arch Dis Child. 1994;71:F75–80.
9. Wenger JD, Hightower AW, Facklam RR, Gaventa S, Broome CV. Bacterial meningitis in the United States, 1986: report of a multistate surveillance study. The Bacterial Meningitis Study Group. J Infect Dis. 1990;162:1316–23.
10. Francis BM, Gilbert GL. Survey of neonatal meningitis in Australia: 1987–1989. Med J Aust. 1992;156:240–3.

11. Levy C, de La Rocque F, Cohen R. [Epidemiology of pediatric bacterial meningitis in France]. Med Mal Infect. 2009;39:419–31.
12. Kalliola S, Vuopio-Varkila J, Takala AK, Eskola J. Neonatal group B streptococcal disease in Finland: a ten-year nationwide study. Pediatr Infect Dis J. 1999;18:806–10.
13. Fluegge K, Siedler A, Heinrich B, Schulte-Moenting J, Moennig MJ, Bartels DB, et al. Incidence and clinical presentation of invasive neonatal group B streptococcal infections in Germany. Pediatrics. 2006;117: e1139–45.
14. Neto MT. Group B streptococcal disease in Portuguese infants younger than 90 days. Arch Dis Child Fetal Neonatal Ed. 2008;93:F90–3.
15. Heath PGT, Balfour G, Weisner AM, Efstratiou A, Lamagni TL, Tighe H, et al. Group B streptococcal disease in UK and Irish infants younger than 90 days. Lancet. 2004;363:292–4.
16. Mulder CJ, Zanen HC. A study of 280 cases of neonatal meningitis in The Netherlands. J Infect. 1984;9:177–84.
17. Franco SM, Cornelius VE, Andrews BF. Long-term outcome of neonatal meningitis. Am J Dis Child. 1992;146:567–71.
18. Bedford H, de Louvois J, Halket S, Peckham C, Hurley R, Harvey D. Meningitis in infancy in England and Wales: follow up at age 5 years. BMJ. 2001;323:533–6.
19. de Louvois J, Halket S, Harvey D. Neonatal meningitis in England and Wales: sequelae at 5 years of age. Eur J Pediatr. 2005;164:730–4.
20. Stoll BJ, Hansen NI, Adams-Chapman I, Fanaroff AA, Hintz SR, Vohr B, et al. Neurodevelopmental and growth impairment among extremely low-birth-weight infants with neonatal infection. JAMA. 2004;292: 2357–65.
21. Palazzi D, Klein J, C B. Bacterial sepsis and meningitis. In: Remington JS, Klein J, editors. Infectious Disease of the Fetus and Newborn Infants 6th ed. Philadelphia: Elsevier Saunders; 2006. p. 247–95.
22. Thompson MJ, Ninis N, Perera R, Mayon-White R, Phillips C, Bailey L, et al. Clinical recognition of meningococcal disease in children and adolescents. Lancet. 2006;367:397–403.
23. Holt DE. Neonatal meningitis in England and Wales: 10 years on. Archives of Disease in Childhood - Fetal and Neonatal Edition. 2001;84:85 F-9.
24. May M. Early onset neonatal meningitis in Australia and New Zealand, 1992–2002. Archives of Disease in Childhood - Fetal and Neonatal Edition. 2005;90:F324-f7.
25. Heath PT, Balfour GF, Tighe H, Verlander NQ, Lamagni TL, Efstratiou A. Group B streptococcal disease in infants: a case control study. Archives of Disease in Childhood. 2009;94:674–80.
26. Merenstein GB. Neonatal sepsis. Current Opinion in Infectious Diseases. 1992;5:553–7.
27. Johnson CE, Whitwell JK, Pethe K, Saxena K, Super DM. Term newborns who are at risk for sepsis: are lumbar punctures necessary? Pediatrics. 1997;99:E10.
28. Fielkow S, Reuter S, Gotoff SP. Cerebrospinal fluid examination in symptom-free infants with risk factors for infection. J Pediatr. 1991;119:971–3.
29. Wiswell TE, Baumgart S, Gannon CM, Spitzer AR. No lumbar puncture in the evaluation for early neonatal sepsis: will meningitis be missed? Pediatrics. 1995;95:803–6.
30. Visser VE, Hall RT. Lumbar puncture in the evaluation of suspected neonatal sepsis. J Pediatr. 1980;96:1063–7.
31. Garges HP. Neonatal Meningitis: What Is the Correlation Among Cerebrospinal Fluid Cultures, Blood Cultures, and Cerebrospinal Fluid Parameters? Pediatrics. 2006;117:1094–100.
32. Ansong AK, Smith PB, Benjamin DK, Clark RH, Li JS, Cotten CM, et al. Group B streptococcal meningitis: cerebrospinal fluid parameters in the era of intrapartum antibiotic prophylaxis. Early Hum Dev. 2009;85:S5–7.
33. Vergnano S, Embleton N, Collinson A, Menson E, Russell BA, Heath P. Missed opportunities for preventing group B streptococcus infection. Archives of Disease in Childhood - Fetal and Neonatal Edition 2009;95:F72–F3.
34. Kestenbaum LA, Ebberson J, Zorc JJ, Hodinka RL, Shah SS. Defining Cerebrospinal Fluid White Blood Cell Count Reference Values in Neonates and Young Infants. Pediatrics. 2010;125:257–64.
35. Sarman G, Moise AA, Edwards MS. Meningeal inflammation in neonatal gram-negative bacteremia. Pediatr Infect Dis J. 1995;14:701–4.
36. Nigrovic LE, Malley R, Macias CG, Kanegaye JT, Moro-Sutherland DM, Schremmer RD, et al. Effect of Antibiotic Pretreatment on Cerebrospinal Fluid Profiles of Children With Bacterial Meningitis. Pediatrics. 2008;122:726–30.
37. Chiba N, Murayama SY, Morozumi M, Nakayama E, Okada T, Iwata S, et al. Rapid detection of eight causative pathogens for the diagnosis of bacterial meningitis by real-time PCR. Journal of Infection and Chemotherapy. 2009;15:92–8.

38. Rajesh NT, Dutta S, Prasad R, Narang A. Effect of delay in analysis on neonatal cerebrospinal fluid parameters. Archives of Disease in Childhood - Fetal and Neonatal Edition. 2009;95:F25–F9.

39. Abo A, Chen L, Johnston P, Santucci K. Positioning for Lumbar Puncture in Children Evaluated by Bedside Ultrasound. Pediatrics. 2010;125:e1149–e53.

40. Heath PT. Neonatal meningitis. Archives of Disease in Childhood - Fetal and Neonatal Edition. 2003;88:173 F–8.

41. Millar M, Philpott A, Wilks M, Whiley A, Warwick S, Hennessy E, et al. Colonization and persistence of antibiotic-resistant Enterobacteriaceae strains in infants nursed in two neonatal intensive care units in East London, United Kingdom. J Clin Microbiol. 2008;46:560–7.

42. Tiskumara R, Fakharee SH, Liu CQ, Nuntnarumit P, Lui KM, Hammoud M, et al. Neonatal infections in Asia. Archives of Disease in Childhood - Fetal and Neonatal Edition. 2009;94:F144–F8.

43. Mitja O, Pigrau C, Ruiz I, Vidal X, Almirante B, Planes AM, et al. Predictors of mortality and impact of amino-glycosides on outcome in listeriosis in a retrospective cohort study. J Antimicrob Chemother. 2009;64:416–23.

44. Klinger G, Chin CN, Beyene J, Perlman M. Predicting the outcome of neonatal bacterial meningitis. Pediatrics. 2000;106:477–82.

45. Fernando AM, Heath PT, Menson EN. Antimicrobial policies in the neonatal units of the United Kingdom and Republic of Ireland. J Antimicrob Chemother. 2008;61:743–5.

46. Rivers E, Nguyen B, Havstad S, Ressler J, Muzzin A, Knoblich B, et al. Early goal-directed therapy in the treatment of severe sepsis and septic shock. N Engl J Med. 2001;345:1368–77.

47. Carcillo JA, Davis AL, Zaritsky A. Role of early fluid resuscitation in pediatric septic shock. JAMA. 1991;266:1242–5.

48. Han YY, Carcillo JA, Dragotta MA, Bills DM, Watson RS, Westerman ME, et al. Early reversal of pediatric-neonatal septic shock by community physicians is associated with improved outcome. Pediatrics. 2003;112:793–9.

49. Nau R, Wellmer A, Soto A, Koch K, Schneider O, Schmidt H, et al. Rifampin reduces early mortality in experimental Streptococcus pneumoniae meningitis. J Infect Dis. 1999;179:1557–60.

50. Grandgirard D, Oberson K, Buhlmann A, Gaumann R, Leib SL. Attenuation of cerebrospinal fluid inflammation by the nonbacteriolytic antibiotic daptomycin versus that by ceftriaxone in experimental pneumococcal meningitis. Antimicrob Agents Chemother. 2010;54:1323–6.

51. McCracken GH, Jr., Mize SG. A controlled study of intrathecal antibiotic therapy in gram-negative enteric meningitis of infancy. Report of the neonatal meningitis cooperative study group. J Pediatr. 1976;89:66–72.

52. McCracken GH, Jr., Mize SG, Threlkeld N. Intraventricular gentamicin therapy in gram-negative bacillary meningitis of infancy. Report of the Second Neonatal Meningitis Cooperative Study Group. Lancet. 1980;1:787–91.

53. Ohlsson A, Lacy JB. Intravenous immunoglobulin for suspected or subsequently proven infection in neonates. Cochrane Database Syst Rev. 2004:CD001239.

54. (http://www.npeu.ox.ac.uk/inis).

55. Bedford Russell AR, Emmerson AJ, Wilkinson N, Chant T, Sweet DG, Halliday HL, et al. A trial of recombinant human granulocyte colony stimulating factor for the treatment of very low birthweight infants with presumed sepsis and neutropenia. Arch Dis Child Fetal Neonatal Ed. 2001;84:F172–6.

56. Carr R, Brocklehurst P, Dore CJ, Modi N. Granulocyte-macrophage colony stimulating factor administered as prophylaxis for reduction of sepsis in extremely preterm, small for gestational age neonates (the PROGRAMS trial): a single-blind, multicentre, randomised controlled trial. Lancet. 2009;373:226–33.

57. Yu JS, Grauaug A. Purulent Meningitis in the Neonatal Period. Arch Dis Child. 1963;38:391–6.

58. Daoud AS, Batieha A, Al-Sheyyab M, Abuekteish F, Obeidat A, Mahafza T. Lack of effectiveness of dexamethasone in neonatal bacterial meningitis. Eur J Pediatr. 1999;158:230–3.

59. Peltola H, Roine I, Fernandez J, Zavala I, Ayala SG, Mata AG, et al. Adjuvant glycerol and/or dexamethasone to improve the outcomes of childhood bacterial meningitis: a prospective, randomized, double-blind, placebo-controlled trial. Clin Infect Dis. 2007;45:1277–86.

60. Baker CJ, Rench MA, McInnes P. Immunization of pregnant women with group B streptococcal type III capsular polysaccharide-tetanus toxoid conjugate vaccine. Vaccine. 2003;21:3468–72.

61. Law MR, Palomaki G, Alfirevic Z, Gilbert R, Heath P, McCartney C, et al. The prevention of neonatal group B streptococcal disease: a report by a working group of the Medical Screening Society. J Med Screen. 2005;12:60–8.

62. Poehling KA, Talbot TR, Griffin MR, Craig AS, Whitney CG, Zell E, et al. Invasive pneumococcal disease among infants before and after introduction of pneumococcal conjugate vaccine. JAMA. 2006;295:1668–74.

63. Slack M, et al. The 7th International Symposium on Pneumococci and Pneumococcal Diseases. Tel Aviv Israel 2010.

64. Sarff LD, Platt LH, McCracken GH, Jr. Cerebrospinal fluid evaluation in neonates: comparison of high-risk infants with and without meningitis. J Pediatr. 1976;88:473–7.
65. Portnoy JM, Olson LC. Normal cerebrospinal fluid values in children: another look. Pediatrics. 1985;75:484–7.
66. Bonadio WA, Stanco L, Bruce R, Barry D, Smith D. Reference values of normal cerebrospinal fluid composition in infants ages 0 to 8 weeks. Pediatr Infect Dis J. 1992;11:589–91.
67. Ahmed A, Hickey SM, Ehrett S, Trujillo M, Brito F, Goto C, et al. Cerebrospinal fluid values in the term neonate. Pediatr Infect Dis J. 1996;15:298–303.

Approaches Towards Avoiding Lifelong Antiretroviral Therapy in Paediatric HIV Infection

Philip J.R. Goulder and Andrew J. Prendergast

1 Introduction

An estimated 2.1 million children are living with HIV worldwide, with 0.43 m new infections each year and 0.28 m paediatric deaths/year [1]. These figures, published in November 2009, refer to the epidemic as it existed in 2008. In that year, WHO guidelines for management of HIV infection in children changed, based principally on data from the Children with HIV Early Antiretroviral Therapy (CHER) study [2]. In this study, infants randomised at 6–12 weeks of age to immediate, compared to deferred, antiretroviral therapy (ART) had significantly lower mortality (4% vs 16%, respectively). WHO therefore recommended that all children diagnosed with HIV infection in the first year of life should start ART as soon after birth as possible, regardless of clinical or immunological disease stage. More recent (2010) WHO guidelines [3] (Table 1) recommend immediate initiation of ART in all infected children diagnosed before 24 months of age, recognising that disease progression remains rapid for most children in the first 2 years of life, and a strategy of 'watching and waiting' for signs of disease progression, even beyond infancy, is risky in many settings. The problem here is that lifelong ART from birth, or as soon after birth as ART can be initiated, is unlikely to be sustainable, for reasons of drug toxicity, resistance and cost. Even in a relatively well-resourced country such as South Africa, where mother-to-child-transmission (MTCT) prevention programmes are well established, high antenatal HIV prevalence rates result in up to 1000 new paediatric infections per week [4, 5]. At current rates of MTCT and ART implementation, the number of HIV-infected children in sub-Saharan Africa will have doubled in 5 years [1]. The prospect of an epidemic of HIV-infected adolescents on salvage therapy regimens is a reality and one that needs to be avoided.

This chapter sets out the options for management of children with HIV infection. It follows that, if lifelong ART from birth is unsustainable, any alternative strategy would incorporate discontinuation of ART at some point. We examine the possibility that initiation of ART from birth, or close to birth, continued for sufficient time to allow maturation of the developing immune system in the setting of undetectably low levels of HIV, followed by ART discontinuation, may be one option that could usefully be adopted in HIV-infected children. A second option would be to induce or boost

P.J.R. Goulder (✉)
Department of Paediatrics, University of Oxford, Oxford, UK
e-mail: philip.goulder@paediatrics.ox.ac.uk

A.J. Prendergast
Centre for Paediatrics, Queen Mary, University of London, London, UK
e-mail: a.prendergast@qmul.ac.uk

N. Curtis et al. (eds.), *Hot Topics in Infection and Immunity in Children VIII*,
Advances in Experimental Medicine and Biology 719, DOI 10.1007/978-1-4614-0204-6_3,
© Springer Science+Business Media, LLC 2011

Table 1 Current World Health Organization guidelines for ART in HIV-infected children

Age	<2 years	2–5 years	≥5 years
CD4 percentage	All	≤25%	NA
Absolute CD4 count	All	≤750 cells/mm^3	<350 cells/mm^3 (as in adults)

specific immune responses effective against HIV at an appropriate time during the period on ART, the aim being that effective HIV-specific immunity generated in infected children on ART could control viral replication once ART was discontinued.

2 How does Paediatric HIV Infection Differ from Adult Infection?

2.1 Rate of Disease Progression, Timing of Infection

Globally, HIV transmission to children occurs predominantly perinatally via MTCT, whilst adult infection is mostly via heterosexual transmission. ART-naïve adults typically progress to AIDS in 10 years [6–8], whereas ART-naïve perinatally infected children most commonly progress to AIDS in 1–2 years [5, 9, 10]. In a recent South African study, undertaken when WHO criteria to initiate ART were based on a CD4% of <25%, as many as 70% of infected infants reached this CD4% criterion by 4 months of age [11]. As with other chronic viral infections [12], such as CMV, HSV and VZV, precise timing of HIV acquisition has a profound impact on disease outcome [10, 13]. Following MTCT, progression is fastest in infants infected in utero, and slowest in those infected post-partum through breastfeeding. In the Zvitambo study in Zimbabwe, before availability of cotrimoxazole prophylaxis and ART, median time to death was 208d for in utero infected infants, 380d for intra-partum infected infants, and >500d for those infected post-partum via breast-feeding [10]. In keeping with the rapid disease progression, in utero-infected infants are also the group most likely to develop HIV encephalopathy [14], one of the most devastating sequelae of advanced HIV disease. Age also affects outcome from adult infection, rate of progression to AIDS increasing with age at infection [6, 8, 13]. Studies of HIV-infected haemophiliacs suggest that progression is slowest in the youngest children [13], although small numbers of HIV-infected haemophiliac children limit further detail. Thus, there appears to be an optimal age at which to become infected with HIV, in terms of disease progression, at which immune responses are fully mature, but not yet in decline.

2.2 Immune Activation and HIV Disease Progression

It is clear from these studies that CD4 decline usually occurs very rapidly after birth in perinatally-infected infants. This is in keeping with the very high viral loads that characterise HIV infection in the first year of life (see Sect. 2.3); however, the precise mechanism by which CD4 depletion occurs remains unclear. It was initially thought that, since HIV is tropic for CD4 cells, the decline in CD4 count is caused by direct viral infection and killing of CD4 cells. However, later studies showed that a minority (<1%) CD4 cells are actually infected with virus [15]; instead, CD4 cells appear to become activated and die by apoptosis [16].

One of the hallmarks of adult HIV infection is therefore a state of generalised immune activation [17], which is characterised by polyclonal B and T cell activation and raised levels of pro-inflammatory cytokines. Chronic activation leads to increased T cell turnover, immune exhaustion and apoptosis. Elaboration of proinflammatory cytokines causes fibrosis and distortion of lymph node architecture, and thymic involution, leading to altered lymphocyte trafficking and reduced lymphocyte renewal. Taken together, these changes have been termed immunosenescence, and parallel the alterations in immune function that occur naturally with ageing. Immune activation is in fact a better indicator of disease progression in adults than viral setpoint [18].

Similarly the phenomenon of immune activation occurs in perinatally infected children; the degree of immune activation present as early as 1–2 months of age can be used to predict which children will become long-term non-progressors [19, 20]. It is likely to be immune activation, rather than the virus itself, which leads to the decline in CD4 count during paediatric as well as adult HIV infection. Clearly immune activation occurs because of the virus, but the exact mechanisms linking the two remain unclear [17]. HIV itself can cause immune activation, particularly through the effects of the envelope glycoprotein gp120 on immune cells; co-infections, such as CMV, may drive activation through induction of a high-magnitude immune response; and damage to the gut during acute infection may allow translocation of gut organisms to the systemic circulation, which can drive peripheral immune activation [21]. In children, the relationship between HIV viral load and immune activation appears to be less clear cut than in adults, in whom a higher level of viraemia is predictably associated with higher levels of activation [18]. Thus the extent to which high viral loads, which are universal in infancy, are linked with differences in immune activation between infants who progress rapidly and those who progress slowly, needs further investigation.

Despite the lack of clarity around the association between the level of viraemia and the degree of immune activation in children, immune control of the virus is an important goal. ART reduces the virus to undetectable levels and thereby slows disease progression; furthermore, the majority of long-term non-progressors, usually diagnosed late in childhood or early in adolescence, maintain unusually low viral loads throughout this period. Maintaining a low viral load is therefore an advantageous, but infrequent, situation in HIV-infected children. For this reason, better understanding the immunological correlates of viral control, and designing interventions to improve containment of viraemia, remain critical to future management approaches.

2.3 Control of HIV viraemia

Rapid control of HIV infection is observed early in adult infection, viraemia declining from a peak of approximately 10^7 HIV copies/ml plasma at 2–3 weeks post infection, to a median viral setpoint of 30,000 copies/ml. This 2-to-3-$\log_{10}$ drop in viral load typically takes 2–4 weeks. In paediatric infection, viraemia declines very little from a similar peak in the first year [11, 22]. Even in the small minority of HIV-infected children whose CD4 counts are sufficiently high not to meet WHO criteria to start ART [11], and who have remained ART-naïve with high CD4 counts for years, a 2-$\log_{10}$ decline in viraemia takes 5–6 years.

The precise reasons for this failure of the paediatric immune system to bring about rapid control of viraemia are unknown. HIV-specific CD8+ T-cell activity appears to play a central role in the rapid reduction of acute viraemia in adults [23, 24] (see below), and although HIV-specific CD8+ T-cell activity is detectable even on the first day of life in many infants infected in utero [25], these responses appear to be relatively ineffective during infancy. A central factor contributing to this is likely to be the delay in a robust HIV-specific T helper (CD4+) response [25–27], critical for the induction and maintenance of effective CD8+ T-cell responses as well as of B cell activity. In contrast to HIV-infected adults in acute infection [28], HIV-specific T helper responses are only detected sporadically and at low levels in perinatally infected infants aged less than

3 months, but appear to improve as they get older. Furthermore, in adults in whom ART is initiated during acute infection, HIV-specific T helper responses increase [28], whereas the presence of HIV-specific T helper activity remains low or undetectable in HIV-infected infants, even when ART has been initiated in the first weeks of life [27]. Similar findings have been made in relation to perinatal HSV or CMV infection, where virus-specific CD4 functional responses are weaker and delayed in comparison to those seen in adult infection [29–33], and indeed the delay in herpesvirus-specific CD4 responses appears to extend beyond the immediate neonatal period well into infancy [12, 33].

Whatever the precise reasons for the delay in HIV-specific T-cell immunity in perinatally infected infants, this is likely to be critical in setting the balance even more heavily in favour of the virus than is the case in adult infection. From the moment of transmission, it is as if a race starts, between the virus and its destruction of the immune system, and the immune system and its containment of the virus (Fig. 1). On the basis of this model, one can propose that if long-term immune control of HIV is to be achieved, as it is in a small minority of infected adults and an even smaller minority of infected children, viral destruction of the immune system needs to be limited early in the course of infection, either through the immune response itself, or through the early initiation of ART or both.

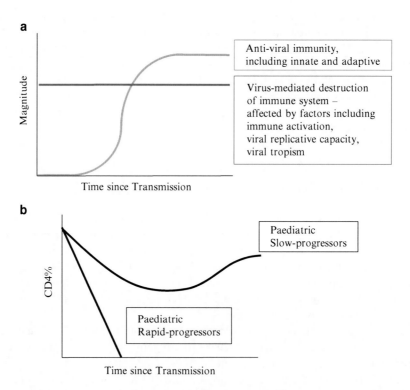

Fig. 1 Race between the virus and immune response. (**a**) Scheme to represent competing elements in operation from transmission: virus-mediated destruction of the immune system on the one hand, and the developing immune response against the virus on the other. (**b**) Scheme to represent diverse resulting outcomes: in paediatric slow-progressors, the anti-viral immune response has emerged sufficiently early to combat virus-mediated destruction of the immune response (and the CD4% of T cells in the blood); in paediatric rapid-progressors, the virus has successfully destroyed the immune system before it can be effective

2.4 Thymic Activity; Immune Reconstitution on ART

In children starting ART, even several years after infection, regeneration of the immune system is substantially more rapid than in adults. This is thought to be because of the contribution of the thymus [34, 35]. Qualitatively, immune reconstitution also differs between children and adults. In practice, adults presenting with HIV disease cannot reconstitute HIV-specific T-helper activity [28], whereas this returns rapidly in HIV-infected children even when ART is initiated 10 or more years after transmission [35]. Furthermore, while high frequency HIV sequence variation provides a highly successful mechanism by which the virus can evade the CD8+ T-cell response [36], the greater diversity of the paediatric CD8+ T cell repertoire may allow children to mount effective variant-specific immune responses to HLA-binding variants, whereas adults typically lack this capacity [37].

3 CD8+ T-Cell Mediated Immune Control of HIV

3.1 CD8+ T-Cell Mediated Immune Control of HIV in Adults

CD8+ T-cells play a central role in control of adult HIV infection [23, 24]. The temporal association between the appearance of HIV-specific CD8+ T-cells and the sharp decline in viraemia in early adult infection [38, 39] suggests the possibility that CD8+ T-cells mediate that decline to a viral 'setpoint', predictive of disease progression. Depletion of CD8+ T-cells in SIV-infected macaques, during acute infection, results in absence of the decline in viraemia from peak to setpoint, and during chronic infection results in a reciprocal rise in viral load [40–42]. That CD8+ T-cells are an important driving force in determining the success or failure of the immune response in controlling HIV infection is most clearly indicated by genetic studies that have shown HIV disease outcome to be strongly influenced by the particular HLA-B alleles expressed [43]. The HLA class I alleles most strongly linked with slow HIV disease progression in Caucasian populations are HLA-B*57 and HLA-B*27 [44–47], and, in African populations, HLA-B*57, HLA-B*5801 and HLA-B*8101 [43, 48–50]. HLA class I alleles most strongly associated with rapid progression to HIV disease are HLA-B*35 in Caucasian populations [51, 52] and HLA-B*5802 and HLA-B*1801 in African populations [43, 53]. Genome-wide association studies (GWAS) provide strong evidence that the MHC region is where single polymorphisms can make the most difference to viral setpoint [54, 55]. Similar GWAS results have recently been published from analysis of adult elite controllers [56].

   HLA class I alleles present fragments of viral proteins ('epitopes') on the surface of virus-infected cells for recognition by CD8+ T-cells that express the cognate T-cell receptor for the given peptide-MHC complex. Thus, one possible explanation for the observed HLA associations with HIV disease outcome could be that the specific HIV epitopes presented by different HLA-B molecules have an important bearing on the effectiveness of the CD8+ T-cell response. A common feature of the immunodominant epitopes presented by HLA alleles that are associated with slow HIV disease progression, HLA-B*27/*57/*5801/*8101, is that these are all within the highly abundant and conserved Gag capsid (p24) protein [45, 57–61]. The immunodominant epitopes presented by disease-susceptible HLA alleles, HLA-B*35/5802/1801, however, are within the Nef and Env proteins [61, 62]. In the case of each of the Gag epitopes, 'escape' mutations are selected by the virus that allow it to evade recognition by CD8+ T-cells, and in each case these escape mutants arise at a significant cost to viral replicative capacity [63–71]. Escape mutants are either not selected as a result of the Nef and Env-specific CD8+ T-cell responses restricted by HLA-B*35/5802/1801, or they do not significantly affect viral replicative capacity [72]. A broad Gag-specific response, comprising CD8+ T-cells targeting several different Gag epitopes, is associated with greater control of viraemia [59, 60, 73–77]. This may be because escape mutations within one epitope are not selected in the presence of Gag-specific CD8+ T-cells targeting a different Gag epitope. Multiple Gag mutations may therefore be required by the virus to evade responses presented by protective MHC class I alleles [78–81], and some of these Gag mutants may sequentially reduce viral replicative capacity to the point where the virus survives only in a relatively crippled state [63, 82].

Gag-specific CD8+ T-cell responses additionally may be more effective than non-Gag-specificities because these cells can recognise and kill virally infected targets before the production of new virions [58, 83], and before *de novo* synthesis of Nef, which downregulates HLA class I, thereby reducing CD8+ T-cell efficacy [84].

Studies in the SIV-macaque model of HIV infection suggest that Gag may not be the only potential source of epitopes that can mediate successful control of viraemia [81, 85, 86]. Several studies describe an important role for polyfunctional CD8+ T-cells, which secrete multiple cytokines, in HIV-infected individuals who successfully control HIV [87–91]. The link between polyfunctionality and expression of protective alleles such as HLA-B*57/5801/8101, however, remains unclear.

HLA Class I has an additional impact on immune control of HIV via the effect of HLA-KIR combinations on antiviral NK activity. So far the principal HLA ligands identified in Caucasian populations that are involved in slower progression are HLA-Bw4-80I alleles that are ligands both for KIR3DL1 and KIR3DS1 [92–94]. These combinations are associated with reduced viral setpoint. The HLA-KIR combinations that have a similar impact on NK-mediated immune control in sub-Saharan African populations are as yet unidentified.

3.2 CD8+ T-Cell Mediated Immune Control of HIV in Children

As stated above, the majority of perinatally-infected children progress rapidly to HIV disease. In the Zvitambo study, 67% had died by 2 years of age [10]. However, approximately 15% of ART-naïve children maintain high and even increasing CD4 counts through the first years of life, in spite of persistently high viral loads, that decline only slowly over this time. It is likely that this group has been underestimated and certainly under-studied, and recent studies suggest that the median survival of these paediatric 'slow-progressors' is as long as 16 years [95].

Studies of HIV-infected children followed from birth in Durban, South Africa, suggest that CD8+ T-cells play a less clear-cut role in control of paediatric HIV infection than in adults [25, 96]. The HLA alleles HLA-B*57/5801/8101 are not associated with slow-progression in children as they are in adults. Similarly, alleles such as HLA-B*5802/1801, associated with high viraemia in adults [43, 53, 97], and therefore predisposing to MTCT [98, 99], are well represented in the paediatric slow-progressor group. In the Durban cohort, paediatric slow-progression was linked with either the mother *or* the child having one of HLA-B*57/5801/8101 [96]. Sharing of 'protective' alleles by mother and child does not benefit the child because allele protectivity is lost if the transmitted virus carries escape mutants in the relevant epitopes [63, 100, 101]. Paediatric slow-progressors expressing HLA-B*57/5801/8101 benefit from these alleles through targeting Gag epitopes [96]. Paediatric slow-progressors whose mothers expressed HLA-B*57/5801/8101 benefited by acquiring a transmitted virus that carried multiple Gag mutants reducing viral replicative capacity [96, 102]. However, most paediatric slow-progressors in Durban [96] and Kimberley (data not published) in South Africa have none of the protective HLA alleles, HLA-B*57/5801/8101, and nor do their mothers. Study of paediatric slow-progressors therefore provides an opportunity to determine how immune control can be achieved in the setting of HLA alleles such as HLA-B*5802 that are not typically associated with immune control in adult infection. This is of direct relevance to HIV vaccine development to protect against adult as well as paediatric HIV disease.

4 ART Interruption Studies

4.1 ART Interruption Studies in HIV-Infected Adults

ART-interruption studies in adults were in vogue 10 years ago but have since been abandoned. One approach was to initiate ART in acute infection, with the rationale that HIV-specific CD4+ T-cell responses would be boosted, and then discontinue ART in the setting of a robust HIV-specific T helper response [28]. Although initially promising, this approach ultimately proved unsuccessful, merely delaying the inevitable recurrence of high-level viraemia and the need for ART to be reinstituted [103, 104]. ART interruption studies in chronically infected adults resulted in an increase in deaths from all causes in those interrupting ART [105]. In the SMART study, in which adults with a CD4 count >350 cells/mL were randomised either to continuous or to episodic use of ART, those taking episodic ART had higher mortality, particularly due to non-AIDS events, such as cardiovascular and renal disease, presumably driven by increased levels of immune activation.

4.2 ART Interruption Studies in HIV-Infected Children

In HIV-infected children the rationale for ART interruption differs from that in adults in several important ways. First, lifelong ART, initiated in the first weeks or months of life, is unsustainable (for the reasons set out above), so an alternative seems essential. In contrast, adults infected in their thirties or later might not require ART for 10 years, and might reasonably expect to experience few problems for decades on ART, especially as new and improved therapies continue to be developed. So whilst it is realistic to envisage a lifetime on ART for adults diagnosed in middle-age, this is unrealistic for infants diagnosed soon after birth. Second, there is a clear benefit to initiating ART in early life following perinatal infection [3], not only to minimise damage to the immune system and reduce mortality, but also to allow time for the developing immune system to reach optimal maturity. In adults, the immune system does not improve with the passage of time that a period of ART would allow. On the contrary, HIV disease progression is more rapid in older adults. Third, the greater thymic activity in children provides immunotherapeutic opportunities that are not available in adult infection. A period of time on ART provides the chance in HIV-infected children to induce immune responses that are not typically developed in natural infection and that may be highly effective against HIV, such as a robust HIV-specific CD4+ T-cell response and a broad Gag-specific CD8+ T-cell response. Induction of these responses might result in better control of HIV following discontinuation of ART than the responses emerging otherwise. In HIV-infected adults, initiation of ART is typically too late in the disease process to rescue HIV-specific CD4+ T-cell responses [28], and the likelihood of altering the character of the HIV-specific CD8+ T-cell response via immunotherapeutic interventions during a period of ART would be low. In contrast, as stated above, even when ART is started several years after perinatal HIV infection, HIV-specific CD4+ T-cell responses are rapidly regenerated [35]. Fourth, HIV-infected children may not be as susceptible to the non-AIDS morbidity and mortality that occurred in the SMART trial [105], since adults have comorbidities that may be compounded by an inflammatory milieu.

Several studies of ART interruption in HIV-infected children are currently in progress and show promise. In the small Durban study undertaken by our group, of 63 infants followed from birth, 43 were randomised to ART from birth (given either as continuous therapy or as viral load-driven structured treatment interruptions), which was then discontinued after 12 months of ART. In children coming off ART, 40% progressed rapidly but overall there was a significant benefit from early ART [106]. The subset of children most likely to benefit from early 12 month ART were those with high CD4% at diagnosis, around the time of birth [106]. A second similar but larger (n = 411) ART interruption study (CHER: Children with Early antiRetroviral therapy) is in progress in South Africa [3]: this study now compares children receiving ART for 12 m vs 24 m. The CHER trial ends in Aug 2011 and first results from that study are expected in 2012 (A Violari, personal communication). A third ART interruption study in children for which data are available is the PENTA-11 (Paediatric European Network for Treatment of AIDS) study [107]. This was a CD4-guided treatment interruption study in 109 children of median 9 years of age, who had high CD4 counts on ART, who were randomised to continuous therapy, or to a period of up to 48 weeks off therapy, until CD4<20%. No children undergoing treatment interruption died or had an AIDS-defining event, showing the intervention was safe; it appeared a particularly promising option for children starting interruption at a younger age and with higher CD4% nadir.

These preliminary data suggest that ART interruption in HIV-infected children is a viable and safe strategy, and the buying of time to allow the immature paediatric immune system time to develop for a period of time, perhaps 1–3 years, before stopping ART, may provide a substantial time off treatment for a significant number of infected children. Improving further on this strategy by vaccination or other immunotherapeutic approaches in order to induce optimally effective HIV-specific immunity in children before ART is discontinued (Fig. 2) would seem to be an option that would be well worth exploring to provide much-needed respite from the current lifetime ART-dependence for all HIV-infected children.

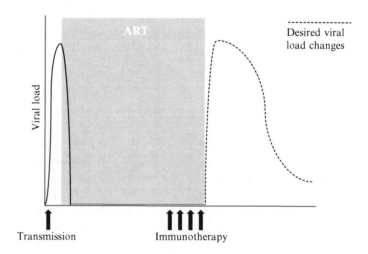

Fig. 2 Goal of early ART and immunotherapy. Scheme to show the desired impact on viral load, first, of early ART to buy time in allowing the immune response to develop in the absence of high viraemia; and, second, of the immune responses when ART is discontinued, that were induced through immunotherapeutic interventions employed at an appropriate time while still under cover of ART

5 Conclusion

The challenges presented by paediatric HIV infection are changing. High mortality rates for HIV-infected children in the first 1–2 years of life, and the unpredictability of deaths in this age group, have led to the current guidelines for initiation of ART in all HIV-infected infants diagnosed at age < 24 months. The rationale for this change in WHO guidelines is therefore strong. As ART accessibility increases in resource-poor settings where the paediatric HIV epidemic is concentrated, the scale of the problem becomes apparent. Even in a relatively well-resourced country such as South Africa, where there are up to 1000 new paediatric infections per week, the expectation of maintaining these children on ART through childhood and adolescence without encountering major problems resulting from drug resistance and/or toxicity seems unrealistic. In addition, the sheer cost of a lifetime of ART from birth, together with that of monitoring CD4 counts and viral loads would be colossal. In parallel with continuing efforts to prevent new paediatric infections arising, a strategy needs to be put in place to avert an epidemic of HIV-infected adolescents on failing salvage therapy regimens in 10 years time.

The strategy proposed here exploits the use of ART initiated in the first weeks or months of life to minimise damage to the immune system caused by HIV infection, and to allow time for the developing immune system to mature sufficiently to engage effectively with HIV. There is evidence that the paediatric immune system can successfully control HIV for some time from the significant numbers of perinatally infected, ART-naïve children who can maintain normal CD4 counts into their teenage years. Current work is focusing on determining at what age HIV-infected children are most likely to generate effective anti-viral immunity, and on developing the vaccines or immunotherapeutic agents to generate optimal efficacy in these responses.

References

1. http://www.avert.org/worldstats.htm.
2. Violari A, Cotton MF, Gibb DM, et al. Children with HIV Early Antiretroviral Therapy (CHER) Trial. Early antiretroviral therapy and mortality among HIV-infected infants. N Eng J Med. 2008;359:2233–44.
3. http://www.who.int/hiv/topics/paediatric/en/index.html.
4. Meyers T, Moultrie H, Naidoo K, Cotton M, Eley B, and Sherman G. Challenges to pediatric HIV Care and Treatment in South Africa. J Infec Dis, 2007;196:S474–S481.
5. Prendergast AJ, Tudor-Williams G, Jeena P, Burchett S and Goulder PJR. International perspectives, progress and future challenges of paediatric HIV infection. Lancet 2007.
6. Munoz A, Sabin CA, Phillips AN. The incubation period of AIDS. AIDS, 1997;11:S69–76.
7. Biggar RJ and Rosenberg PS. HIV infection/AIDS in the US during the 1990s. Clin Infec Dis 1993;17: S219–223.
8. Collaborative Group on AIDS Incubation and HIV Survival including the CASCADE EU Concerted Action on SeroConversion to AIDS and Death in Europe. Time from HIV-1 seroconversion to AIDS and death before widespread use of highly-active antiretroviral therapy: a collaborative re-analysis. Lancet 2000;355(9210): 1131–7.
9. Spira R, Lepage P, Msellati P et al. Mother-to-child HIV-1 Transmission Study Group. Natural history of HIV-1 infection in children: a five-year prospective study in Rwanda. Pediatrics 1999;104: e56.
10. Marinda E, Humphrey JH, Iliff PJ, et al. Child mortality according to maternal and infant HIV status in Zimbabwe. Ped Infec Dis J 2007;26:519–526.
11. Mphatswe W, Blanckenberg N, Tudor-Williams G, et al. High frequency of rapid immunological progression in African infants infected in the era of perinatal HIV prophylaxis. AIDS 2007;19;21:1253–61.
12. Lewis DB and Wilson CB. Developmental immunology and role of host defences in fetal and neonatal susceptibility to infection. Chapter 4 in: Remington JS, Klein JO, Wilson CB and Baker CJ eds. 7th edition. 2010.
13. SC Darby, DW Ewart, PL Giangrande, RJ Spooner and CR Rizza, Importance of age at infection with HIV-1 for survival and development of AIDS in UK haemophilia population, Lancet 1996;347:1573–1579.
14. Mayaux MJ, Burgard M, Teglas JP, Cottalorda J, Krivine A, Simon F, Puel J, Tamalet C, Dormont D, Masquelier B, Doussin A, Rouzioux C, Blanche S. Neonatal characteristics in rapidly progressive perinatally acquired HIV-1 disease. The French Pediatric HIV Infection Study Group. JAMA 1996;275(8):606–10.
15. Brinchmann JE, Albert J, Vartdal F. Few infected CD4+ T cells but a high proportion of replication-competent provirus copies in asymptomatic human immunodeficiency virus type 1 infection. J Virol 1991;65(4): 2019–23.
16. Finkel TH, Tudor-Williams G, Banda NK, Cotton MF, Curiel T, Monks C, Baba TW, Ruprecht RM, Kupfer A. Apoptosis occurs predominantly in bystander cells and not in productively infected cells of HIV- and SIV-infected lymph nodes. Nat Med 1995;1(2):129–34.
17. Appay V, Sauce D. Immune activation and inflammation in HIV-1 infection: causes and consequences. J Pathol 2008;214(2):231–41. Review.
18. Giorgi JV, Hultin LE, McKeating JA, Johnson TD, Owens B, Jacobson LP, et al. Shorter survival in advanced human immunodeficiency virus type 1 infection is more closely associated with T lymphocyte activation than with plasma virus burden or virus chemokine coreceptor usage. J Infect Dis 1999;179:859–870.
19. Mekmullica J, Brouwers P, Charurat M, Paul M, Shearer W, Mendez H, et al. Early immunological predictors of neurodevelopmental outcomes in HIV-infected children. Clin Infect Dis 2009;48:338–346.

20. Paul ME, Mao C, Charurat M, Serchuck L, Foca M, Hayani K, et al. Predictors of immunologic long-term nonprogression in HIV-infected children: implications for initiating therapy. J Allergy Clin Immunol 2005;115: 848–855.
21. Brenchley JM, Price DA, Schacker TW, Asher TE, Silvestri G, Rao S, Kazzaz Z, Bornstein E, Lambotte O, Altmann D, Blazar BR, Rodriguez B, Teixeira-Johnson L, Landay A, Martin JN, Hecht FM, Picker LJ, Lederman MM, Deeks SG & Douek DC. Microbial translocation is a cause of systenic immune activation in chronic HIV infection. Nature Medicine 2006;12:1365–1371.
22. Shearer WT, Quinn TC, LaRussa P et al. Viral load and disease progression in infants infected with HIV-1. Women and Infants Transmission Study Group. N Eng J Med 1997;336:1337–42.
23. Goulder PJR and Watkins DI. Impact of MHC class I diversity on immune control of immunodeficiency virus replication. Nature Rev Immunol 2008;8:619–630.
24. McMichael AJ, Borrow P, Tomaras GD, Goonetilleke N, Haynes BF. The immune response during acute HIV infection: clues for vaccine development. Nature Reviews Immunology 2010;10:11–23.
25. Thobakgale CF, Ramduth D, Reddy S, et al. HIV-specific CD8+ T cell activity is detectable from birth in the majority of in utero infected infants. J Virol 2007;81:12775–84.
26. Ramduth D, Thobakgale CF, Mkhwanazi NP, et al. Detection of HIV-1 Gag-specific CD4+ T cell responses in acutely infected infants AIDS Res Hum Retr 2008;24:265–270.
27. Luzuriaga K, et al. Early therapy of vertial HIV-1 infection: control of viral replication and absence of persistent HIV-specific immune responses. J Virol 2000;74:6984–6991.
28. Rosenberg ES, Altfeld M, Poon SP, et al. Immune control of HIV-1 after early treatment of acute infection. Nature 2000;407:523–526.
29. Sullender WM, et al. Humoral and cell-mediated immunity in neonates with herpes simplex virus infection. J Infec Dis 1987;155:28–37.
30. Burchett SK, et al. Diminished IFN-gamma and lymphocyte proliferation in neonatal and postpartum primary herpes simplex virus infection. J Infec Dis 1992;165:813–818.
31. Starr SE, et al. Impaired cellular immunity to cytomegalovirus in congenitally infected children and their mothers. J Infec Dis 1979;140:500–505.
32. Pass, RF, et al. Specific cell-mediated immunity and the natural history of congenital infection with cytomegalovirus. J Infec Dis 1983;148:953–961.
33. Tu, W, et al. Persistent and selective deficiency of CD4+ T cell immunity to cytomegalovirus in immunocompetent young children. J Immunol 2004;172:3260–3267.
34. Mackall CL, Fleischer TA, Brown MR, et al. Age, thymopoiesis, and CD4 T-lymphocyte regeneration after intensive chemotherapy. New Eng J Med 1995;332:143–9.
35. Feeney ME, Draenert R, Roosevelt KA, et al. Reconstitution of virus-specific CD4 proliferative responses in pediatric HIV-1 infection. J Immunol 2003;171:6968–6975.
36. Goulder PJR and Watkins DI. HIV and SIV CTL Escape: Implications for Vaccine design. Nature Reviews Immunology 2004;4:630–640.
37. Feeney ME, Tang Y, Pfafferott KJ, et al. HIV-1 viral escape in infancy followed by emergence of a variant-specific CTL response. J Immunol 2005;174:7524–30.
38. Borrow P, Lewicki H, Hahn BH, Shaw GM and Oldstone MBA. Virus-specific CD8+ cytototoxic T lymphocyte activity associated with control of viraemia in primary HIV infection. J Virol 1994;68:6103–6110.
39. Koup RA, Safrit JT, Cao Y, Andrew CA, et al. Temporal association of cellular immune responses with the initial control of viremia in primary HIV infection. J Virol 1994;68:4650–5.
40. Schmitz JE, Kuroida MJ, Santra S, Sasseville VG, Simon MA et al. Control of viremia in simian immunodeficiency virus infection by CD8+ lymphocytes. Science 1999;283:857–60.
41. Matano T, Shibata R, Siemon C et al. Administration of anti-CD8 monoclonal antibody interferes with the clearance of chimeric SIV/HIV during primary infections of rhesus macaques. J Virol 1998;72:164–9.
42. Jin X, Bauer DE, Tuttleton SE, Lewin S, Gettie A, Blanchard J, Irwin CE, Safrit JT, Mittler J, Weinberger L, Kostrikis LG, Zhang L, Perelson AS, Ho DD: Dramatic rise in plasma viremia after CD8(+) T cell depletion in simian immunodeficiency virus-infected macaques. J Exp Med 1999;189:991–8.
43. Kiepiela P, Leslie AJ, Honeyborne I, et al. Dominant influence of HLA-B in mediating the potential co-evolution of HIV and HLA. Nature 2004;432:769–774.
44. Kaslow RA, Carrington M, Apple R, et al. Influence of combinations of human major histocompatibility complex genes on the course of HIV infection. Nat Med 1996;2:405–11.
45. Goulder PJR, Phillips RE, Colbert R, et al. Late escape from an immunodominant cytotoxic T lymphocyte response associated with progression to AIDS. Nature Medicine 1997;3:212–217
46. Migueles SA, Sabbaghian MS, Shupert WL et al. HLA-B*5701 is highly associated with restriction of virus replication in a subgroup of HIV-infected long-term non-progressors. PNAS 2000;97:2709–14.
47. O'Brien SJ, Gao X and Carrington M. HLA and AIDS: a cautionary tale. Trends in Molecular Medicine 2001;7: 379–81.

48. Lazaryan A, Lobashevsky E, Mulenga J, Karita E, Allen S, Tang J et al. Human leukocyte antigen B58 super-type and HIV infection in native Africans. J Virol 2006;80:6056–60.

49. Shrestha S, Aissani B, Song W, Wilson CM, Kaslow R and Tang J. Host genetics and HIV viral set-point in African-Americans. AIDS 2009;23:673–7.

50. Koehler, R. N., A. M. Walsh, E. Saathoff, S. Tovanabutra, M. A. Arroyo, J. R. Currier, L. Maboko, M. Hoelsher, M. L. Robb, N. L. Michael, F. E. McCutchan, J. H. Kim, and G. H. Kijak. 2010. Class I HLA-A*7401 Is Associated with Protection from HIV-1 Acquisition and Disease progression in Mbeya, Tanzania. J Infec Dis 2010;202:1562–6.

51. HLA and HIV-1: heterozygote advantage and B*35-Cw*04 disadvantage. Carrington M, Nelson GW, Martin MP et al. Science 1999;283:1748–52.

52. Gao X, Nelson GW, Karacki P, et al. Effect of a single amino acid change in MHC class I molecules on the rate of HIV disease progression. New Eng J Med 2001;344:1668–75.

53. Ngumbela KC, Day CL, Mncube Z, et al. Targeting of a CD8 T Cell Env Epitope Presented by HLA-B*5802 Is Associated with Markers of HIV Disease Progression and Lack of Selection Pressure. AIDS Res Hum Retroviruses 2008;24(1):72–82.

54. Fellay J, Shianna KV, Ge D, et al. A whole-genome study of major determinants for host control of HIV-1. Science 317:944–947.

55. Fellay J, Ge D, Shianna KV et al. Common genetic variation and the control of HIV-1 in humans. PLoS Genet 2009,5:e1000791.

56. Pereyra F, et al, The International Controllers Study. The major determinants of HIV-1 control affect HLA class I peptide presentation. Science 2010;Nov 4. [epub ahead of print].

57. Goulder PJR, Bunce M, Krausa P et al. Novel, cross-restricted, conserved and immunodominant cytotoxic T lymphocyte epitopes in slow p[ropgressors in HIV-1 infection. AIDS Res Hum Retr 1996;10:1691–8.

58. Payne RP, Kløverpris H, Sacha JB, et al. Efficacious early antiviral activity of HIV Gag- and Pol-specific HLA-B*2705-restricted CD8+ T-cells. J Virol 2010;84:10543–57.

59. Kiepiela P, Ngumbela K, Thobakgale C, et al. CD8+ T cell responses to different HIV proteins have discordant associations with viral load. Nature Medicine 2007;13:46–54.

60. Honeyborne I, Prendergast A, Pereyra F, et al. Control of human immunodeficiency virus type 1 is associated with HLA-B*13 and targeting of multiple Gag-specific CD8+ T-cell epitopes J Virol 2007;81:3667–72.

61. Streeck H, Lichterfeld M, Alter G et al. Recognition of a defined region within p24 Gag by CD8+ T cells during primary HIV-1 infection in individuals expressing protectiove HLA class I alleles. J Virol 2007;81:7725–31.

62. Jin X, Gao X, Ramanathan M Jr, et al. HIV-specific CD8+_ T-cell responses for groups of HIBV-infected individuals with different HLA-B*35 genotypes. J Virol 2002;76:12603–10.

63. Crawford H, Lumm W, Leslie A, et al. Evolution of HLA-B*5703 HIV-1 escape mutations in HLA-B*5703-positive individuals and their transmission recipients. J Exp Med 2009;206:909–21.

64. Leslie A, Pfafferott KJ, Chetty P, et al. HIV evolution: CTL escape mutation and reversion after transmission. Nature Medicine 2004;10:282–9.

65. Martinez-Picado J, Prado JG, Fry EE, et al. Fitness cost of escape mutation in p24 Gag in association with control of HIV-1. J Virol 2006;80:3617.

66. Brockman MA, Schneidewind A, Lahaie M, et al. Escape and compensation from early HLA-B57-mediated cytotoxic T-lymphocyte pressure on human immunodeficiency virus type 1 Gag alter capsid interactions with cyclophilin A. J Virol 2007;81:12608–18.

67. Schneidewind A, Brockman MA, Yang R, et al. Escape from the Dominant HLA-B27 Restricted CTL Response in Gag is Associated with a Dramatic Reduction in HIV-1 Replication. J Virol 2007;81:12382–93.

68. Crawford H, Prado JG, Leslie A, et al. Compensatory mutation partially restores fitness and delays reversion of escape mutation with the immunodominant HLA-B*5703-restricted Gag epitope in HIV-1 infection. J Virol 2007;81:8346–51.

69. Prado J, Honeyborne I, Brierley I, et al. Functional consequences of HIV-1 escape from an HLA-B*13-restricted CD8+ T-cell Epitope in the p1 Gag protein. J Virol 2009;83:1018–1025.

70. Wright JK, Brumme ZL, Carlson JM, et al. Gag-Protease-Mediated Replication Capacity in HIV-1 Subtype C Chronic Infection: Associations with HLA Type and Clinical Parameters. J Virol 2010;84:10820–31.

71. Matthews PC, Prendergast AJ, Leslie AJ, et al. Central role of reverting mutations in HLA associations with HIV viral setpoint. J Virol 2008;82:8548–59.

72. Troyer RM, McNevin J, Liu Y, et al. Variable fitness impact of HIV-1 escape mutations to cytotoxic T lympho-cyte response. PLoS Pathogens 2009;5(4):e1000365.

73. Julg B, Williams KL, Reddy S, et al. Enhanced anti-HIV functional activity associated with Gag-specific CD8 T-cell responses. J Virol 2010;84:5540–9.

74. Pereyra F, Addo MM, Kaufmann DE, et al. Genetic and immunologic heterogeneity among persons who control HIV infection in the absence of therapy. J Infect Dis 2008;197(4):563–71.

75. Riviere Y, McChesney MB, Porrot F, et al. Gag-specific cytotoxic responses to HIV type 1 are associated with a decreased risk of progression to AIDS-related complex or AIDS. AIDS Res Hum Retr 1995;11(8):903–7.

76. Klein MR, van Baalen CA, Holwerda AM, et al. Kinetics of Gag-specific cytotoxic T lymphocyte responses during the clinical course of HIV-1 infection: a longitudinal analysis of rapid progressors and long-term asymptomatics. J Exp Med 1995181(4):1365–72.

77. Edwards, BH, Bansal A, Sabbaj S, et al. Magnitude of functional CD8+ T-cell responses to the gag protein of human immunodeficiency virus type 1 correlates inversely with viral load in plasma. J Virol 2002;76:2298–305.

78. Friedrich, TC, Doods EJ, Yant LJ, et al. Reversion of CTL escape-variant immunodeficiency viruses in vivo. Nat Med 2004;10:275–81.

79. Matano T, Kobayashi M, Igarashi H, et al. Cytotoxic T lymphocyte-based control of simian immunodeficiency virus replication in a preclinical AIDS vaccine trial. J Exp Med 2004;199(12):1709–18.

80. Kobayashi M, Igarashi H, Takeda A, et al. Reversion in vivo after inoculation of a molecular proviral DNA clone of simian immunodeficiency virus with a cytotoxic-T-lymphocyte escape mutation. J Virol 2005;79(17):11529–32.

81. Maness NJ, Yant LJ, Chung C, et al. Comprehensive immunological evaluation reveals surprisingly few differences between elite controller and progressor Mamu-B*17-positive SIV-infected Rhesus macaques. J Virol 200882:5245–54.

82. Friedrich TC, Frye CA, Yant LJ, O'Connor DH et al. Extraepitopic compensatory substitutions partially restore fitness to SIV variants that escape from an immunodominant CTL response. J Virol 2004;78:2581–5.

83. Sacha JB, Chung C, Loffredo JT, et al. Gag-specific CD8+ T lymphocytes recognize infected cells before AIDS-virus integration and viral protein expression. J Immunol 2007;178:2746–54.

84. Collins KL, Chen BK, Kalms SA, et al. HIV-1 Nef protein protects infected primary cells against killing by cytotopxic T lymphocytes. Nature 1998;39:397–401.

85. Maness NJ, Valentine LE, May GE, Reed J, Piaskowski SM, Soma T, Furlott J, Rakasz EG, Friedrich TC, Price DA, Gostick E, Hughes AL, Sidney J, Sette A, Wilson NA and Wakins DI. AIDS virus specific CD8+ lymphocytes against an immunodominant cryptic epitope select for viral escape. J Exp Med 2007;204:2505–12.

86. Loffredo JT, Friedrich TC, Leon EJ, Stephany JJ, Rodriguez DS, Spencer SP, Bean AT, Beal DR, Burwitz BJ, Rudersdorf RA, Wallace LT, Piaskowski SM, May GE, Sidney J, Gostick E, Wilson NA , Price DA, Kallas EG, Sette A, and Wakins DI. CD8+ T cells from SIV elite controller macaques recognize Mamu-B*08 bound epitoes and select for widespread viral variation. PLoS ONE 2007;2:e1152.

87. Mackedonas G, Hutnick N, Haney D et al. Perforin and IL-2 upregulation define qualitative differences among highly functional virus-specific human T-cells. PLos Path 2010;5:e1000798.

88. Hersperger AR, Pereyra F, Nason M, et al. Perforin expression directly ex vivo by HIV-specific CD8 T-cells is a correlate of HIV elite control. PLoS Pathog. 2010;27;6(5):e1000917.

89. Betts MR, Nason MC, West SM, et al. HIV non-progressors preferentially maintain highly functional HIV-specific CD8+ T-cells. Blood 2006;107:4781–9.

90. Almeida JR, Sauce D, Price DA, et al. Antigen sensitivity is a major determinant of CD8+ T-cell polyfunctionality and HIV suppressive activity. Blood 2009;113:6351–60.

91. Migueles SA, Osborne CM, Royce C et al. Lytic granule loading of CD8+ T- cells is required for HIV-infected cell elimination associuated with immune control. Immunity 2008;29:1009–21.

92. Martin MP, Gao X, Lee JH, et al. Epistatic interaction between KIR3DS1 and HLA-B delays the progression to AIDS. Nature Genetics 2002;31:429–434.

93. Martin MP, Qi Y, Gao X, et al. Innate partnership of HLA-B and KIR3DL1 subtypes against HIV-1. Nature Genetics 2007;39:1114–9.

94. Altfeld M and Goulder PJR. Unleashed natural killers hinder HIV. Nature Genetics 2007;39:708–710.

95. Ferrand RA, Corbett EL, Wood R, et al. AIDS among older children and adolescents in Southern Africa: projecting the time course and magnitude of the epidemic. AIDS 2009;23:2039–46.

96. Thobakgale CF, Prendergast A, Crawford H, et al. Impact of HLA in Mother and Child on Paediatric HIV-1 disease progression. J Virol 2009;83:10234–44.

97. Leslie A, Matthews PC, Listgarten J, et al. Additive contribution of HLA class I alleles in the immune control of HIV-1 infection. J Virol 2010;84:9879–88.

98. Sperling RS, Shapiro DE, Coombs RW et al. Maternal viral load, zidovudine treatment, and the risk of transmission of HIV-1 from mother to infant. New Eng J Med 1996;335:1621–1629.

99. Cao Y, Krogstad P, Korber BT et al. Maternal HIV-1 viral load and vertical transmission of infection: The Ariel Project and prevention of HIV transmission form mother to infant. Nat Med 1997;3:549–552.

100. Goulder PJR, Brander C, Tang Y, et al. Evolution and transmission of stable CTL escape mutants in HIV infection. Nature 2001;412:334–8.

101. Goepfert P, Lumm W, Farmer P, et al. Transmission of HIV-1 Gag immune escape mutations is associated with reduced viral load in linked recipients. J Exp Med 2008;205:1009–17.

102. Prado JG, Prendergast A, Thobakgale C, et al. Replicative capacity of human immunodeficiency virus type 1 transmitted from mother to child is associated with pediatric disease progression rate. J Virol 2010;84:492–502.
103. Kaufmann DE, Lichterfeld M, Altfeld M et al. Supervised treatment interruption fails to control HIV infection. PLoS Medicine 2004;1:e36.
104. Markowitz M, Jin X, Hurley A, Simon V, Ramratnam B, et al. (2002) Discontinuation of antiretroviral therapy commenced early during the course of human immunodeficiency virus type 1 infection, with or without adjunctive vaccination. J Infect Dis 186:634–643.
105. El-Sadr WM, Lundgren JD, Neaton JD, et al. The Strategies for Management of Antiretroviral Therapy (SMART) Study Group. CD4+ count-guided interruption of antiretroviral treatment. N Eng J Med 2006;355: 2283–2296.
106. Prendergast AJ, Oral presentation, 14th Conference on Retroviruses and Opportunistic Infections 2008.
107. Paediatric European Network for Treatment of AIDS. Response to planned treatment interruptions in HIV infection varies across childhood. AIDS 2010;24:231–41.

How Short Is Long Enough for Treatment of Bone and Joint Infection?

Markus Pääkkönen and Heikki Peltola

1 Introduction

"We decided to continue antimicrobials just to be on the safe side" is a phrase one often hears when a patient with a severe disease is being treated. The words are not always those of an inexperienced physician in training, but those of a senior consultant. The flaw in the rationale behind this is the presumption that a longer course will benefit the patient more than a short course. Sometimes this holds true, but usually it does not. A spectacular example is meningococcal meningitis in which more than one prospective study shows that a single injection of long-acting penicillin, chloramphenicol, or in more recent studies, cephalosporin cures the great majority of patients [1, 2]. However, this regimen should certainly not be tried for other types of bacterial meningitis.

Historically acute osteoarticular infections of childhood – osteomyelitis (OM), septic arthritis (SA), and OM with adjacent SA (OMSA) – used to be fatal or otherwise devastating diseases. Before the era of antimicrobials, the attending physician (usually a surgeon) had a dilemma: if you operated on the patient immediately, mortality was higher but sequelae in survivors developed less frequently – and vice versa: if you waited for a week or so, the risk of death decreased but the child's chances of being left crippled increased [3]. The same destructive effect has been shown in animal models [4]. An elegant analysis of the pathogenesis of OM almost a century ago [5] was realistic when commenting as follows on the role of operation (which was virtually the only thing a doctor could do) in the treatment: "However much one may dislike it, it is necessary to appreciate the extent to which, in appropriate instances, operation can and frequently does do more harm than good."

2 Role of Antimicrobials

The advent of sulphonamides in the late 1930s was a major step forward in the treatment of osteoarticular as well as other bacterial infections. The American, Frank Dickson, commenting on the article by Penberthy and Weller, 1941 [6], stated: "If the results of the use of chemotherapy continue to be as promising as they are at the present time, I believe that we may look upon it as a boon in

M. Pääkkönen (✉)
Turku University Hospital, Turku, Finland
e-mail: markus.paakkonen@helsinki.fi

H. Peltola
Children's Hospital, Helsinki University Central Hospital, University of Helsinki, Helsinki, Finland
e-mail: heikki.peltola@hus.fi

N. Curtis et al. (eds.), *Hot Topics in Infection and Immunity in Children VIII*,
Advances in Experimental Medicine and Biology 719, DOI 10.1007/978-1-4614-0204-6_4,
© Springer Science+Business Media, LLC 2011

the treatment of this disease that … has, heretofore, left behind it a trail of more or less incapacitated and crippled individuals." In the same paper we find that sulphonamides, as an adjuvant therapy, were combined with surgery which was carried out to all patients soon after arrival in hospital. Of note was the observation that medication could mostly be administered orally. The effect of sulphonamides was so dramatic that mortality rates which had been exceeding 20% in 1934–1936, fell to 9% in 1936 and nil in 1939.

Also, the duration of medical treatment used was typically very short (Table 1). One 6 year-old girl with proven *Staphylococcus aureus* OM in the tibia received sulphapyridine for just 3 days and "healed in 6 weeks" [6]. Another contemporary article [7] recommended penicillin discontinuation 5 days after the defervescence of the patient. This early experience from the dawn of the era of antimicrobials, when they were used in small quantities to conserve scarce supplies, showed that the treatment course need not always be prolonged, although just how to select the patients for whom a short course would be sufficient remained unclear. When facilities to produce larger quantities of antimicrobials were established in the 1950s, scarcity ceased to be an important issue in most instances.

With this experience, it is a little surprising that then, for several decades, antimicrobials were administered routinely for several weeks or even months [8]. Only recently have questions been raised as to whether this is really necessary. Lengthy courses of antimicrobials are costly, especially when expensive anti-staphylococcal agents are used, and they are not without risks [9]. As illustrated in Fig. 1 (Panels a and b), there is a moment when all the advantages of treatment have been obtained while all the negative effects have not yet accumulated. This is the optimal time to stop treatment. Unfortunately, identifying this moment is not easy in the real-life situation. There is no doubt that there has long been a great need for sufficiently-powered, prospective studies examining the ideal duration of antimicrobial treatment [10].

Still, acknowledging this need, all the information that has been gained over the decades has not yet been put into practice. For example, in many Commonwealth countries, a 6 week intravenous course for childhood OM is still a routine – despite compelling evidence showing that it is not necessary. Table 2 summarizes the current practice in nine countries from various parts of the world [11–20].

3 Risk of Reoccurrence

One spectre in the treatment of osteoarticular infections is their tendency to reoccur. This may be a true *relapse*, a new episode caused by the same agent, or *late re-infection* in which the same anatomical site undergoes infection due to different agents [21]. This second category cannot be deemed a "true" treatment failure, as once infected, an osteoarticular site remains prone to further infections for some time. An analogy with endocarditis is evident. No data exist suggesting that any prolonged medication would prevent all these late re-infections. The incidence of relapses is not known, there may be some association with the length of history and very short treatments, but 95% of relapses occur within 12 months [22]. Overall, these events are very rare and their occurrence does not justify long treatment courses for all cases. This said, in osteoarticular infections, like endocarditis, the treatment should be taylored to the needs of each patient.

An article in 1960 [23] examined relapses of acute osteomyelitis and identified two risk factors for failure: history of at least 3 days from the onset of symptoms to presentation and an antimicrobial course of 5–10 days. It came to be believed that long treatment courses and use of normalization of the erythrocyte sedimentation rate (ESR) as the yardstick for the discontinuation of antimicrobials were the lines to be followed in future. However since then numerous antimicrobials have arrived on the market, and much has been learned of their very heterogeneous pharmacokinetics.

Table 1 Duration of antimicrobial treatment (total) for childhood staphylococcal osteomyelitis after the arrival of sulphonamides (Adapted from: Penberthy CG, Weller CN. Ann Surg 1941: 129–146 [6])

Age (years), gender	History (Days)	Focus	Staphylococcus aureus isolated from	Operation	Medication[a] (Days)	Days in hospital	Outcome
11, M	14	Ilium	Site	Resection[b]	14	28	Good
8, F	6	Tibia	Site & Blood	Fenestration	6	23	Small sequestrum
11, F	3	Tibia	Site & Blood	Fenestration	8	8	Good
7, M	7	Tibia	Site & Blood	Fenestration	8	18	Good
2, F	7	Femur	Site & Blood	Arthrotomy	19	36	"Lesion treated at 1 year"[c]
4, F	3	Femur	Site & Blood	Fenestration	20	23	At 11 months, intermittent wound discharge
6, F	14	Tibia	Site & Blood	Fenestration	9	11	Good after sequestrectomy at 2 months
6, F	1	Tibia	Site	Fenestration	3	12	Good[d]
13, F	6	Tibia	Site	Fenestration	9	16	Good, "little discharge"[c]

[a] Sulphanilamide, sulphapyridine, or sulphathiazole every 4–6 h
[b] Subperiosteal resection of one-third of the ilium
[c] No more details
[d] "Healed in 6 weeks"

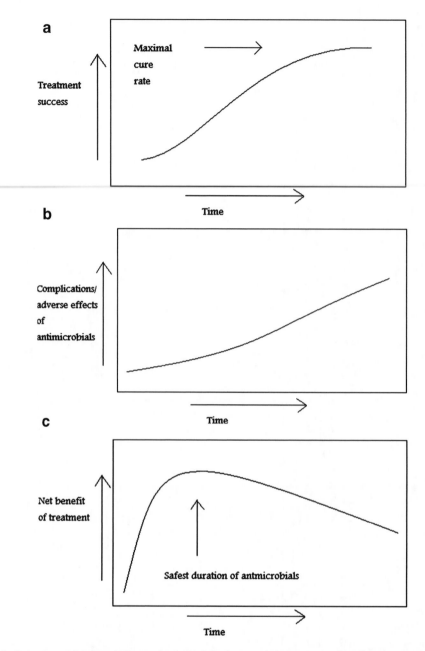

Fig. 1 As in treatments in general, the osteoarticular infections not being an exemption, there is a moment when all advantages have been obtained but not all adverse effects have yet arrived. The treatment should be stopped then. Panels A, B and C

4 Monitoring the Course of OM, SA, and OMSA

Regarding ESR, Fig. 2 [24] shows how arbitrarily ESR behaves in childhood OM and OMSA. A more reliable laboratory index is needed.

We have been measuring the serum C-reactive protein (CRP) levels sequentially now for three decades [25] and are very happy with this practice [26]. However, a necessary condition is that the

Table 2 Duration of antimicrobial treatment (total) for childhood osteoarticular infections in the 2000s

Country	Duration	Reference
Australia	3 weeks	Vinod et al. 2002 [11]
Chile	4–6 weeks	Prado et al. 2010 [12]
Finland		
Septic arthritis (SA)	10–14 days[a]	Peltola et al. 2009 [13]
Osteomyelitis w/w-out SA	3 weeks[a]	Peltola et al. 2010 [14]
France	5–6 weeks	Abuamara et al. 2004 [15]
India	3–4 weeks	Shetty and Gedalia. 2004 [16]
Iran	5 weeks	Jaberi et al. 2002 [17]
Nigeria	2–4 weeks	Eyichukwu et al. 2010 [18]
Taiwan	5 weeks	Kao et al. 2003 [19]
United States	4 weeks	Ballock et al. 2010 [20]

[a] only 2–4 days i.v., all the rest orally

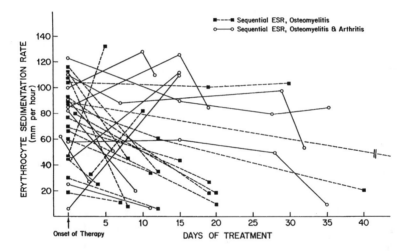

Fig. 2 Sequential measurement of the erythrocyte sedimentation rate (ESR) is a suboptimal laboratory index in the monitoring of osteoarticular infections of childhood (From: Dich, V.Q., et al. (1975) *Am J Dis Child* (129),1273–1278. Copyright © (1975) American Medical Association. All rights reserved)

test result is quantitative and arrives on the day the sample is taken since only then does it serve the patient and clinician by adding usefully to the monitoring of the course of disease. We usually discontinue antimicrobial once the level of 20 mg/L has been reached. If the patient recovers uneventfully, this normally occurs within a fortnight in SA and OMSA, and after about a week in OM [26]. The CRP measurements are currently easy to perform, as a finger/heel prick whole-blood sample suffices and the test result can be ready within minutes. We do not believe that more costly measurements of serum procalcitonin levels would add anything to the information obtained by CRP.

5 Towards Shorter Antimicrobial Courses

A clinical trial of 62 patients from the 1960s [27] was one of the first using a β-lactam, cloxacillin, in the treatment of osteoarticular infections (OM). Cloxacillin was given "arbitrarily for a period of 5 weeks". For many, this arbitrarily chosen length of antimicrobial soon became the "clinical standard" from which not much deviation occurred.

Two decades later, a large retrospective analysis from Texas [28] concluded that a course of 3–4 weeks would suffice for most cases of OM. For SA, a little less, around 3 weeks, would be fine, as earlier suggested by Tetzlaff et al. [29]. Still, an early switch from i.v. to oral administration was deemed risky, and was "allowed" only if serum concentrations were measured and good compliance was guaranteed. The i.v. period was supposed to last 1–2 weeks. Since then, antimicrobial concentrations have not much been measured, but good compliance is, of course, a continuous requirement.

6 Lessons Learned from Prospective Studies in OM, SA, and OMSA

The first sufficiently-powered, randomized and prospective study seeking for optimum durations of treatment for OM, SA, and OMSA was completed recently in Finland [13, 14]. Paucity of cases in that country was a problem, as it would have been elsewhere in the industrialized world, in contrast to resource-poor countries [16–18]. As a consequence the enrolment of patients took more than two decades. Nevertheless, more than 250 cases, all culture-positive, were randomized and ultimately analyzed. The comprehensiveness of the series permitted conclusions which were better founded than ever before. The lessons learned are summarized below, but a few *caveats* should be recognized.

First, all information derives from two antimicrobials only, clindamycin and first generation cephalosporins. Caution should be entertained if this experience is extrapolated to other agents.

Second, although *Staphylococcus aureus* was overwhelmingly the most common causative organism, no methicilllin resistance (MRSA) was encountered. Fortunately, most MRSA strains (e.g. in the USA) have retained their susceptibility to clindamycin [30].

Third, we cannot comment on treatment of, say, salmonella infections which may require considerably longer administration of medication.

Eight major lessons were learned:

1. Large doses are probably needed [29]; for clindamycin 40 mg/kg/day divided in 4 equal doses *(qid)*, and for 1st generation cephalosporins no less than 150 mg/kg/day *qid*. The same dose can be used both for i.v. and oral administration. As these agents are "time-dependent" – effectiveness correlates with the time that their plasma concentration exceeds the minimal inhibitory concentration – thus their *qid* (not *tid*) dosing is probably important.
2. Regardless of whether OM, SA, or OMSA is being treated, the initial i.v. course (if needed at all) can be short, 2–4 days or so. Supporting this, 3–5 days was also sufficient in another study [31].
3. Most cases of OM heal without a notable risk of recrudescence or sequelae with a course which lasts for 3 weeks in total – provided the clinical response is good and serum C-reactive protein (CRP) decreases to <20 mg/L in 1 or 2 weeks or so [13, 14]. The same principles apply to OMSA.
4. In "pure" SA (adjacent bone not involved), a 10–14 day total course of high-dose clindamycin or first generation cephalosporins is enough, with the same requirements as above [13].
5. Cases caused by *S. aureus* do not need special treatment [13, 14].
6. Open surgery is needed less than has been thought before. In hip SA, there is no need for routine arthrotomy [32].
7. Sequential CRP measurements provide a useful guide to clinicians [13, 14, 32]. If no decrease is observed on day four of treatment in OM, one may encounter some troubles [33]. After diagnosis, following the erythrocyte sedimentation rate is of little value [13, 14, 26].
8. Although more than 200 culture-positive cases of OM, SA or OMSA have been treated along these lines without a single recrudescence or an increased frequency of sequelae, no treatment

guarantees a 100% cure. In clinical medicine, exceptions always exist, and in those cases, you have to cope with the facts. If the signs and symptoms do no subside within days, fever continues, and CRP remains high for more than 10 days, the treatment might warrant modification. In those cases, surgery should also be considered.

Studies aiming at shortening treatments raise little interest from industry, which is one reason why research in this field has been meagre. Public resources are mostly directed towards more common diseases than OM, SA, or OMSA, which, when they occur, require considerable investment of time and resources. Fortunately, the traditional month-long i.v. course has now shrunk into a few days with no disadvantage to the patient. Clearly, our aim, as seen in panel C of Fig. 1, was in the wrong place. Consequently, we have not previously been offering maximal benefit to our patients. As clinicians and scientists, we need to meet the challenge to optimize treatment in all situations.

References

1. MacFarlane JT, Anjorin FI, Cleland PG, Hassan-King M, Tor-Agbidye S, Wali SS et al. Single injection treatment of meningococcal meningitis. 1. Long-acting penicillin. Transact R Soc Trop Med Hyg 1979;73(6):693–697.
2. Nathan N, Borel T, Djibo A, Evans D, Djibo S, Corty JF, et al. Ceftriaxone as effective as long-acting chloramphenicol in short-course treatment of meningococcal meningitis during epidemics: a randomised non-inferiority study. Lancet 2005;366(9482):308–313.
3. Kenney W. The prognosis in acute hematogenous osteomyelitis with and without chemotherapy. Surgery 1944;16:477–484.
4. Emslie KR, Nade S. Acute hematogenous staphylococcal osteomyelitis: the effects of surgical drilling and curettage in an animal model. Pathology 1986;18 (2):227–233.
5. Wilensky AO. The pathogenesis of the end results of the lesions of acute osteomyelitis. Ann Surgery 1926;84(5):651–662.
6. Penberthy GC, Weller CN. Chemotherapy as an aid in the management of acute osteomyelitis. Ann Surg 1941;114(1):129–146.
7. Compere EL, Schnute WJ, Cattell LM. The use of penicillin in the treatment of acute hematogenous osteomyelitis in children: report of twelve consecutive cases. Ann Surg 1945;122(6):954–962.
8. Carek PJ, Dickerson LM, Sack JL. Diagnosis and management of osteomyelitis. Am Fam Physician 2002;63(12):2413–2420.
9. Ceroni D, Regusci M, Pazos JM, Saunders CT, Kaelin A. Risks and complications of prolonged parenteral antibiotic treatment in children with acute osteoarticular infections. Acta Othop Belg 2003;69(5):400–404.
10. Gillespie WJ, Mayo KM. The management of acute haematogenous osteomyelitis in the antibiotic era: a study of the outcome. J Bone Joint Surg Br 1981;63-B(1):126–131.
11. Vinod MB, Matusske J, Curtis N, Graham HK, Carapetis JR. Duration of antibiotics in children with osteomyelitis and septic arthritis. J Paediatr Childh Health 2002;38(4):363–367.
12. Prado SMA, Lizama CM, Pena DA, Valenzuela MC, Viviani ST. Short duration of initial intravenous treatment in 70 pediatric patients with osteoarticular infections. Rev Chilena Infectol 2008;25(1):30–36.
13. Peltola H, Pääkkönen M, Kallio P, Kallio MJ. Prospective, randomized trial of 10 days versus 30 days of antimicrobial treatment, including a short-term course of parenteral therapy, for childhood septic arhtritis. Clin Infect Dis 2009;48(9):1201–1210.
14. Peltola H, Pääkkönen M, Kallio P, Kallio MJT. The OM-SA Study Group. Twenty versus 30 Days of antimicrobial including a short course of parenteral therapy for acute hematogenous osteomyeltiis of childhood. A randomized, controlled trial on 131 culture-positive cases. Pediatr Infect Dis J 2010;29(7):716–719.
15. Abuamara S, Louis JS, Quyard MF, Barbier-Frebourg N, Lechevallier J. Osteoarticular infection in children: evaluation of a diagnostic and management protocol. Rev Chir Orthop Reparatrice Appar Mot 2004;90(8):703–713.
16. Shetty AK, Gedalia A. Management of septic arthritis. Indian J Pediatr 2004;71(9):819–824.
17. Jaberi FM, Shahcheragni GH, Ahadzadeh M. Short-term intravenous antibiotic treatment of acute hematogenous bone and joint infections: a prospective randomized trial. J Pediatr Orthop 2002;22(3):317–320.
18. Eyichykwu GO, Onyemaechi NO, Onyegbule EC. Outcome of management of non-gonococcal septic arthritis at National Orthopaedic Hospital, Enugu, Nigeria. Niger J Med 2010;19(1):69–76.

19. Kao HC, Huang YC, Chiu CH, Chang LY, Lee ZL, Chung PW, et al. Acute hematogenous osteomyelitis and septic arthritis in children. J Microbio Infect 2003;36(4):260–265.
20. Ballock RT, Newton PO, Evans SJ, Estabrook M, Farnsworth CL, Bradley JS. A comparison of early versus late conversion from intravenous to oral therapy in the treatment of septic arthritis. J Pediatr Orthop 2009;29(6):636–642.
21. Uçkay I, Assal M, Legout L, Rohner P, Stern R, Lew D et al. Recurrent osteomyelitis caused by infection with different bacterial strains without obvious source of reinfection. J Clin Micro 2006;44(3):194–196.
22. Tice AD, Hoaglund PA, Shoultz DA. Risk factors and treatment outcomes in osteomyelitis. J Antim Chemother 2003;51(5):1261–1268.
23. Harris NH. Some problems in the diagnosis and treatment of acute osteomyelitis. J Bone Joint Surg B. 1960;42-B:535–41.
24. Dich VQ, Nelson JD, Haltalin KC. Osteomyelitis in infants and children. A review of 163 cases. Am J Dis Child 1975;129(11):1273–1278.
25. Peltola H, Räsänen JA. Quantitative C-reactive protein in relation to erythrocyte sedimentation rate, fever, and duration of antimicrobial therapy in bacteraemic diseases of childhood. J Infection 1982;5:257–267.
26. Pääkkönen M, Kallio MJT, Kallio PE, Peltola H. Sensitivity of erythrocyte sedimentation rate and C-reactive protein in childhood bone and joint infections. Clin Orthoped Relat Res 2010;468(3):861–866.
27. Green JH. Cloxacillin in treatment of acute osteomyelitis. BMJ 1967;2(5549):414–416.
28. Syrogiannopoulos GA, Nelson JD. Duration of antibiotic therapy for acute suppurative osteoarticular infections. Lancet 1988;1(8575–6):37–40.
29. Tetzlaff TR, McCracken GH Jr, Nelson JD. Oral antibiotic therapy for skeletal infections of children II. Therapy of osteomyelitis and suppurative arthritis. J Pediatr 1978;92(3):485–490.
30. Martínez-Aguilar G, Hammerman WA, Mason EO Jr, Kaplan SL. Clindamycin treatment of invasive infections caused by community-acquired, methicillin-resistant and methicillin-susceptible Staphylococcus aureus in children. J Pediatr Infect Dis 2003;22(7):593–598.
31. Jagodzinski NA, Kanwar R, Graham K, Bache CE. Prospective evaluation of a shortened regimen of treatment for acute osteomyelitis and septic arthritis in children. J Pediatr Orthop 2009; 29(5):518–525.
32. Pääkkönen M, Kallio MJT, Peltola H, Kallio PE. Pediatric septic hip with or without arthrotomy retrospective analysis of 62 consecutive nonneonatal culture-positive cases. J Pediatr Orthop B 2010;19(3):264–269
33. Roine I, Faingezicht I, Arguedas A, Herrera JF, Rodríguez F. Serial serum C-reactive protein to monitor recovery from acute haematogenous osteomyelitis in children. Pediatr Infect Dis J 1995;14(1):40–44.

What's New in Diagnostic Testing and Treatment Approaches for *Mycoplasma pneumoniae* Infections in Children?

Ken B. Waites

1 Introduction

Mycoplasma pneumoniae infections are among the most common bacterial infections that occur in children; they can sometimes be severe, spread systemically, and evoke autoimmune reactions affecting multiple organ systems. Despite the potential for significant illness in some children, a microbiological confirmation is often not sought because of the perception among some physicians that there are no satisfactory diagnostic tests suitable for the acute care setting, that treatment with macrolide antibiotics is always curative, and that the majority of infections, even if untreated, will eventually resolve without sequelae. While these beliefs that are held by many primary care physicians and reinforced in medical education programs to some extent, there is much that has been learned about mycoplasmal respiratory infections in the past several years that warrants reconsidering these concepts.

An illustrative example of how mycoplasmal respiratory infection can cause problems both from the diagnostic as well as the therapeutic standpoints comes from a case that was encountered in Alabama, USA in 2009 [1].The patient was a ten year-old boy who developed fever, sore throat, a bullous rash and stomatitis, prompting hospitalization with the diagnosis of Stevens-Johnson Syndrome. Despite initial treatment with vancomycin and later with ceftriaxone and azithromycin, he continued to have high fever with worsening cutaneous and oral lesions over a 1 month period. Chest radiographs indicated pneumonia with pleural effusion. Upon transfer to a tertiary care medical center, IgM, IgG and IgA antibodies and a polymerase chain reaction (PCR) assay from a throat swab were positive for *M. pneumoniae*. Additional PCR assays confirmed the presence of a point mutation in the ribosomal RNA gene known to confer high-level macrolide resistance. He was then treated with levofloxacin and recovered. The upcoming discussion addresses some pertinent matters regarding clinical recognition and diagnostic testing that should be considered by primary care physicians who encounter children with *M. pneumoniae* infections and the emerging problem of macrolide resistance that potentially complicates management.

K.B. Waites (✉)
Department of Pathology, University of Alabama at Birmingham, Alabama, USA
e-mail: waiteskb@uab.edu

N. Curtis et al. (eds.), *Hot Topics in Infection and Immunity in Children VIII*,
Advances in Experimental Medicine and Biology 719, DOI 10.1007/978-1-4614-0204-6_5,
© Springer Science+Business Media, LLC 2011

2 Epidemiology of *Mycoplasma* Respiratory Infections

M. pneumoniae infection is not a reportable disease and is infrequently confirmed by laboratory testing on a routine basis, except in seriously ill patients requiring hospitalization in locations where diagnostic testing is available. Therefore, data concerning its prevalence are somewhat sparse for many parts of the world and are limited mainly to information obtained from studies that are conducted to evaluate a specific antimicrobial treatment regimen or targeted surveillance for specific conditions such as community acquired pneumonia (CAP). Since *M. pneumoniae* can produce diseases in the upper respiratory tract or the lower tract as well as produce a wide array of extrapulmonary manifestations without overt respiratory disease, patients may have a highly variable presentation [2]. Thus, depending on the criteria for patient inclusion (i.e., age, type of illness), severity, and the method of diagnostic confirmation (i.e., culture, serology, PCR), surveillance studies may yield quite divergent results when considering the contribution of *M. pneumoniae* to human infections. Mycoplasmal infection is endemic in many areas, becoming epidemic from time to time, and can be transmitted by respiratory droplets from person to person, with a 1–3 week incubation period. Thus, in some years there will be more cases than others. Localized outbreaks often occur in schools, military bases, summer camps, prisons and other locations where large numbers of people are housed in close proximity to one another.

While most epidemiological data are from the United States, Europe and Japan, there does not seem to be any particular geographic or climatic relationship with *M. pneumoniae* infection and it can be considered a worldwide pathogen. The cyclical nature that is well known for *M. pneumoniae* infections is at least partly due to the fact that a primary infection does not normally provide complete protective immunity from future infections, although in adults who have had multiple mycoplasmal infections over time tend to be more likely to have illness of more modest severity than young children who are experiencing the infection for the first time. The organism may be carried in the respiratory tract for variable periods in asymptomatic persons [3].

A study from the USA conducted in the mid–1990s detected *M. pneumoniae* in 23% of CAP among children aged 3–4 years [4], whereas a study of Finnish children [5] reported its occurrence in 30% of pediatric CAP overall and in more than 50% of children aged 5 years or older, making it the single most common pathogen encountered. A French surveillance program aimed at characterization of adults and children with upper respiratory influenza-like manifestations which was conducted over several years detected *M. pneumoniae* in 2–20% of nasal swabs by PCR, often concomitantly with influenza viruses. In some time periods *M. pneumoniae* was detected in 50% of the samples [6]. Using a combination of serology and PCR, Italian investigators [7] determined that *M. pneumoniae* was present in 23% of nasopharyngeal aspirates from children with pharyngitis that were negative for *Streptococcus pyogenes*. A subsequent study by the same authors [8] found *M. pneumoniae* to be more common than *S. pyogenes,* second only to various respiratory viruses as a cause of pharyngitis and it tended to cause more severe lower respiratory illness if inadequately treated.

The overall message of the many studies conducted over several years to assess the relative contribution of *M. pneumoniae* as an infectious agent is that this bacterium can cause infections which range from very mild or even subclinical to severe and that they can affect persons of any age, beginning in infancy all the way to elderly adults. An important epidemiological link between mycoplasmal infections in children and elderly adults is that the latter often provide childcare for younger children and likely acquire infections as a result of that interaction.

3 Clinical Presentation and Basis for Pathogenesis

3.1 Respiratory Disease

Recognition of *M. pneumoniae* infection in children can be challenging for the pediatrician who should always have a keen suspicion for it in a child of any age with a respiratory tract infection for which another microbial etiology is not readily apparent. The key to recognition of potential cases of *M. pneumoniae* respiratory infection is to understand that illness is usually indolent, there are often other family members with current or recent illnesses of a similar nature, and there may be variable severity. The infection can involve the upper respiratory tract manifesting as sinonasal congestion, coryza, hoarseness, and sore throat, as well as the lower tract with dry hacking cough, wheezing, and pneumonia. Combined involvement of the upper and lower tract is common as are other nonspecific findings such as fever, chills, headache and earache. Chest radiographs may reveal consolidation, usually unilateral, with bibasilar streaky infiltrates, and sometimes pleural effusion, even though the child may not appear significantly ill [9].

Particular attention should be paid to children with chronic asthma who are suspected of having *M. pneumoniae* infection because of the growing body of evidence that the two conditions can be related and concern that prolonged airway dysfunction may result in children who are already at risk for lung problems even without supervening infection. Considerable work has been done in recent years to investigate the role of *M. pneumoniae* in asthma to elucidate whether it may be of primary etiologic significance or merely act as an exacerbating agent in an individual already predisposed to the condition. There are several lines of evidence supporting a role for this mycoplasma, both in underlying pathogenesis and as an exacerbator of asthma. The role in asthma has been discussed in more depth in recent publications [9, 10].

The basis of the respiratory manifestations associated with *M. pneumoniae* infection can be explained to a great extent by two inter-related pathogenic features of the organism. *M. pneumoniae* possesses a specialized terminal attachment organelle which enables it to bind to molecules on the surface of the respiratory epithelium. The cytadherence process occurs as a result of the interactions of the P1 adhesin and numerous other well characterized proteins which have been described in detail elsewhere [9, 10]. Attachment to the respiratory mucosa places the organism in close association with the host cells and thereby enables a second major process to occur involving the secretion of various chemical mediators that include hydrogen peroxide, superoxide radicals and an exotoxin, aptly named the Community Acquired Respiratory Distress Syndrome (CARDS) toxin [11]. This toxin has similar sequence homology with the pertussis toxin S1 subunit which carries out ADP ribosylation and ultimately, in conjunction with other chemical mediators, causes vacuolation, ciliostasis and exfoliation of mucosal cells. Stimulation of host cell production of proinflammatory cytokines and lymphocyte activation complete the pathogenic process of damaging the respiratory mucosa and eliciting acute inflammation which results in the characteristic symptoms of *M. pneumoniae* infection such as the dry hacking cough and the infiltrates associated with pneumonia. The potential for intracellular replication which has been documented in vitro but not in vivo, may also help explain the chronicity of mycoplasmal respiratory infections, difficulty of organism eradication, and circumvention of the host immune response [10, 12].

3.2 Extrapulmonary Manifestations

Children with hypogammaglobulinemia seem especially prone to invasive and prolonged mycoplasmal infections that may disseminate to the joints and other organs, indicating the importance of an intact humoral immune system in host defense against this organism, but children with an intact immune system may also develop severe and disseminated disease [13]. The reasons why some immunocompetent children have mild or only modest illness when infected with *M. pneumoniae* while others develop severe pneumonia, sometimes with dissemination and extrapulmonary complications, are complex and incompletely understood. In some cases, such as in the child described earlier, nonpulmonary manifestations may be more prominent than a respiratory component, which may be minimal or even nonexistent. Cytokine formation and lymphocyte activation may either minimize disease through enhancement of host defenses or exacerbate disease through development of hypersensitivity. The more vigorous the immune response and cytokine stimulation, the more severe the clinical illness and organ damage [14]. Thus, it appears that to a large degree the course and outcome of mycoplasmal infection rests in the host response.

Table 1 summarizes some of the most common extrapulmonary manifestations of *M. pneumoniae* infection and information concerning possible mechanisms. It is important to understand that extrapulmonary complications can in some instances be due to direct spread of the organisms to distant sites, presumably through the bloodstream where they elicit a local inflammatory response

Table 1 Extrapulmonary complications of *Mycoplasma pneumoniae* infection[a]

System	Clinical manifestation	Additional information and proposed mechanisms
Hematologic	Cold agglutinin IgM antibodies, aplastic and hemolytic anemia, thrombotic thrombocytopenic purpura, disseminated intravascular coagulation/shock, mononucleosis-like syndrome	Cross-reactive antibodies against erythrocyte I antigens and/or plasma proteases as well as vasculitic and/or thrombotic occlusion have been implicated in pathogenesis.
Neurologic/ocular	Ascending paralysis (Guillain-Barré Syndrome), cerebellar syndrome, polyradiculitis, cranial nerve palsy, peripheral neuropathy, coma, aseptic meningitis, encephalitis, coma, optic neuritis, anterior uveitis, iritis	Neurologic complications are common, can be severe, and may occur in absence of overt respiratory illness. Autoantibodies against myelin and leukoencephalopathy suggest autoimmune mechanisms for some conditions, but organisms have also been detected in CSF and neural tissue.
Cardiovascular	Pericarditis, endocarditis, myocarditis, pericardial effusion, cardiac tamponade, Kawasaki Disease	Autoimmunity and direct invasion have been documented in various CV diseases. Organisms have been detected in pericardial fluid.
Dermatologic	Erythematous maculopapular and vesicular-bullous rashes, Stevens-Johnson Syndrome, ulcerative stomatitis,	These are the most common nonrespiratory complications. Autoimmunity and/or direct invasion may be involved. Organisms have been detected in skin lesions.
Musculoskeletal	Polyarthropathy, myalgias, arthritis, osteomyelitis, acute rhabdomyolysis	Invasive joint infections occur in persons with antibody deficiency and normal hosts. There is a possible association with rheumatoid arthritis. Autoantibodies, vascular occlusion and direct invasion of joints have been implicated. Organisms have been detected in synovial fluid.
Gastrointestinal	Nausea, vomiting, diarrhea, cholestatic hepatitis, pancreatitis	Nonspecific GI complaints often occur in association with respiratory disease, but mechanisms have not been investigated carefully.
Urogenital	IgA nephropathy, acute glomerulonephritis, tubulointerstitial nephritis, renal failure, priapism	Autoimmunity with immune complex formation is likely predominant, but direct invasion may also occur as mycoplasmal antigens have been detected in kidney tissue by immunofluorescence.
Bloodstream	Bacteremia	Organisms are sometimes detectable by PCR in blood of patients with pneumonia.

[a] Information in table was derived from material discussed in references [2, 14, 15, 41]

through elaboration of various soluble mediators such as interleukins, interferon γ, and tumor necrosis factor α. Direct spread is proven by numerous reports in which the organisms or their DNA have been detected directly in the organs involved [2]. Though direct spread outside of the respiratory tract certainly occurs, it is likely that autoimmune reactions are equally, if not more important as causative factors in extrapulmonary diseases of some types. This includes production of autoantibodies that cross react with host antigens, generation of circulating immune complexes, and elaboration of other substances such as IgE and stimulation of mast cell degranulation [2, 15]. The presence of any of these clinical conditions should warrant consideration for a possible mycoplasmal etiology, even though they are non-specific and might be due to other infectious and sometimes non-infectious causes.

4 Diagnostic Approach

Given the mild to modest severity of many mycoplasmal respiratory infections, the numerous limitations of laboratory diagnostic procedures, their costs, turnaround time, and lack of widespread availability in many countries, it is not surprising that many physicians rely solely on clinical suspicion and empiric treatment despite the inherent limitations of this approach. The types of diagnostic tests available and their costs will vary from one country to another. As a general rule, it is reasonable to pursue a microbiological diagnosis when respiratory illness is sufficient to warrant hospitalization, if there is an unsatisfactory clinical response to empiric treatment, if the patient has underlying co-morbid conditions or immunodeficiency that would make severe and disseminated disease more likely, and when there are significant extrapulmonary symptoms present.

There are three broad categories of laboratory testing that can be applied to aid in the diagnosis of *M. pneumoniae* infection: culture, serology and PCR. This topic has been reviewed recently so only a few basic comments need to be reiterated [9, 10, 16].

4.1 Culture

Culture was developed in the 1960s but is rarely used for clinical purposes due to the fact that it requires several days to weeks for the organisms to become visible on specialized enriched media such as SP4 agar. Even in hands of the most experienced microbiologists, false negatives are common. The main value for performing cultures is that it will provide clinical isolates that can be studied further for genotyping and tested for antimicrobial susceptibilities if desired.

4.2 Serology

Serology has been the most widely used method for detection of *M. pneumoniae* infection for many years. Some assays to measure antibodies became available fairly soon after the organism was first described. The serologic approach was logical at that time since culture was so time consuming, complex, and insensitive. Serum is easy to obtain and store and the basic principles of complement fixation to measure antibody response had already been developed for some of the common viral infections when *M. pneumoniae* was first characterized. The older non-specific cold agglutinin test and labor intensive complement fixation assay have now been replaced with a variety of newer commercially sold products that include enzyme-linked immunosorbent assays (EIAs), immuno-fluorescence, and particle agglutination tests. The EIA format is the most widely used among these various methods and has been adapted to technologies that allow precise quantitation of IgM, IgG and IgA antibodies in microtiter plate format as well as single use qualitative point-of-care assays that have become popular with pediatricians in some countries.

Despite its widespread use for so many years and its general acceptance by clinicians, the development of commercial assays and publications comparing them with one another and with detection of the organism by PCR, the limitations of serology for detection of acute *M. pneumoniae* infection have become more apparent. It is likely that some assays are better than others, but their relative merits are difficult to discern because so many publications simply compare several different tests against one another or with complement fixation without really knowing what constitutes a "true positive". To appreciate fully the problems inherent to serology it is necessary to understand some basic facts concerning the host immune response to the presence of mycoplasmas in the respiratory tract.

Within about a week after primary infection with *M. pneumoniae* antibody to protein and glycolipid antigens form, peaking at 3–6 weeks and then gradually declining. In children and young adults IgM is the first antibody to develop, typically preceding IgG by about 2 weeks. As a result of reinfections over time, many adults beyond 40 years of age never develop IgM in an acute infection and there may be a very high background of IgG as well as IgM in the general population without acute mycoplasmal infection [16]. The problem of using IgM as a marker of acute infection in children is further complicated because infants under 6 months of age, often do not produce IgM. Older children may mount an IgM response erratically at variable periods after infection, if at all, and it may persist for months afterwards [16]. Another obvious limitation for some patient populations is that a functional humoral immune system is prerequisite for mounting a measureable antibody response and those whose immune systems are impaired for one reason or another appear to be at greatest risk for invasive mycoplasmal disease. Reliance on the analysis of a single acute phase serum for IgM alone or with IgG using commercial assays has unacceptable performance as demonstrated by some large studies that compared several commercial products with one another, one of which used patients with PCR-proven infections as a reference standard [17, 18]. The ability of serology to detect acute infection is improved when acute and convalescent sera collected 2–4 weeks apart are evaluated. In a study of Japanese children with pneumonia, using the Meridian ImmunoCard (Meridian Biosciences, Cincinnati, OH, USA), a rapid qualitative IgM assay suitable for point of care diagnosis, the sensitivity was 31.8% when basing results on a single test, but it rose to 88.6% when paired sera were analyzed [19]. Thus, from the numerous evaluations of some of the most commonly used serological tests in Europe and North America, it may be concluded that the main value of serology is for epidemiological, rather than diagnostic purposes since awaiting results for seroconversion from tests on paired sera collected at two separate clinic visits is not practical from a patient management standpoint.

IgA has been purported to be a useful marker of acute *M. pneumoniae* infection in children as well as in adults who do not mount an IgM because it may rise more quickly and decline sooner than either of the other classes, but very few commercial assays include IgA. A recent evaluation (unpublished) of a commercial IgA test in our laboratory that investigated serological response in children and adults with radiologically proven pneumonia who were PCR-positive for *M. pneumoniae* found that there were no IgA-positive specimens that were not also positive for IgM.

4.3 Nucleic Acid Amplification Tests

Given the shortcomings of culture and serology for diagnosis of acute infection, descriptions of PCR assays for detection of *M. pneumoniae* were one of the very first applications of this technology when it first became available in the late 1980s. Since then many assays have been described with several different gene targets that include 16 S rRNA, 16 S-23 S rRNA spacer, P1 adhesin, ATPase operon, *tuf*, *parE*, *dnaK, pdhA*, repMp1 noncoding element, and CARDS toxin gene, [10, 16]. Considerable interest has arisen to develop multiplex assays to detect other respiratory pathogens including *Legionella pneumophila* and *Chlamydophila pneumoniae* and several of these have been described [10]. During the past decade, the emphasis of nucleic acid amplification testing has shifted towards real-time PCR assays that can be completed in a single day on a single specimen that does not require viable mycoplasmas and which may be collected from a patient earlier in the course of illness than would be possible for measuring antibodies.

These real-time PCR assays can use a variety of different instrumentations and assay conditions. Real-time PCR has numerous advantages over traditional gel-based PCR assays. One advantage is improved sensitivity for detection of systemic spread of *M. pneumoniae* infection. A real-time PCR found that 15 of 29 (52%) patients with serological evidence of *M. pneumoniae* infection had a positive assay versus 0% with conventional PCR [20].

Just as the literature concerning serology is fraught with a large number of studies of questionable design in which various commercial EIAs were compared with one another and/or with complement fixation without a clear understanding of what constitutes a "true positive", such is also the case for PCR. In the past 20 years there have been more than 200 indexed publications describing the use of PCR in association with some aspect of infection with *M. pneumoniae*. The main problems in drawing conclusions about this growing body of literature is that there are so many different assay formats, targets, and assay conditions for which the accuracy has never been validated. In some cases assays of unknown sensitivity and specificity are simply compared to one another, to serology, or to culture, sometimes using patient populations that are not well-defined.

There has been no consensus of which specimen type is optimum for PCR testing (i.e. lower respiratory tract specimens or throat swabs), though all specimen types used for culture may be used [16]. In pediatric patients, throat swabs or nasopharyngeal specimens may be the only specimen available if sputum is not being produced and the patient is not sick enough to justify more invasive sampling.

Despite the obvious advantages and superior diagnostic sensitivities of PCR assays for detection of acute M. *pneumoniae* infections, their utilization has not become as widespread as one might expect, though this varies from country to country. In the United States as of 2011 there are no complete PCR assays for *M. pneumoniae* cleared for diagnostic use by the U.S. Food and Drug Administration (FDA) that are sold commercially. Some larger hospital clinical laboratories and reference laboratories offer PCR assays using their own in-house assays, or their adaptations of published PCR primers, probes, and assay conditions. The relatively high cost and necessity of shipping specimens to an off-site laboratory which may lengthen turnaround time for results may lessen enthusiasm for their use in ambulatory care settings. Lack of FDA-approved assays that are commercially manufactured significantly limits offering PCR testing in the USA since most diagnostic laboratories will be unable to perform the necessary research to develop their own assays. Even though some laboratories in the USA offer PCR testing for *M. pneumoniae,* there is no organized proficiency testing program in place so the relative quality of patient results they provide may not be possible to ascertain. In Europe there are several companies that market their products for diagnostic testing of *M. pneumoniae.* Some of these products have been compared to in-house assays with generally favorable results with regards to analytical sensitivity [21].

Loens and coworkers [16] have provided a comprehensive critical discussion of nucleic acid amplification tests used for detection of *M. pneumoniae,* including several of the most recent comparative studies of in-house as well as commercial assays used mainly in Europe. As might be expected for extremely sensitive molecular based assays for which there are no universally applied standards for performance, there has been significant discordance for organism detection, even among laboratories that use the same tests. One very promising development in Europe is the production of a PCR proficiency test panel for detection of *M. pneumoniae* produced by Quality Control Center for Molecular Diagnostics [22].

Many problems remain with the current PCR technology as mentioned above since there are no mandated standards and most assays with some exceptions in various European countries consist of in-house assays that have undergone little if any comparative validations and may not include acceptable controls to determine whether assay inhibition, contamination, or nonspecific reactions with other microorganisms exist. Since PCR can theoretically detect one colony forming unit, a relevant question is whether PCR detection of a very low number of genomes denotes clinically significant infection since it is known that some people can carry *M. pneumoniae* in absence of disease caused by it for variable periods. It has been suggested that a combination of PCR and serology may provide more accurate results and help distinguish colonization from true infection. This approach may not be realistic since it would increase the costs of testing and interpretation of results would potentially be hindered by a high background positivity of both IgG and IgM antibodies, even in children.

The future of *M. pneumoniae* diagnostics utilizing PCR technology looks promising and it should be considered the test of choice whenever microbiological confirmation is needed for patient management and the test is available. A pediatrician who has access to a diagnostic laboratory that offers PCR tests for *M. pneumoniae* should inquire about the nature of test, the gene target, how the test was validated, and whether the laboratory subscribes to any type of external proficiency testing or specimen exchange to verify the accuracy of patient results. Performing PCR assays on patients who do not fit basic criteria on which to suspect mycoplasmal infection is not recommended.

5 Macrolide Resistance and Treatment Alternatives

Clinical studies from as far back as the 1960s indicated that antibiotic treatment of children with mycoplasmal pneumonia reduces morbidity and shortens the duration of illness [23], so a course of antibiotics should be given when mycoplasmal disease is proven or strongly suspected. However, as illustrated in the case study above, treatment of *M. pneumoniae* infection is not always straightforward. The majority of mycoplasmal infections can be successfully managed on an outpatient basis. Considering that microbiological diagnosis may not be available and children with mycoplasmal respiratory infection will rarely have a sufficiently unique clinical presentation to distinguish their illness from other common bacterial and viral infections, pediatricians must provide empiric treatment in many situations when mycoplasmal infection is suspected without ever knowing the true infectious etiology. Historically, the treatment of choice for pediatric patients has been drugs in the macrolide class which are typically the most potent antimycoplasmal agents in terms of MIC values [2].

Clinical studies demonstrating the efficacy of newer macrolides such as azithromycin and clarithromycin were published in the 1990's and these agents have largely supplanted older drugs such as erythromycin due to improved tolerability, convenience of less frequent dosages and shorter treatment durations [4, 24]. The widespread use of macrolides for pediatric respiratory infections that intensified over the past decade on a worldwide basis largely aimed at management of penicillin-resistant pneumococcal infections, has likely caused collateral damage by affecting the susceptibility of *M. pneumoniae* to this class of drugs through selective pressure.

Although macrolide resistance in *M. pneumoniae* has been known to occur for more than 30 years, reports were sporadic and most often were associated with prior macrolide-treatment. Everything changed in about 2000 when reports from Japan [25–28] and later China [29, 30], Europe [31] and the USA [32] documented increased occurrence of macrolide resistance in *M. pneumoniae*. Recent reports from China indicate from 80% to more than 90% of clinical isolates now have high-level macrolide resistance with erythromycin MICs often exceeding 64 μg/ml [29, 30]. While the majority of macrolide-resistant *M. pneumoniae* have occurred in children, adults may also be affected, but to a lesser extent [33]. This disparity could possibly be related to more use of fluoroquinolones and tetracyclines as alternatives to macrolides in adults than in children, thereby causing less selective pressure for mutations.

The first reports of macrolide resistance utilized culture followed by in vitro susceptibility testing which is practical only for epidemiological surveillance purposes since several days to weeks are required to obtain the isolates and measure the MICs. A significant advantage for studying the epidemiology of macrolide resistance in *M. pneumoniae* that may also have potential benefit for management of individual cases has been the development of rapid molecular-based assays that can detect the major ribosomal RNA gene mutations that are known to confer macrolide resistance through reducing drug affinity to the 50 S ribosomal subunit [31, 32, 34]. Molecular-based assays to detect antibiotic resistance are not available on a widespread basis since none is commercially sold, but the technology is relatively straightforward, the necessary oligonucleotide primers, probes, reagents, and assay conditions have been published and the assays can be performed in any laboratory that has the necessary equipment and experience in molecular diagnostics. In our mycoplasma reference laboratory at the University of Alabama at Birmingham, USA, we routinely perform real-time PCR targeting the repMp1 gene [35] to detect *M. pneumoniae* in clinical specimens of children who have pneumonia, especially if the illness is of sufficient severity to warrant hospitalization. If this initial PCR assay is positive, a second assay [34] is performed reflexively to detect the macrolide resistance mutations, thereby providing the physician with immediate guidance for initiation of the most appropriate treatment.

Determining the true clinical significance of macrolide resistance is difficult and can be complicated by the fact that many infections are of modest severity and may improve without aggressive antibiotic treatment. Since macrolides are known to have substantial immunomodulatory activities that can reduce inflammation, there could also be a beneficial effect independent of antibacterial activity [36]. Reports from Japan described prolonged febrile illnesses necessitating substitution of drugs from other antibiotic classes due to persistent cough or worsening of chest radiographs. Two Alabama children described in another publication had severe pneumonias requiring hospitalization with supplemental oxygen requirements. Both failed azithromycin treatment and were later proven to have macrolide-resistant *M. pneumoniae* infections [34]. These children and the case study described earlier provide ample evidence that this newly emergent resistance does indeed matter in some instances.

Pediatricians faced with selection of treatment alternatives when macrolide-resistant *M. pneumoniae* infection is suspected based on poor response to macrolides or confirmed by MIC or molecular-based tests, have few therapeutic options. Tetracyclines or fluoroquinolones may be used in these settings, but neither class is deemed appropriate for young children under normal circumstances. There is no documented naturally occurring resistance to either of these antimicrobial classes in *M. pneumoniae*, so multiple representatives of either class should be effective. Levofloxacin is bactericidal against *M. pneumoniae* [37], giving it and similar drugs in this class a potential advantage over bacteriostatic drugs such as minocycline or other tetracyclines. Whether fluoroquinolones will eradicate infection sooner or reduce organism shedding more than other drug classes is not known with certainty. Although clinical experience with levofloxacin in pediatrics is limited, a pharmacokinetic study recommended that children 5 years or older need a daily dose of 10 mg/kg, whereas children 6 months up to 5 years should receive 10 mg/kg every 12 h in order to obtain exposures associated with clinical effectiveness and safety in adults [38]. Recommended minocycline dosages for children 9 years or older are 4 mg/kg initially followed by 2 mg/kg every 12 h.

Due to the nature of their mechanisms of action, none of the beta lactams, sulfonamides, trimethoprim or rifampin may be considered active against *M. pneumoniae*. Although lincosamides such as clindamycin may appear active somewhat active in vitro, previous experience is that it does not work in vivo and should not be considered as a therapeutic option [39].

Our recent experience with an investigational fluoroketolide, CEM 101 (Cempra Pharmaceuticals) indicate it has potential for use in treatment of macrolide-resistant *M. pneumoniae*. Even though its affinity to ribosomes with mutated rRNA is reduced in comparison to wild-type, its in vitro potency is so much greater overall, that it is reasonable to expect it might be useful clinically. Two *M. pneumoniae* clinical isolates with azithromycin MICs of ≥ 32 µg/ml had MICs for CEM 101 of 0.5 µg/ml [40]. In view of the worldwide emergence of macrolide resistance in *M. pneumoniae* that mainly affects children; the pharmaceutical industry should certainly be evaluating new agents suitable for pediatric use. Use of adjunctive treatments for management of extrapulmonary disease, (i.e., steroids for neurological complications) has been discussed elsewhere [2] and need not be reiterated here.

References

1. Atkinson TP, Boppana S, Theos A, CLements LS, Xiao L, Waites K. Stevens-Johnson Syndrome in a patient with macrolide-resistant *Mycoplasma pneumoniae*. Pediatrics 2011;127:e1605–1609.
2. Waites KB, Talkington DF. *Mycoplasma pneumoniae* and its role as a human pathogen. Clin Microbiol Rev 2004;17(4):697–728.
3. Foy HM. Infections caused by Mycoplasma pneumoniae and possible carrier state in different populations of patients. Clin Infect Dis 1993;17 Suppl 1:S37–46.
4. Block S, Hedrick J, Hammerschlag MR, Cassell GH, Craft JC. *Mycoplasma pneumoniae* and *Chlamydia pneumoniae* in pediatric community-acquired pneumonia: comparative efficacy and safety of clarithromycin vs. erythromycin ethylsuccinate. Pediatr Infect Dis J 1995;14(6):471–7.
5. Korppi M, Heiskanen-Kosma T, Kleemola M. Incidence of community-acquired pneumonia in children caused by *Mycoplasma pneumoniae*: serological results of a prospective, population-based study in primary health care. Respirology 2004;9(1):109–14.
6. Layani-Milon MP, Gras I, Valette M, Luciani J, Stagnara J, Aymard M, et al. Incidence of upper respiratory tract *Mycoplasma pneumoniae* infections among outpatients in Rhone-Alpes, France, during five successive winter periods. J Clin Microbiol 1999;37(6):1721–6.
7. Esposito S, Blasi F, Bosis S, Droghetti R, Faelli N, Lastrico A, et al. Aetiology of acute pharyngitis: the role of atypical bacteria. J Med Microbiol 2004;53(Pt 7):645–51.
8. Esposito S, Bosis S, Begliatti E, Droghetti R, Tremolati E, Tagliabue C, et al. Acute tonsillopharyngitis associated with atypical bacterial infection in children: natural history and impact of macrolide therapy. Clin Infect Dis 2006;43(2):206–9.
9. Atkinson TP, Balish MF, Waites KB. Epidemiology, clinical manifestations, pathogenesis and laboratory detection of *Mycoplasma pneumoniae* infections. FEMS Microbiol Rev 2008;32(6):956–73.
10. Waites KB, Balish MF, Atkinson TP. New insights into the pathogenesis and detection of *Mycoplasma pneumoniae* infections. Future Microbiol 2008;3(6):635–48.
11. Kannan TR, Baseman JB. ADP-ribosylating and vacuolating cytotoxin of *Mycoplasma pneumoniae* represents unique virulence determinant among bacterial pathogens. Proc Natl Acad Sci U S A 2006;103(17):6724–9.
12. Dallo SF, Baseman JB. Intracellular DNA replication and long-term survival of pathogenic mycoplasmas. Microb Pathog 2000;29(5):301–9.
13. Roifman CM, Rao CP, Lederman HM, Lavi S, Quinn P, Gelfand EW. Increased susceptibility to Mycoplasma infection in patients with hypogammaglobulinemia. Am J Med 1986;80(4):590–4.
14. Waites KB, Simecka JW, Talkington DF, Atkinson TP. Pathogensis of *Mycoplasma pneumoniae* infections: adaptive immunity, innate immuinity, cell biology and virulence factors. In: Suttorp N, Welte T, Marre R, editors. Community-acquired pneumonia. Basel, Switzerland: Burkhauser Verlag; 2007. p. 183–99.
15. Narita M. Pathogenesis of extrapulmonary manifestations of *Mycoplasma pneumoniae* infection with special reference to pneumonia. J Infect Chemother 2010;16(3):162–9.
16. Loens K, Goossens H, Ieven M. Acute respiratory infection due to *Mycoplasma pneumoniae*: current satus of diagnostic methods. Eur J Clin Microbiol Infect Dis 2010;29:1055–69.
17. Beersma MF, Dirven K, van Dam AP, Templeton KE, Claas EC, Goossens H. Evaluation of 12 commercial tests and the complement fixation test for *Mycoplasma pneumoniae*-specific immunoglobulin G (IgG) and IgM antibodies, with PCR used as the "gold standard". J Clin Microbiol 2005;43(5):2277–85.
18. Talkington DF, Shott S, Fallon MT, Schwartz SB, Thacker WL. Analysis of eight commercial enzyme immunoassay tests for detection of antibodies to *Mycoplasma pneumoniae* in human serum. Clin Diagn Lab Immunol 2004;11(5):862–7.

19. Ozaki T, Nishimura N, Ahn J, Watanabe N, Muto T, Saito A, et al. Utility of a rapid diagnosis kit for *Mycoplasma pneumoniae* pneumonia in children, and the antimicrobial susceptibility of the isolates. J Infect Chemother 2007;13(4):204–7.

20. Daxboeck F, Khanakah G, Bauer C, Stadler M, Hofmann H, Stanek G. Detection of *Mycoplasma pneumoniae* in serum specimens from patients with mycoplasma pneumonia by PCR. Int J Med Microbiol 2005;295(4): 279–85.

21. Dumke R, Jacobs E. Comparison of commercial and in-house real-time PCR assays used for detection of *Mycoplasma pneumoniae*. J Clin Microbiol 2009;47(2):441–4.

22. Loens K, Mackay WG, Scott C, Goossens H, Wallace P, Ieven M. A multicenter pilot external quality assessment programme to assess the quality of molecular detection of *Chlamydophila pneumoniae* and *Mycoplasma pneumoniae*. J Microbiol Methods 2010;82(2):131–5.

23. McCracken GH, Jr. Current status of antibiotic treatment for *Mycoplasma pneumoniae* infections. Pediatr Infect Dis 1986;5(1):167–71.

24. Harris JA, Kolokathis A, Campbell M, Cassell GH, Hammerschlag MR. Safety and efficacy of azithromycin in the treatment of community-acquired pneumonia in children. Pediatr Infect Dis J 1998;17(10):865–71.

25. Morozumi M, Iwata S, Hasegawa K, Chiba N, Takayanagi R, Matsubara K, et al. Increased Macrolide Resistance of *Mycoplasma pneumoniae* in Pediatric Patients with Community-Acquired Pneumonia. Antimicrob Agents Chemother 2008;52(1):348–50.

26. Suzuki S, Yamazaki T, Narita M, Okazaki N, Suzuki I, Andoh T, et al. Clinical evaluation of macrolide-resistant *Mycoplasma pneumoniae*. Antimicrob Agents Chemother 2006;50(2):709–12.

27. Matsuoka M, Narita M, Okazaki N, Ohya H, Yamazaki T, Ouchi K, et al. Characterization and molecular analysis of macrolide-resistant *Mycoplasma pneumoniae* clinical isolates obtained in Japan. Antimicrob Agents Chemother 2004;48(12):4624–30.

28. Morozumi M, Hasegawa K, Kobayashi R, Inoue N, Iwata S, Kuroki H, et al. Emergence of macrolide-resistant *Mycoplasma pneumoniae* with a 23 S rRNA gene mutation. Antimicrob Agents Chemother 2005;49(6):2302–6.

29. Xin D, Mi Z, Han X, Qin L, Li J, Wei T, et al. Molecular mechanisms of macrolide resistance in clinical isolates of *Mycoplasma pneumoniae* from China. Antimicrob Agents Chemother 2009;53(5):2158–9.

30. Liu Y, Ye X, Zhang H, Xu X, Li W, Zhu D, et al. Antimicrobial susceptibility of *Mycoplasma pneumoniae* isolates and molecular analysis of macrolide-resistant strains from Shanghai, China. Antimicrob Agents Chemother 2009;53(5):2160–2.

31. Peuchant O, Menard A, Renaudin H, Morozumi M, Ubukata K, Bebear CM, et al. Increased macrolide resistance of *Mycoplasma pneumoniae* in France directly detected in clinical specimens by real-time PCR and melting curve analysis. J Antimicrob Chemother 2009;64(1):52–8.

32. Wolff BJ, Thacker WL, Schwartz SB, Winchell JM. Detection of macrolide resistance in *Mycoplasma pneumoniae* by real-time PCR and high-resolution melt analysis. Antimicrob Agents Chemother 2008;52(10):3542–9.

33. Miyashita N, Maruyama T, Kobayashi T, Kobayashi H, Taguchi O, Kawai Y, et al. Community-acquired macrolide-resistant *Mycoplasma pneumoniae* pneumonia in patients more than 18 years of age. J Infect Chemother.

34. Li X, Atkinson TP, Hagood J, Makris C, Duffy LB, Waites KB. Emerging macrolide resistance in *Mycoplasma pneumoniae* in children: detection and characterization of resistant isolates. Pediatr Infect Dis J 2009;28(8): 693–6.

35. Dumke R, Schurwanz N, Lenz M, Schuppler M, Luck C, Jacobs E. Sensitive detection of *Mycoplasma pneumoniae* in human respiratory tract samples by optimized real-time PCR approach. J Clin Microbiol 2007;45(8):2726–30.

36. Lopez-Boado YS, Rubin BK. Macrolides as immunomodulatory medications for the therapy of chronic lung diseases. Curr Opin Pharmacol 2008;8(3):286–91.

37. Duffy LB, Crabb DM, Bing X, Waites KB. Bactericidal activity of levofloxacin against *Mycoplasma pneumoniae*. J Antimicrob Chemother 2003;52(3):527–8.

38. Chien S, Wells TG, Blumer JL, Kearns GL, Bradley JS, Bocchini JA, Jr., et al. Levofloxacin Pharmacokinetics in Children. J Clin Pharmacol 2005;45(2):153–60.

39. Smilack JD, Burgin WW, Jr., Moore WL, Jr., Sanford JP. *Mycoplasma pneumoniae* pneumonia and clindamycin therapy. Failure to demonstrate efficacy. Jama 1974;228(6):729–31.

40. Waites KB, Crabb DM, Duffy LB. Comparative in vitro susceptibilities of human mycoplasmas and ureaplasmas to a new investigational ketolide, CEM-101. Antimicrob Agents Chemother 2009;53(5):2139–41.

41. Talkington DF, Waites KB, Schwartz SB, Besser RE. Emerging from obscurity: understanding pulmonary and exrrapulmonary syndromes, pathogenesis, and epidemiology of human *Mycoplasma pneumoniae* infections. In: Scheld WM, Craig WA, Hughes JM, editors. Emerging Infections 5. Washington, D.C.: American Society for Microbiology; 2001. p. 57–84.

Pigs, Poultry, and Pandemic Influenza: How Zoonotic Pathogens Threaten Human Health

Thijs Kuiken, Ron Fouchier, Guus Rimmelzwaan, Judith van den Brand, Debby van Riel, and Albert Osterhaus

1 Introduction

Emerging infections have an enormous impact on human health, food supply, economics, and the environment. Examples of pathogens causing emerging infections in humans in recent decades are Chikungunya virus, human immunodeficiency virus type 1, Ebola virus, hantavirus pulmonary syndrome virus, Hendravirus, highly pathogenic avian influenza (HPAI) H5N1 virus, Nipah virus, and SARS-coronavirus [1]. Animals, and wild animals in particular, are considered to be the source of more than 70% of all emerging infections in humans [2]. Therefore, understanding how these zoonotic pathogens—pathogens of nonhuman vertebrate animals that may be transmitted to humans under natural circumstances—emerge in the human population is critical for containing and eradicating these infections.

Zoonotic pathogens need to cross the host species barrier in order to become capable of infecting and being maintained in the human population. The host species barrier is not a simple concept: it consists of the interaction of factors that collectively limit the transmission of an infection from a donor species to a recipient species [3]. The process of crossing the host species barrier can be divided into four phases (Fig. 1): interspecies host-host contact between donor species and recipient species; pathogen-host interactions within an individual host of the recipient species, allowing replication and shedding of the pathogen; intraspecies host-host contact in the recipient species, allowing pathogen spread; and persistence in the recipient species population even during epidemic troughs [3].

A zoonotic pathogen that has successfully crossed the host species barrier to humans is avian influenza A virus. Wild waterbirds are considered to be the original reservoir for all influenza A viruses in poultry and mammals, including humans [4] (Fig. 2). From this reservoir, novel strains or even novel subtypes of influenza A virus may cross the host species barrier to humans. This may occur either directly from birds or indirectly via an intermediate host such as domestic swine. If such an influenza virus adapts sufficiently to its new human host to be efficiently transmitted, a pandemic may occur.

Until 2009, the two influenza A viruses circulating endemically in the human population were human influenza viruses of the subtypes H1N1 and H3N2, which caused annual epidemics every winter season. Two influenza viruses that successfully spread from animals to humans are HPAI H5N1 virus, originating from poultry, and pandemic H1N1 influenza (H1N1) virus, thought to originate from domestic swine. However, these two influenza viruses differ in a critical aspect: their

T. Kuiken (✉) • R. Fouchier • G. Rimmelzwaan • J. van den Brand • D. van Riel • A. Osterhaus
Department of Virology, Erasmus Medical Centre, Rotterdam, The Netherlands
e-mail: t.kuiken@erasmusmc.nl

N. Curtis et al. (eds.), *Hot Topics in Infection and Immunity in Children VIII*,
Advances in Experimental Medicine and Biology 719, DOI 10.1007/978-1-4614-0204-6_6,
© Springer Science+Business Media, LLC 2011

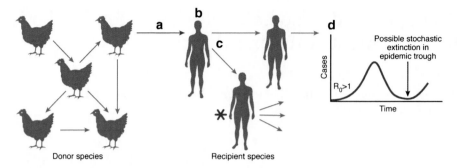

Fig. 1 Schematic illustrating phases in overcoming species barriers. (**a**) Interspecies host-host contact must allow transmission of virus from donor species to recipient species. (**b**) Virus-host interactions within an individual of recipient species affect the likelihood of the virus replicating and being shed sufficiently to infect another individual of recipient species. (**c**) Intraspecies host-host contact in recipient species must allow viral spread ($R_0 > 1$) in the presence of any pre-existing immunity. Superspreader events (*red* asterisk) early in the transmission chain can help this process. (**d**) The pathogen must persist in the recipient species population even during epidemic troughs (after most susceptible individuals have had the disease) so that subsequent epidemics can be seeded: If few susceptibles are left, the virus may (stochastically) go extinct in epidemic troughs. Viral variation and evolution can aid invasion and persistence, particularly by affecting host-virus interactions. Reprinted from *Science* [3] with permission from the American Association for the Advancement of Science

Fig. 2 Schematic representation of known events involving cross-species transmission of avian influenza viruses to mammals besides humans. Cross-species transmission of avian influenza viruses to swine, horses, harbour seals, whales and mink. The source of infection is not precisely known but is thought to be wild bird reservoirs (Anseriformes, such as ducks, and Charadriiformes, such as gulls). Poultry can become infected with avian influenza viruses and may transmit the viruses to swine and horses, when reared together. Horses have transmitted equine influenza H3N8 virus to domestic dogs. Reprinted from *Scientific and Technical Review* [11] with permission from the World Organisation for Animal Health (OIE)

human-to-human transmissibility. Between 1997 and 2010, HPAI H5N1 virus has caused 507 documented infections in humans, of whom 302 have died (www.who.int/csr/disease/avian_ influenza/country/cases_table_2010_10_18/en/index.html), but is not transmitted efficiently among humans [5]. In contrast, human-to-human transmission of pH1N1 virus is efficient: within months of the emergence of pH1N1 virus in Mexico at the beginning of 2009, it had been reported world-wide, resulting in the first human influenza pandemic of the twenty-first century.

In the past 14 years, we have studied different aspects of the host species barrier for HPAI H5N1 virus and pH1N1 virus. In this review, we will discuss the host species barrier for these viruses, concentrating on three questions: how does HPAI H5N1 virus transmit from birds to humans; what are the within-host dynamics of HPAI H5N1 virus and pH1N1 virus in humans and other mammals; and what determines transmission of influenza viruses among humans.

2 How does HPAI H5N1 Virus Transmit from Birds to Humans?

Before HPAI H5N1 virus emerged in humans in 1997 in Hong Kong, it was generally thought that avian influenza viruses had little affinity for the human respiratory tract [6]. This was based on experimental avian influenza virus infections in humans and on studies of the human trachea. How, then, could human infection with HPAI H5N1 virus, which was genetically confirmed to be completely of avian origin, be explained?

Attachment of influenza virus to its host cell is the first step in the virus replication cycle. Because virus attachment is an important determinant of the host range that a virus can infect, we determined the pattern of HPAI H5N1 virus attachment to the human lower respiratory tract (LRT). The LRT starts at the trachea, divides progressively into bronchi and bronchioles, and ends at the pulmonary alveoli. We used a technique which we coined "virus histochemistry": we incubated formalin-fixed, paraffin-embedded tissue sections with formalin-inactivated fluorescein isothiocyanate (FITC)-labeled influenza virus and detected virus with a peroxidase-labeled rabbit antibody to FITC that was amplified with a tyramide signal amplification system [7].

As expected from previous research, HPAI H5N1 virus did not attach to epithelial cells of the human trachea [7] (Fig. 3). However, HPAI H5N1 virus did attach to epithelial cells of bronchi and bronchioles. The virus also attached to type II pneumocytes and alveolar macrophages in the

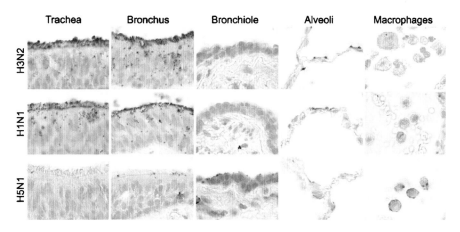

Fig. 3 Attachment of human (H3N2 and H1N1) and avian (highly pathogenic H5N1) influenza viruses in human lower respiratory tract. Reprinted from *American Journal of Pathology* [8] with permission from the American Society for Investigative Pathology

pulmonary alveoli. The pattern of virus attachment of HPAI H5N1 virus contrasted with that of the seasonal human influenza viruses of the subtypes H1N1 and H3N2 [8]. These viruses attached abundantly to epithelial cells of the human trachea and bronchi, and less abundantly to epithelial cells of bronchioles. Within the pulmonary alveoli, they attached predominantly to type I pneumocytes and not to alveolar macrophages.

The results from these studies demonstrated for the first time that an avian influenza virus was able to attach to cells of the human respiratory tract, a prerequisite for virus infection. Also, the difference in pattern of virus attachment between HPAI H5N1 virus and seasonal human H1N1 and H3N2 influenza viruses fits with the difference in primary disease presentation: tracheo-bronchitis for seasonal human influenza viruses, and diffuse alveolar damage for HPAI H5N1 virus infection.

3 What are the Within-Host Dynamics of HPAI H5N1 Virus and pH1N1 Virus in Humans and Other Mammals?

Knowledge of the within-host dynamics of an emerging pathogen in a new host species are not only important to understand virus shedding, but also to improve diagnosis and to assess pathogenicity. The within-host dynamics of HPAI H5N1 virus infection were of specific interest because of the high case fatality rate and unusual clinical presentation of human cases, as well as the wide range of mammalian host species in which HPAIV H5N1 virus infection causes severe disease. The within-host dynamics of pH1N1 virus infection were of specific interest to assess the severity of the influenza pandemic.

HPAI H5N1 virus infection in humans is unusual because it can cause different clinical symptoms than expected from a virus traditionally associated with respiratory tract infections. This is perhaps best illustrated by a case report of a child who died of HPAI H5N1 virus infection in Vietnam [9]. This child presented with severe diarrhea but no apparent respiratory illness, followed by rapidly progressive coma, leading to a clinical diagnosis of acute encephalitis. HPAI H5N1 virus was isolated from cerebrospinal fluid, fecal, throat, and serum specimens. The sibling of this child died of a similar illness, although the lack of clinical specimens did not allow diagnosis. Separately, another patient with HPAI H5N1 virus was described with an initial presentation of fever and diarrhea alone [10]. These cases emphasize that clinical surveillance of HPAI H5N1 virus infections should focus not only on respiratory illnesses, but also on clusters of unexplained deaths or severe illnesses of any kind [9].

HPAI H5N1 virus infection in mammal species other than humans also cause extra-respiratory disease. Other mammal species that have been reported dead from natural HPAI H5N1 virus infection include tigers (*Panthera tigris*), leopards (*P. pardus*), domestic cats, domestic dogs, Owston's palm civets (*Chrotogale owstoni*), a stone marten (*Mustela foina*), and an American mink (*M. vison*) [11]. At necropsy of these animals, evidence for lesions associated with HPAI H5N1 virus infection were found not only in the lungs but also in multiple extra-respiratory organs, including brain and liver [11]. In order to examine the extra-respiratory spread of HPAI H5N1 virus infection in mammals more carefully, we experimentally infected domestic cats with HPAI H5N1 virus and examined them by virological and pathological assays 7 days after inoculation [12, 13]. Severe necrosis and inflammation were present in lungs, brain, heart, kidneys, liver, and adrenal glands (Fig. 4). The presence of these pathological changes co-localized with the expression of influenza virus antigen in parenchymal cells of these organs (Fig. 4). (Interestingly, experimental infections of domestic cats with other influenza viruses, HPAI H7N7 virus [14] or pH1N1 virus [15], did not extend beyond the respiratory tract.) This study demonstrates that HPAI H5N1 virus causes systemic disease in domestic cats, with results corresponding in part to the findings of the above-described non-typical cases of fatal HPAI H5N1 virus infection in humans.

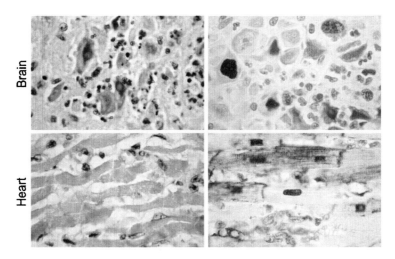

Fig. 4 Cats infected with influenza A virus (H5N1) have lesions associated with virus replication in multiple tissues, including brain and heart. Necrotizing and inflammatory changes are seen (*left* column). Serial sections of these tissues (*right* column) show that these lesions are closely associated with the expression of influenza virus antigen. Tissues were stained either with hematoxylin and eosin (*left* column) or by immunohistochemistry using a monoclonal antibody against the nucleoprotein of influenza A virus as a primary antibody (*right* column). Reprinted from *American Journal of Pathology* [13] with permission from the American Society for Investigative Pathology

When pH1N1 virus emerged in Mexico early in 2009 and started spreading in humans across the world, there was major concern about the level of disease burden and mortality that it would cause. In order to assess the pathogenesis and pathogenicity of pH1N1 virus, we performed experimental infections in ferrets, which are considered one of the most suitable laboratory animal species to model influenza in humans. In these ferrets, we compared the within-host dynamics of pH1N1 virus with that of seasonal human H1N1 virus, which is well-adapted to its human host, and of HPAI H5N1 virus, which is known to cause a high case fatality rate in humans [16]. Our results showed seasonal human H1N1 virus and pH1N1 virus were restricted mainly to the respiratory tract, while HPAI H5N1 virus also replicated extensively in extra-respiratory tissues. Also, we found that the severity of pneumonia and cumulative mortality rate of ferrets infected with pH1N1 virus was intermediate between that for seasonal H1N1 virus and HPAI H5N1 virus infection. Interestingly, influenza virus antigen expression in the pH1N1 virus group was high at all 3 levels of the lower respiratory tract (alveoli, bronchioles, and bronchi) (Fig. 5). This was associated with histopathological changes of diffuse alveolar damage, bronchiolitis, and bronchitis, respectively (Fig. 5). This contrasted with the influenza virus antigen expression in the HPAI H5N1 virus group, which was highest in alveoli, and lower in bronchioles and bronchi; and with that in the seasonal H1N1 virus group, which was low at all three levels. The results of this study suggest that pH1N1 virus has the intrinsic ability to cause more severe pneumonia than seasonal H1N1 virus. This corresponds with the results of other experimental studies in mice, ferrets, and macaques [17–19].

The pathogenesis of pH1N1 virus infection as described in above studies with laboratory animals corresponded with the pathogenesis of pH1N1 virus infection in humans. In a pathological study of 100 fatal human cases of pH1N1 virus infection, marked differences in viral tropism and tissue damage were observed compared with seasonal influenza virus infection and HPAI H5N1 virus infection [20]. In that study, patients with fatal pH1N1 virus infection not only had diffuse alveolar damage associated with the presence of viral antigen in the alveoli, a pattern somewhat similar to patients with fatal HPAI H5N1 virus infection, but also had viral localization along with inflammation and other histopathological changes in trachea, bronchi, and bronchioles, a pattern more commonly seen in severe or fatal cases of seasonal influenza.

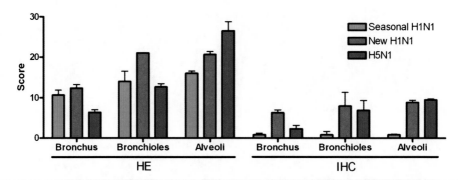

Fig. 5 Histological and immunohistochemical scoring in the lungs of ferrets inoculated with different influenza viruses. Histological scoring of samples stained with hematoxylin and eosin (HE) showed that the alveolar lesions in the new H1N1 virus (synonym for pH1N1 virus) group were intermediate in severity between those of the seasonal H1N1 virus group and the highly pathogenic avian influenza (HPAI) H5N1 virus group, and that the bronchiolar lesions in the new H1N1 virus group were the most severe of all three groups. Scoring of the immunohistochemical analysis (IHC) showed that influenza virus antigen expression in the new H1N1 virus group was high in alveoli, bronchioles, and bronchi. In the HPAI H5N1 virus group, the scores were highest for alveoli and lower in bronchioles and bronchi. In the seasonal H1N1 virus group, scores were low at all three levels. Reprinted from *Journal of Infectious Diseases* [16] with permission from Oxford University Press

4 What Determines Transmission of Influenza Viruses Among Humans?

To cause a pandemic, a zoonotic influenza virus must not only be able to cross the species barrier from animals to humans, but also to transmit efficiently among humans. pH1N1 virus, which is thought to originate from domestic swine, is transmitted efficiently among humans; in contrast, HPAIV H5N1, originating from poultry, is not. The factors determining these differences in transmission are poorly understood. One factor that may be important is tropism of the influenza virus for the human upper respiratory tract (URT) [8, 21, 22]. However, there is no consensus on this [23].

To address this question, we used virus histochemistry to compare the pattern of attachment to human URT of influenza viruses with inefficient or efficient human-to-human transmission. We used avian influenza viruses, including HPAI H5N1 virus, to represent inefficiently transmitted viruses; and seasonal human H1N1 virus, seasonal human H3N2 virus, and pH1N1 virus to represent efficiently transmitted viruses [24]. We found that the seasonal human influenza viruses and pH1N1 virus attached abundantly to epithelial cells throughout the human URT. In contrast, the avian influenza viruses, including HPAI H5N1 virus, attached only rarely [24] (Fig. 6). These results indicate that the ability of an influenza virus to attach to human URT epithelium is a critical factor for efficient transmission in the human population.

5 Conclusions

Populations of both wild and domestic animals form a vast reservoir of influenza A viruses, and provide ample opportunity for these viruses to reassort and mutate. The human population is therefore permanently at risk of becoming infected with new variants of influenza virus from this animal reservoir. To contain and eradicate such infections requires not only strategic virus surveillance of both animal and human populations, but also a better understanding of the hurdles that a zoonotic influenza virus needs to jump over in order to cross the species barrier and cause a human pandemic. Advances in these two areas will allow us to better predict the risk of emergence of zoonotic influenza viruses in the human population.

Fig. 6 Attachment of
pandemic influenza H1N1
virus (*upper* panel) and
highly pathogenic avian
influenza H5N1 virus (*lower*
panel) to ciliated epithelial
cells in the human upper
respiratory tract. Reprinted
from *American Journal of
Pathology* [24] with
permission from the
American Society for
Investigative Pathology

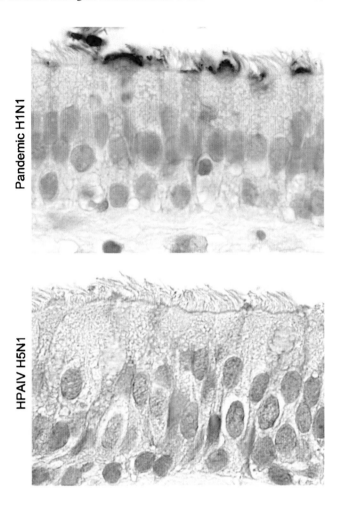

References

1. Kuiken T, Leighton FA, Fouchier RA, LeDuc JW, Peiris JS, Schudel A, Stohr K, Osterhaus AD. Public health. Pathogen surveillance in animals. Science 2005;309:1680–81.
2. Taylor LH, Latham SM, Woolhouse ME. Risk factors for human disease emergence. Philos Trans R Soc Lond B 2001;356:983–89.
3. Kuiken T, Holmes EC, McCauley J, Rimmelzwaan GF, Williams CS, Grenfell BT. Host species barriers to influenza virus infections. Science 2006;312:394–97.
4. Olsen B, Munster VJ, Wallensten A, Waldenstrom J, Osterhaus AD, Fouchier RA. Global patterns of influenza A virus in wild birds. Science 2006;312:384–88.
5. Beigel JH, Farrar J, Han AM, Hayden FG, Hyer R, de Jong MD, Lochindarat S, Nguyen TK, Nguyen TH, Tran TH, Nicoll A, Touch S, Yuen KY. Avian influenza A (H5N1) infection in humans. N Engl J Med 2005;353:1374–85.
6. Baigent SJ, McCauley JW. Influenza type A in humans, mammals and birds: determinants of virus virulence, host-range and interspecies transmission. Bioessays 2003;25:657–71.
7. van Riel D, Munster VJ, de Wit E, Rimmelzwaan GF, Fouchier RAM, Osterhaus ADME, Kuiken T. H5N1 virus attachment to lower respiratory tract. Science 2006;311:399.
8. van Riel D, Munster VJ, de Wit E, Rimmelzwaan GF, Fouchier RA, Osterhaus AD, Kuiken T. Human and avian influenza viruses target different cells in the lower respiratory tract of humans and other mammals. Am J Pathol 2007;171:1–9.

9. de Jong MD, Bach VC, Phan TQ, Vo MH, Tran TT, Nguyen BH, Beld M, Le TP, Truong HK, Nguyen VV, Tran TH, Do QH, Farrar J. Fatal avian influenza A (H5N1) in a child presenting with diarrhea followed by coma. N Engl J Med 2005;352:686–91.

10. Apisarnthanarak A, Kitphati R, Thongphubeth K, Patoomanunt P, Anthanont P, Auwanit W, Thawatsupha P, Chittaganpitch M, Saeng-Aroon S, Waicharoen S, Apisarnthanarak P, Storch GA, Mundy LM, Fraser VJ. Atypical avian influenza (H5N1). Emerg Infect Dis 2004;10:1321–24.

11. Reperant LA, Rimmelzwaan GF, Kuiken T. Avian influenza viruses in mammals. Rev Sci Tech 2009;28:137–59.

12. Kuiken T, Rimmelzwaan G, van Riel D, van Amerongen G, Baars M, Fouchier R, Osterhaus A. Avian H5N1 influenza in cats. Science 2004;306:241.

13. Rimmelzwaan G, van Riel D, Baars M, Bestebroer TM, van Amerongen G, Fouchier RAM, Osterhaus ADME, Kuiken T. Influenza A virus (H5N1) infection in cats causes systemic disease with potential novel routes of virus spread within and between hosts. Am J Pathol 2006;168:176–83.

14. van Riel D, Rimmelzwaan GF, van Amerongen G, Osterhaus AD, Kuiken T. Highly pathogenic avian influenza virus H7N7 isolated from a fatal human case causes respiratory disease in cats but does not spread systemically. Am J Pathol 2010;177:2185–90.

15. van den Brand JM, Stittelaar KJ, van Amerongen G, van de Bildt MW, Leijten LM, Kuiken T, Osterhaus AD. Experimental pandemic (H1N1) 2009 virus infection of cats. Emerg Infect Dis 2010;16:1745–47.

16. van den Brand JM, Stittelaar KJ, van Amerongen G, Rimmelzwaan GF, Simon J, de Wit E, Munster V, Bestebroer T, Fouchier RA, Kuiken T, Osterhaus AD. Severity of pneumonia due to new H1N1 influenza virus in ferrets is intermediate between that due to seasonal H1N1 virus and highly pathogenic avian influenza H5N1 virus. J Infect Dis 2010;201:993–99.

17. Munster VJ, de Wit E, van den Brand JM, Herfst S, Schrauwen EJ, Bestebroer TM, van de Vijver D, Boucher CA, Koopmans M, Rimmelzwaan GF, Kuiken T, Osterhaus AD, Fouchier RA. Pathogenesis and transmission of swine-origin 2009 A(H1N1) influenza virus in ferrets. Science 2009;325:481–83.

18. Itoh Y, Shinya K, Kiso M, Watanabe T, Sakoda Y, Hatta M, Muramoto Y, Tamura D, Sakai-Tagawa Y, Noda T, Sakabe S, Imai M, Hatta Y, Watanabe S, Li C, Yamada S, Fujii K, Murakami S, Imai H, Kakugawa S, Ito M, Takano R, Iwatsuki-Horimoto K, Shimojima M, Horimoto T, Goto H, Takahashi K, Makino A, Ishigaki H, Nakayama M, Okamatsu M, Takahashi K, Warshauer D, Shult PA, Saito R, Suzuki H, Furuta Y, Yamashita M, Mitamura K, Nakano K, Nakamura M, Brockman-Schneider R, Mitamura H, Yamazaki M, Sugaya N, Suresh M, Ozawa M, Neumann G, Gern J, Kida H, Ogasawara K, Kawaoka Y. In vitro and in vivo characterization of new swine-origin H1N1 influenza viruses. Nature 2009;460:1021–25.

19. Herfst S, van den Brand JM, Schrauwen EJ, de Wit E, Munster VJ, van Amerongen G, Linster M, Zaaraoui F, van Ijcken WF, Rimmelzwaan GF, Osterhaus AD, Fouchier RA, Andeweg AC, Kuiken T. Pandemic 2009 H1N1 influenza virus causes diffuse alveolar damage in cynomolgus macaques. Vet Pathol 2010;47:1040–47.

20. Shieh WJ, Blau DM, Denison AM, Deleon-Carnes M, Adem P, Bhatnagar J, Sumner J, Liu L, Patel M, Batten B, Greer P, Jones T, Smith C, Bartlett J, Montague J, White E, Rollin D, Gao R, Seales C, Jost H, Metcalfe M, Goldsmith CS, Humphrey C, Schmitz A, Drew C, Paddock C, Uyeki TM, Zaki SR. 2009 pandemic influenza A (H1N1): pathology and pathogenesis of 100 fatal cases in the United States. Am J Pathol 2010;177:166–75.

21. Shinya K, Ebina M, Yamada S, Ono M, Kasai N, Kawaoka Y. Influenza virus receptors in the human airway. Nature 2006;440:435–36.

22. Thompson CI, Barclay WS, Zambon MC, Pickles RJ. Infection of human airway epithelium by human and avian strains of influenza A virus. J Virol 2006;80:8060–68.

23. Nicholls JM, Chan MCW, Chan WY, Wong HK, Cheung CY, Kwong DL, Wong MP, Chui WH, Poon LL, Tsao SW, Guan Y, Peiris JS. Tropism of avian influenza A (H5N1) in the upper and lower respiratory tract. Nat Med 2007;13:147–49.

24. van Riel D, den Bakker MA, Leijten LM, Chutinimitkul S, Munster VJ, de Wit E, Rimmelzwaan GF, Fouchier RA, Osterhaus AD, Kuiken T. Seasonal and pandemic human influenza viruses attach better to human upper respiratory tract epithelium than avian influenza viruses. Am J Pathol 2010;176:1614–18.

Kingella kingae Infections in Children: An Update

Inbal Weiss-Salz and Pablo Yagupsky

1 Introduction

For most of the four decades that have elapsed since the first description of *Kingella kingae*, this Gram-negative ß-hemolytic member of the *Neisseriaceae* family was considered exceptional rare cause of human disease, infrequently isolated from infected joints, bones and cardiac valves [1–3]. The serendipitous discovery that inoculation of synovial fluid and bone exudates into blood culture vials (BCV) significantly improved detection of the organism, resulted in the appreciation of *K. kingae* as an emerging invasive pathogen in young children [4–7]. Since the last time this topic was covered in this series [8], increasing adoption of the BCV technique for culturing joint and bone aspirates and growing familiarity of clinical microbiology laboratories with the identification of the organism, coupled with the development of nucleic acid amplification techniques (NAAT) [9–11], has considerably increased our knowledge of *K. kingae*. The present review summarizes recent advances in the detection, epidemiology, clinical presentation, pathogenesis, immunology, and treatment of pediatric infections caused by the organism.

2 Microbiology

K. kingae is a facultative anaerobic, ß-hemolytic, Gram-negative member of the *Neisseriaceae* family. The organism shows a characteristic morphology and appears as pairs or short chains of plump bacilli with tapered ends and tends to resist decolorization, sometimes resulting in its misidentification as Gram-positive [12] (Fig. 1). *K. kingae* is non-motile and oxidase positive and yields negative catalase, urease, and indole reactions. It produces acid from glucose and maltose (but not from other sugars), contains alkaline and acid phosphatases [12], and can be readily identified by commercial kits such as the API NH or VITEK 2 cards (bioMerieux, Marcy-l'Etoile, France).

I. Weiss-Salz (✉)
Department of Health Services Research, Ministry of Health, Jerusalem, Israel
e-mail: dr.isalz@gmail.com

P. Yagupsky
Clinical Microbiology Laboratory, Soroka University Medical Center, Ben-Gurion
University of the Negev, Beer-Sheva, Israel
e-mail: yagupsky@bgu.ac.il

N. Curtis et al. (eds.), *Hot Topics in Infection and Immunity in Children VIII*,
Advances in Experimental Medicine and Biology 719, DOI 10.1007/978-1-4614-0204-6_7,
© Springer Science+Business Media, LLC 2011

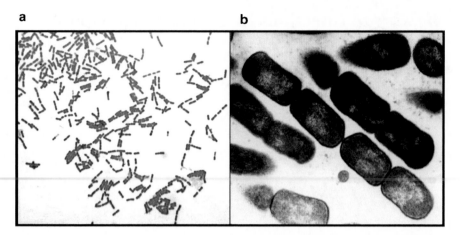

Fig. 1 (**a**) Gram stain of *K. kingae* showing pairs and short chains of Gram-negative coccobacilli; (**b**) typical morphology of *K. kingae* organisms as observed with electron microscope (original magnification × 10,000)

K. kingae grows on routine sheep blood agar and chocolate agar but fails to grow on MacConkey agar or Krigler media [12]. A CO2-enriched atmosphere enhances growth [13]. When the organism is cultured on solid media, three different colony types can be observed: a spreading/corroding type characterized by a small raised central colony surrounded by a large fringe, a non-spreading/non-corroding morphology characterized by a flat colony and a small fringe, and a dome shaped colony with no fringe [14]. These colony morphologies correlate with the level of piliation: spreading/corroding colonies are associated with dense piliation, non-spreading/non-corroding colonies are associated with sparse piliation, and dome shaped colonies are associated with no pili [14].

3 Carriage and Transmission of *K. kingae*

Asymptomatic colonization of the upper respiratory tract is the common strategy adopted by many important bacterial pathogens, such as *Streptococcus pneumoniae*, *Neisseria meningitidis*, and *Haemophilus influenzae*, to survive and spread in the human population. Organisms colonizing respiratory surfaces disseminate by contaminated respiratory secretions, establishing chains of person-to-person transmission, and also are the source of mucosal as well as invasive infections.

Recent studies have demonstrated that *K. kingae* is a normal component of the commensal oropharyngeal flora in young children [15]. Children first acquire *K. kingae* after the age of six months and the colonization rate increases to 9–12% between the ages of 12 and 24 months. Carriage of the bacterium gradually declines during the third year of life and is low in school-aged children and adults, suggesting the development of a mounting immune response that eliminates the organism from the pharynx in older individuals [15–17].

To investigate the dissemination of *K. kingae* in the community and the potential clinical significance of respiratory colonization, pharyngeal isolates from a large cohort of young healthy carriers living side-by-side in southern Israel were studied. Colonizing organisms were analyzed by pulsed field gel electrophoresis and compared to isolates recovered over a 15-year period from carriers and children with bacteremia or osteoarticular infections [16]. Significant geographic clustering of isolates exhibiting indistinguishable genotyping profiles was found in households and neighborhoods, indicating child-to-child spread of *K. kingae* between siblings and playmates. Interestingly, some isolates were identical to *K. kingae* strains recovered in the past from asymptomatic carriers and

Fig. 2 (**a**) Pulsed-field electrophoresis analysis of *K. kingae* strains (represented by colored stars) isolated in a southern Israel neighborhood. Each color represents a distinct genotype. Lanes without stars represent molecular markers and unrelated control strains. (**b**) Geographic clustering of the *K. kingae* strains in households. Each polygon represents a household and each colored star a positive isolation (Color code as in Fig. 2a)

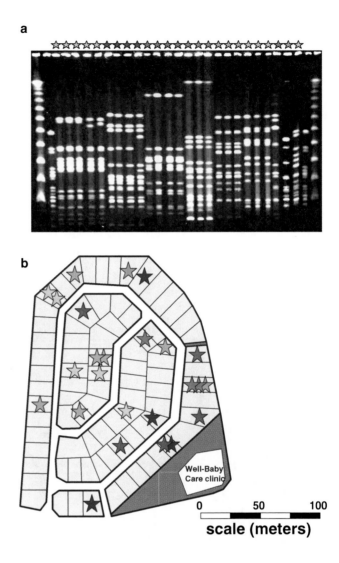

patients with invasive diseases, indicating long-term persistence and potential virulence of some of the colonizing strains [16] (Fig. 2).

In a recent report the potential implications of mucosal colonization in the pathogenesis of *K. kingae* disease was studied in three children [18]. Genotypically identical *K. kingae* organisms were isolated from pharyngeal and blood cultures in the three patients, supporting the concept that colonization of the pharynx represents the first step in the invasion of the bloodstream [18].

4 Epidemiology of Invasive *K. kingae* Infections

Based on data originated at the Soroka University Medical Center (SUMC), the only hospital serving the entire population of southern Israel and where the BCV technique has been routinely used since 1988, the annual incidence of invasive *K. kingae* disease in children aged 0–4 years living in the region was 9.4/100,000 [19]. During this 23-year period, a total of 128 patients were identified of whom 127 were children (all but one younger than 4 years), and one was an adult. The age distribution of affected

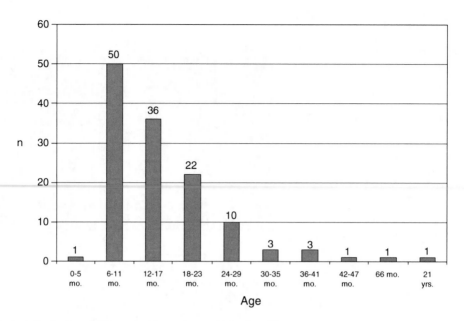

Fig. 3 Age distribution of 128 consecutive patients with invasive *K. kingae* infections diagnosed in southern Israel in the 1988–2010 period

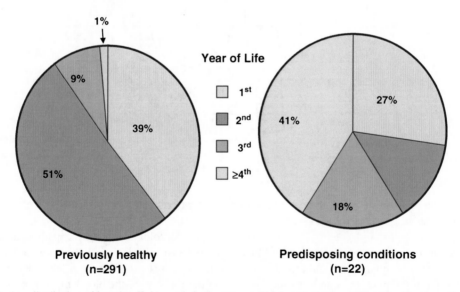

Fig. 4 Age distribution of invasive *K. kingae* infection among 313 Israeli children with and without predisposing conditions

children show that occurrence of disease below the age of 6 months is exceptional (Fig. 3). Cases rapidly accumulate thereafter reaching a peak in children aged 6–11 months. Morbidity gradually decreases between the second and forth years of life. Most children younger than 4 years who develop *K. kingae* disease are otherwise healthy. In contrast, older children and adults often suffer from underlying chronic diseases, immunodeficiency conditions, malignancy, or cardiac valve pathology [7, 13, 19] (Fig. 4).

The male-to-female ratio among 320 Israeli pediatric patients with invasive disease enrolled in a nationwide collaborative study was 1.3 (181 males and 139 females) [19]. The monthly distribution of cases showed increase occurrence between July and December with a nadir in the 3 months period February through April [19].

5 Clinical Presentation of *K. kingae* Infections

Approximately two-thirds of children with culture-proven invasive disease had an acute illness (upper respiratory tract infections, pharyngitis, aphthous stomatitis, and/or diarrhea) within the week prior to the invasive *K. kingae* infection [19]. With the exception of patients with endocarditis, most children with invasive *K. kingae* infections present in good general condition, 61% have a body temperature ≥39.0°C, and up to one quarter are afebrile [7, 13, 19]. Peripheral blood leukocyte count, C-reactive protein level, and erythrocyte sedimentation rate are generally mildly to moderately elevated, but may be entirely normal [7, 13, 19].

The old medical literature on *K. kingae* infections consisted mostly of case reports of patients in which unusual clinical manifestations such as meningitis or endocarditis were probably overrepresented. Based on prospective data systematically collected at the SUMC over more than two decades, a more precise picture can be drawn (Fig. 5). Among the 127 infections in children, 56% involved the skeletal system, followed by bacteremia with no apparent focus in 40%, bacteremia with lower respiratory tract infection in 4%, and endocarditis in 1%. Meningitis [20], ocular infections [21], peritonitis [22], and pericarditis [23] have not been detected in southern Israel, suggesting that these clinical presentations of *K. kingae* disease are rather rare.

5.1 Skeletal Infections

Among the 71 children with skeletal system involvement diagnosed in southern Israel between 1988 and 2010, 56 had septic arthritis, 6 had osteomyelitis, 1 had septic arthritis and contiguous osteomyelitis, 2 had tenosynovitis, 1 had dactylitis, and 5 presented with self-limited *K. kingae* bacteremia and transient skeletal system symptoms.

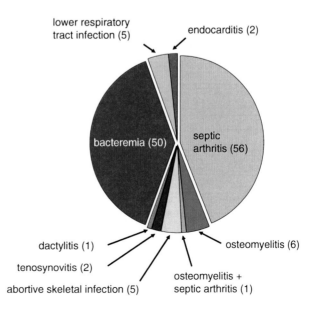

Fig. 5 Clinical presentation of 128 patients with culture-proven *K. kingae* infections

lower respiratory tract infection (5)

endocarditis (2)

bacteremia (50)

septic arthritis (56)

dactylitis (1)

tenosynovitis (2)

abortive skeletal infection (5)

osteomyelitis (6)

osteomyelitis + septic arthritis (1)

Septic arthritis caused by *K. kingae* generally involves the large joints, knee, ankle, hip, wrist, shoulder, and elbow [9, 13, 19]. However, the small metacarpophalangeal, sternoclavicular, and tarsal joints are over-represented in *K. kingae* disease compared with infections caused by other bacteria [9, 13, 19]. In the Israeli multicenter study, the mean leukocyte count in the synovial fluid of 78 children with culture-proven *K. kingae* arthritis was 130,000/mm^3, with a range of ~5,000/mm^3 to ~300,000/mm^3, and 18 (23%) had <50,000 WBC/mm^3 [19].

K. kingae osteomyelitis usually involves the long bones, including femur, humerus or tibia, although involvement of the calcaneus, talus, sternum, or clavicle, which are infrequently invaded by other skeletal system pathogens, may also occur [13, 19, 24]. Onset of *K. kingae* osteomyelitis is generally more insidious than *K. kingae* septic arthritis, frequently resulting in a considerable delay in diagnosis. Nevertheless, the overall prognosis is favorable, and chronic osteomyelitis and orthopedic sequelae are exceptional [13, 25, 26].

Nowadays, *K. kingae* is the most common etiology of hematogenous spondylodidscitis in the 6–48 months age group [27–30]. The disease usually involves the lumbar discs, and patients present with limping, low back pain, refusal to sit or walk, or neurological symptoms. X-ray or MRI studies demonstrate narrowing of the intervertebral space. Patients with *K. kingae* spondylodiscitis respond favorably to appropriate antibiotic treatment and recover without long-term complications [28, 29].

Although bone and joint infections are not considered self-limiting diseases, transient involvement of the skeletal system during an episode of *K. kingae* bacteremia may occur [24, 31, 32]. Children present with limping or refusal to walk or bear weight but without objective signs of osteoarthritis. By the time blood cultures became positive (usually after 2–4 days), the fever and skeletal complaints had resolved without antimicrobial therapy, suggesting an abortive infection. Despite the favorable experience, caution is recommended, and adequate antibiotics should be probably administered to all patients in whom the organism is recovered from a normally sterile body fluid.

5.2 Bacteremia

K. kingae bacteremia without focal infection (occult bacteremia) is the second most common presentation of *K. kingae* disease in children [7, 13, 19, 33–35]. In the study by Dubnov-Raz et al. [19], the maximal temperature measured in children with this condition was 38.8±0.8°C, only half had a body temperature ≥39°C and one-third had a leucocyte count >15,000 WBC/mL: therefore the current guidelines for managing young febrile children with no apparent focus, which rely on the height of fever and leukocyte count results for obtaining blood cultures [36], may be not sensitive enough for detecting *K. kingae* bacteremia. The clinical course of the disease is benign and patients respond favorably to short antibiotic courses [13, 33].

5.3 Endocarditis

K. kingae is included in the HACEK group of organisms that is collectively responsible for up to 5% of cases of bacterial endocarditis. In contrast to other *K. kingae* infections, endocarditis is also diagnosed in older children and adults [13]. In approximately one-half of patients the condition affects a native valve [13]. Predisposing cardiac malformations or rheumatic disease are commonly observed, but some patients have previously healthy valves [13, 37–42]. Typically the left side of the heart is involved, usually the mitral valve [13]. Despite the remarkable benign course observed in other *K. kingae* infections and the susceptibility of the organism to antibiotics, occurrence of cardiac failure, septic shock, mycotic aneurisms, pulmonary infarctions, meningitis, cerebrovascular accidents, and other life-threatening complications is common in patients with endocarditis, and the mortality rate is unusually high [13, 37–40]. Because of the potential severity of endocarditis in this setting, routine echocardiographic evaluation of all individuals from whom *K. kingae* is isolated from a normally sterile site is recommended [43].

5.4 Lower Respiratory Tract Infections

K. kingae has been isolated from the blood or respiratory secretions of previously healthy and immunocompromized adult and pediatric patients with epiglottitis, laryngotracheobronchitis, pneumonia, or pyothorax [13], suggesting that the organism may also cause lower respiratory tract infections.

6 Daycare Center Attendance as a Risk Factor for Carriage and Morbidity

These days, a growing proportion of children attend daycare outside the home. This trend has substantial public health significance because the risk of infection, mostly caused by respiratory pathogens such as pneumococci that spread horizontally by direct contact or through fomite transmission, is significantly increased in daycare center (DCC) attendees [44, 45]. Daycare attendance also appears to increase the risks for *K. kingae* colonization and transmission. In a longitudinal study, 35 of 48 (73%) children attending a DCC facility in southern Israel carried the organism at least once during an 11-month follow-up period and, on average, 28% of the children harbored the organism in the pharynx at any given time [15]. This carriage rate is significantly higher than that observed in the general population of the same age. Molecular typing of the isolates showed that *K. kingae* strains simultaneously or successively colonized multiple daycare attendees, showing person-to-person transmission of the bacterium in the DCC facility [46]. Two different genotypes represented 28% and 46% of all isolates respectively, indicating that some strains are particularly successful mucosal colonizers. DDC attendees frequently harbored the same strain continuously or intermittently for weeks or months and then replaced it by a new strain, demonstrating that carriage of *K. kingae* is a dynamic process with frequent turnover of colonizing organisms, as observed in other respiratory pathogens [46].

DCC classes comprise large numbers of children of approximately the same age and, therefore, with similar degrees of immunological immaturity and susceptibility to infectious agents. Under these circumstances, introduction of a virulent bacterium in a crowded DCC attended by immunologically naïve youngsters may result in prompt dissemination of the organism and cause outbreaks of disease. Recently, clusters of invasive *K. kingae* infections have been detected in DCCs, in the USA [37, 47] and Israel [48]. Affected attendees presented with a wide spectrum of diseases including arthritis [47], osteomyelitis [48], spondylodiscitis [37], and endocarditis [37].

When surveillance cultures were obtained from asymptomatic attendees of the DCC where the outbreaks occurred, it was found that many children were colonized by *K. kingae* organisms [37, 47, 48]. Typing of recovered *K. kingae* isolates by molecular methods demonstrated that organisms causing invasive infections disseminated widely among the healthy daycare population, indicating that the invasive outbreak strains were also highly transmissible [47, 48].

7 Immunity to *K. kingae*

In a longitudinal serological study, titers of IgG antibodies against *K. kingae* outer membrane proteins were high at 2 months of age, reached a nadir at the age of 6–7 months, remained low until the age of 18 months, and increased at 24 months [49]. Serum IgA levels against *K. kingae* outer membrane proteins were lowest at 2 months of age and were similarly increased in 24-month-old children [49]. The low attack rate of disease, absence of pharyngeal carriage, and high levels of IgG but no IgA antibodies detected in the first 6 months of life suggest that immunity to colonization and disease in early infancy is probably conferred by maternal antibodies. The high prevalence of *K. kingae* in the pharynx and the increase incidence of invasive infections among 6–24-month-old children coincide with the age at which antibody levels are lowest, suggesting that young children are unable to mount an effective immune response to the organism. Higher antibody levels among older children presumably reflect immunological maturation and cumulative exposure to *K. kingae* or with cross-reacting bacterial antigens of other species, resulting in a decline in the carriage rate and burden of clinical disease [49]. Because of the relative rare occurrence of invasive *K. kingae* infections, it is believed that asymptomatic pharyngeal colonization is the 'immunizing' event. Exposed epitopes, however, are polymorphic and the immune response elicited by carriage appears to be strain-specific and, thus, does not prevent colonization by an antigenically-different strain [50].

8 Pathogenesis of Disease

The pathogenesis of disease due to *K. kingae* is believed to begin with asymptomatic colonization of the oropharyngeal mucosa. The process involves adherence to respiratory epithelial cells, which is mediated *in-vitro* by polymeric hair-like fibers referred to as pili [51]. *K. kingae* pili are classified as type IV pili and are composed primarily of a major pilin subunit called PilA1 [51, 52]. The PilA1 protein shares homology with the major pilin subunit in type IV pili expressed by other bacterial pathogens, including *Pseudomonas aeruginosa* PilA, *N. meningitidis* PilE, and *Neisseria gonorrhoeae* PilE. In addition, *K. kingae* pili contain two minor pilins referred to as PilC1 and PilC2, which appear to play complementary roles and influence the efficiency of adherence. Similar to type IV pili expressed by other pathogens, *K. kingae* pili are regulated by the σ54 transcription factor and the PilS/PilR two-component sensor/regulator system [52]. Mutations in the PilS sensor result in reduced density of pili, while mutations in the PilR response regulator result in elimination of piliation altogether [52]. Examination of isolates of *K. kingae* demonstrated that the majority of pharyngeal isolates are piliated and express abundant pili [14].

Following colonization of the posterior pharynx, *K. kingae* must breach the epithelium to enter the bloodstream. Patients with invasive *K. kingae* disease frequently present with symptoms of a viral respiratory infection, evidence of herpetic gingivostomatitis, or concomitant buccal aphthous ulcers, suggesting that viral induced damage to the respiratory mucosa facilitates *K. kingae* penetration into the bloodstream [13, 33, 53]. Beyond the role of viral co-infection, *K. kingae* produces a potent extracellular toxin that belongs to the RTX family, capable of lysing epithelial, synovial, and macrophage cells [54]. This RTX toxin probably plays an important role in the pathogenesis of disease facilitating disruption of the respiratory epithelium, and promoting invasion of skeletal tissues [54].

After invading the bloodstream, *K. kingae* must be able to evade multiple host defense mechanisms. Most invasive pediatric pathogens produce an extracellular polysaccharide capsule that facilitates resistance to serum killing activity and phagocytosis. Analysis of the genome of *K. kingae* strain 269–492 reveals open reading frames with homology to the *ctrABCD* operon involved in polysaccharide export and encapsulation in *N. meningitidis* (St Geme JW 3 rd, unpublished data). In addition, the genome of *K. kingae* contains homologs of genes required for the biosynthesis of polysialic acid, a common bacterial capsule component (St Geme JW 3 rd, unpublished data). These preliminary observations suggest that *K. kingae* elaborates a polysaccharide capsule as a mechanism to promote intravascular survival and dissemination to distant sites, and may explain the increased incidence of disease among young children lacking maturation of the T-cell independent arm of the immune system [55].

In patients in whom *K. kingae* is able to persist in the bloodstream, the organism may produce non-focal bacteremia or instead disseminate to joints, bones, or the endocadium. Recent analysis of a large collection of *K. kingae* clinical isolates demonstrated that strains recovered from the blood of patients with uncomplicated bacteremia were generally piliated but typically expressed relatively few pili. In contrast, strains recovered from joint fluid, bone aspirates, or the blood of patients with endocarditis were generally non-piliated [14]. At this point, it is unclear what environmental factors influence σ54, PilS, and PilR activity and control density of *K. kingae* piliation and the reason for selection against high-density piliation of *K. kingae* organisms invading skeletal tissues or causing endocarditis.

9 Diagnosis

9.1 Culture Detection

In the late 1980s it was discovered that seeding of synovial fluid exudates into BCV resulted in isolation of *K. kingae* from joint fluid and bone exudates of young children in whom routine cultures of the specimens on solid media were negative [4]. When positive BCV were subcultured onto the same solid media, *K. kingae* grew readily, suggesting that exudates exert an inhibitory effect on the bacterium and that dilution of purulent material in a large volume of broth decreases the concentration of inhibitory factors, improving the recovery of this fastidious organism [4]. Use of the BCV technique employing Isolator 1.5 Microbial Tubes (Wampole Laboratories, Cranbury, N.J.) or aerobic blood culture bottles from a variety of automated blood culture systems such as BACTEC (Becton Dickinson, Cockeysville, MD), BacT/Alert (Organon Teknika Corporation, Durham, N.C.), and Hemoline DUO (Biomerieux, Paris, France) resulted in the recognition of the organism as an important etiology of pediatric skeletal infections [13]. In studies conducted in Israel and France in which the BCV method was routinely used for culturing synovial fluid aspirates, *K. kingae* was isolated in almost half of young children with culture-proven joint infections [6, 56].

9.2 Molecular Detection

Because of the risk for long-term orthopedic sequelae following bone and joint infections, prompt bacteriological confirmation of the diagnosis and early initiation of appropriate antibiotic therapy are of critical importance. However, definitive identification of the causative agent by culture and determination of antibiotic susceptibility testing require, on average, 2–3 days.

In recent years, use of conventional and real-time PCR has enabled identification of the etiology of septic arthritis and osteomyelitis within 24 h [57]. This approach involves extracting DNA from clinical samples, incubation of the DNA with broad range oligonucleotide primers that anneal to constant regions of the 16 s ribosomal RNA (rRNA) gene, and amplification of the intervening sequence, which varies according to bacterial species [57]. The resulting amplification products are either sequenced and compared with sequences in the Genbank database or hybridized with organism-specific probes [58–63]. As an example, Verdier et al. described 171 children with presumptive osteoarticular infections who had cultures of synovial fluid or bone samples using solid media and blood culture vials [62]. Among the culture-positive specimens, *K. kingae* was isolated in blood culture vials in 9 of 64 (14%). When the remaining 107 culture-negative specimens were further tested using conventional PCR and universal 16 s rRNA primers, specific *K. kingae* sequences were found in additional 15 children. In this study no other pathogen was missed by culture and detected by PCR, confirming the intrinsic difficulties of recovering *K. kingae* by culture methods.

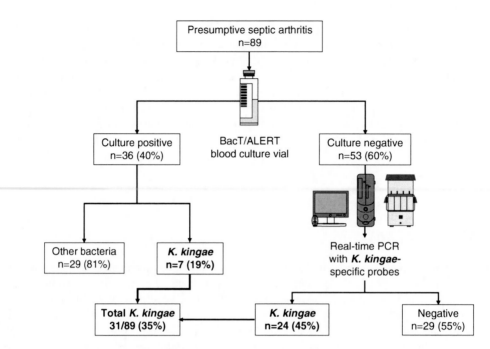

Fig. 6 Detection of *K. kingae* in synovial fluid exudates by combined use of the BCV method and real-time PCR [11]

In addition, PCR assays that amplify *K. kingae*-specific targets such as *cpn60* or the RTX toxin genes have been developed and have been associated with high reliability [9–11, 63, 64]. In a study by Ilharreborde et al. of 89 children with suspected septic arthritis, joint fluid was cultured using blood culture vials and was examined using real-time PCR reaction and highly specific primers that amplify a 169-bp fragment of the *K. kingae cpn60* gene [11] (Fig. 6). The diagnosis of septic arthritis was confirmed by culture in 36 (40%) specimens, including seven that grew *K. kingae*. Real-time PCR identified another 24 cases of *K. kingae* among the 53 culture-negative patients. Of note, real-time PCR identified *K. kingae* in all seven samples that grew the bacterium and was negative in all other children in whom other microorganisms were identified. All together, *K. kingae* was present in 31 (52%) of the 60 microbiologically documented cases [11].

These results clearly demonstrate that recovery of *K. kingae* by culture remains unsatisfactory, even when samples are inoculated into BCV. Nucleic acid amplification methods improve sensitivity and shorten time-to-detection, and clearly demonstrate that *K. kingae* is a leading bacterial etiology of septic arthritis and osteomyelitis in children aged 6–48 months and responsible for a large fraction of culture-negative infections in this age group.

10 Treatment

10.1 Antibiotic Susceptibility

K. kingae is almost always highly susceptible to penicillins and cephalosporins drugs that are empirically administered to young febrile children with suspected bacteremia, septic arthritis or osteomyelitis, and β-lactamase production has been rarely reported [13, 65–67]. With few exceptions, *K. kingae* is also susceptible to aminoglycosides, macrolides, trimethoprim-sulfamethoxazole, fluoroquinolones, tetracycline, and chloramphenicol. The organism exhibits decreased susceptibility to oxacillin and is resistant to trimethoprim and glycopeptide antibiotics such as vancomycin and teicoplanin [13, 67]. In a large study comprising isolates from carriers and patients with a variety of invasive *K. kingae* infections, nearly 40% of isolates were resistant to clindamycin [67].

10.2 Treatment of Invasive Infections

The empiric antibiotic therapy for skeletal infections in children usually consists of intravenous administration of a β-lactamase-resistant penicillin or a second or third-generation cephalosporin, pending culture results [13]. In areas where community associated methicillin-resistant *Staphylococcus aureus* is prevalent, a combination of a β-lactam antibiotic and vancomycin or clindamycin are recommended [68]. In children in whom *K. kingae* is detected, the initial antibiotic regimen is usually changed to ampicillin, once production of β-lactamase is excluded, or cefuroxime. Clinical response and a CRP values are used to guide switching to oral antibiotics and defining duration of therapy. Antibiotic treatment has generally varied from 2 to 3 weeks for *K. kingae* septic arthritis, from 3 to 6 weeks for *K. kingae* osteomyelitis, and from 3 to 12 weeks for *K. kingae* spondylodiscitis [13]. Although some children with *K. kingae* septic arthritis have been managed with repeat joint aspirations and lavage [32], invasive surgical procedures are unnecessary in most cases [13]. Children with *K. kingae* bacteremia are generally treated initially with an intravenous β-lactam antibiotic and subsequently with an oral β-lactam drug once the clinical condition has improved. In most cases, the total duration of therapy ranges from 1 to 2 weeks [13, 33].

Patients with *K. kingae* endocarditis are usually treated with an intravenous β-lactam antibiotic alone or in combination with an aminoglycoside for 4 to 7 weeks. Early surgical intervention is necessary for life-threatening complications unresponsive to medical therapy [13, 39–42].

References

1. Henriksen SD, Bøvre K. *Moraxella kingii* sp. nov., a hemolytic saccharolytic species of the genus *Moraxella*. J Gen Microbiol 1968;51:377–385.
2. Henriksen SD, Bøvre K. Transfer of *Moraxella kingii* Henriksen and Bøvre to the genus *Kingella* gen. nov. in the family *Neisseriaceae*. J Syst Bacteriol 1976;26:447–450.
3. Graham DR, Band JD, Thornsberry C, Hollis DG, Weaver RE. Infections caused by *Moraxella, Moraxella urethralis, Moraxella*-like groups M-5 and M-6, and *K. kingae* in the United States, 1953–1980. Rev Infect Dis 1990;12:423–431.
4. Yagupsky P, Dagan R, Howard CW, Einhorn M, Kassis I, Simu A. High prevalence of *Kingella kingae* in joint fluid from children with septic arthritis revealed by the BACTEC blood culture system. J Clin Microbiol. 1992 30:1278–1281.
5. Moylett EH, Rossmann SN, Epps HR, Demmler GJ. Importance of *Kingella kingae* as a pediatric pathogen in the United States. Pediatr Infect Dis J 2000;19:263–265.
6. Moumile K, Merckx J, Glorion C, Pouliquen JC, Berche P, Ferroni A. Bacterial aetiology of acute osteoarticular infections in children. Acta Paediatr 2005;94:419–422.
7. Dubnov-Raz G, Scheuerman O, Chodick G, Finkelstein Y, Samra Z, Garty BZ. Invasive *Kingella kingae* infections in children; clinical and laboratory characteristics. Pediatrics 2008; 122:1305–1309.
8. Yagupsky P. *Kingella kingae*: an emerging pediatric pathogen. *In:* Pollard A. J., Finn A. (editors). Advances in Experimental Medicine and Biology. Hot Topics in Infection and Immunity in Children III. Springer, New York, 2006, Vol. 582, pp. 179–190.
9. Ceroni D, Cherkaoui A, Ferey S, Kaelin A, Schrenzel J. *Kingella kingae* osteoarticular infections in young children: clinical features and contribution of a new specific real-time PCR assay to the diagnosis. J Pediatr Orthop 2010;30:301–304.

10. Chometon S, Benito Y, Chaker M, Boisset S, Ploton C, Bérard J, Vandenesch F, Freydiere AM. Specific real-time polymerase chain reaction places *Kingella kingae* as the most common cause of osteoarticular infections in young children. Pediatr Infect Dis J 2007;26:377–381.

11. Ilhaerreborde B, Bidet P, Lorrot M, Even J, Mariani-Kurkdjian P, Ligouri S, Vitoux C, Lefevre Y, Doit C, Fitoussi F, Pennecot G, Bingen E, Mazda K, Bonacorsi S. A new real-time PCR-based method for *Kingella kingae* DNA detection: application to a prospective series of 89 children with acute arthritis. J Clin Microbiol 2009;47:1837–1841.

12. Von Graevenitz A, Zbinden R, Mutters R. *Actinobacillus, Capnocytophaga, Eikenella, Kingella, Pasteurella*, and other fastidious or rarely encountered Gram-negative rods. In Murray PR, Baron EJ, Jorgensen JH, Pfaller MA, Yolken RH (eds.) Manual of Clinical Microbiology 8th edition. Washington DC. American Society for Microbiology 2003; pp. 614–615.

13. Yagupsky P. *Kingella kingae*: from medical rarity to an emerging paediatric pathogen. Lancet Infect Dis 2004;4:358–367.

14. Kehl-Fie TE, Porsch EA, Yagupsky P, Grass EA, Obert C, Benjamin DK Jr, St Geme JW 3 rd. Examination of type IV pilus expression and pilus-associated phenotypes in *Kingella kingae* clinical isolates. Infect Immun 2010;78:1692–1699.

15. Yagupsky P, Dagan R, Prajgrod F, Merires M. Respiratory carriage of *Kingella kingae* among healthy children. Pediatr Infect Dis J 1995;14:673–678.

16. Yagupsky P, Weiss-Salz I, Fluss R, Freedman L, Peled N, Trefler R, Porat N, Dagan R. Dissemination of *Kingella kingae* in the community and long-term persistence of invasive clones. Pediatr Infect Dis J 2009;28:707–710.

17. Yagupsky P, Peled N, Katz O. Epidemiological features of invasive *Kingella kingae* infections and respiratory carriage of the organism. J Clin Microbiol 2002;40:4180–4184.

18. Yagupsky P, Porat N, Pinco E. Pharyngeal colonization by *Kingella kingae* in children with invasive disease. Pediatr Infect Dis J 2009;28:155–157.

19. Dubnov-Raz G, Ephros M, Garty BZ, Schlesinger Y, Maayan-Metzger A, Hasson J, Kassis I, Schwartz-Harari O, Yagupsky P. Invasive pediatric *Kingella kingae* infections: a nationwide collaborative study. Pediatr Infect Dis J 2010;29; 639–643.

20. Van Erps J, Schmedding E, Naessens A, Keymeulen B. *Kingella kingae*, a rare cause bacterial meningitis. Clin Neurol Neurosurg 1992;94:173–175.

21. Carden SM, Colville DJ, Gonis G, Gilbert GL. *Kingella kingae* endophtalmitis in an infant. Aust N Z J Ophtalmol 1991;19:217–220.

22. Bofinger JJ, Fekete T, Samuel R. Bacterial peritonitis caused by *Kingella kingae*. J Clin Microbiol 2007;45:3118–3120.

23. Matta M, Wermert D, Podglajen I, Sanchez O, Buu-Hoï A, Gutmann L, Meyer G, Mainardi JL. Molecular diagnosis of *Kingella kingae* pericarditis by amplification and sequencing of the 16 S rRNA gene. J Clin Microbiol 2007;45:3133–3134.

24. Baticle E, Courtivron B, Baty G, Holstein A, Morange V, Mereghetti L, Goudeau A, Lanotte P. Pediatric osteoarticular infections caused by *Kingella kingae* from 1995 to 2006 at CHRU de Tours. Ann Biol Clin 2008;66:454–458.

25. Lundy DW, Kehl DK. Increasing prevalence of *Kingella kingae* in osteo-articular infections in young children. J Pediatr Orthop 1998;18:262–267.

26. Luhmann JD, Luhmann SJ. Etiology of septic arthritis in children: an update for the 1990s. Pediatr Emerg Care 1999;15:40–42.

27. Fuursted K, Arpi M, Lindblad BE, Pedersen LN. Broad-range PCR as a supplement to culture for detection of bacterial pathogens in patients with a clinically diagnosed spinal infection. Scand J Infect Dis 2008;40:772–777.

28. Chanal C, Tiget F, Chapuis P, Campagne D, Jan M, Sirot J. Spondilytis and osteomyelitis caused by *Kingella kingae* in children. J Clin Microbiol 1987;25:2407–2409.

29. Garron E, Viehweger E, Launay F, Guillaume JM, Jouve JL, Bollini G. Nontuberculous spondylodiscitis in children. J Pediatr Orthop 2002;22:321–328.

30. Ceroni D, Cherkaoui A, Kaelin A, Schrenzel J. *Kingella kingae* spondylodiscitis in young children: toward a new approach for bacteriological investigations? A preliminary report. J Child Orthop 2010;4:173–175.

31. Yagupsky P, Press J. Unsuspected *Kingella kingae* infections in afebrile children with mild skeletal symptoms: the importance of blood cultures. Eur J Pediatr 2004;163:563–564.

32. Lebel E, Rudensky B, Karasik M, Itzchaki M, Schlesinger Y. *Kingella kingae* infections in children. J Pediatr Orthop B 2006;15:289–292.

33. Yagupsky P, Dagan R. *Kingella kingae* bacteremia in children. Pediatr Infect Dis J 1994;13:1148–1149.

34. Caballero Rabasco MA, Gonzalez Cuevas A, Martinez Roig A. Isolated bacteremia caused by *Kingella kingae*. An Pediatr (Barc) 2010;72:89–90.

35. Pavlovsky M, Press J, Peled N, Yagupsky P. Blood culture contamination in pediatric patients: young children and young doctors. Pediatr Infect Dis J 2006;25:611–614.
36. Baraff LJ, Bass JW, Fleisher GR, Klein JO, McCracken GH, Powell KR, Schriger DL. Practice guideline for the management of infants and children 0 to 36 months of age with fever without source. Agency for Health Policy and Research. Ann Emerg Med 1993;22:1198–210.
37. Seña AC, Seed P, Nicholson B, Joyce M, Cunningham CK. *Kingella kingae* endocarditis and a cluster investigation among daycare attendees. Pediatr Infect Dis J 2010;29:86–88.
38. Berkun Y, Brand A, Klar A, Halperin E, Hurvitz H. *Kingella kingae* endocarditis and sepsis in an infant. Eur J Pediatr 2004;163:687–688.
39. Wells L, Rutter N, Donald F. *Kingella kingae* endocarditis in a sixteen-month-old-child. Pediatr Infect Dis J 2001;20:454–455.
40. Rotstein A, Konstantinov IE, Penny DJ. *Kingella*-infective endocarditis resulting in a perforated aortic root abscess and fistulous connection between the sinus of Valsalva and the left atrium in a child. Cardiol Young 2010;20:332–333.
41. Youssef D, Henaine R, Di Filippo S. Subtle bacterial endocarditis due to *Kingella kingae* in an infant: a case report. Cardiol Young 2010;20: 448–450.
42. Sarda H, Ghazali D, Thibault M, Leturdu F, Adams C, Le Loc'h H. Infection multifocale invasive à *Kingella kingae*. Arch Pediatr 1998;5:159–62.
43. Dayan A, Delclaux B, Quentin R, et al. The isolation of *Kingella kingae* by hemoculture must always suggest the diagnosis of endocarditis. Presse Med 1989;18:1340–1341.
44. Nafstad P, Hagen JA, Oie L, Magnus P, Jaakola JK. Day care and respiratory health. Pediatrics 1999;103:753–758.
45. Robinson J. Infectious diseases in schools and child care facilities. Pediatr Rev 2001;22:39–45.
46. Slonim A, Walker ES, Mishori E, Porat N, Dagan R, Yagupsky P. Person-to-person transmission of *Kingella kingae* among day care center attendees. J Infect Dis 1998;178:1843–1846.44.
47. Kiang KM, Ogunmodede F, Juni BA, Boxrud DJ, Glennen A, Bartkus JM, Cebelinski EA, Harriman K, Koop S, Faville R, Danila R, Lynfield R. Outbreak of osteomyelitis/septic arthritis caused by *Kingella kingae* among child care center attendees. Pediatrics 2005;116:e206-213.
48. Yagupsky P, Erlich Y, Ariela S, Trefler R, Porat N. Outbreak of *Kingella kingae* skeletal system infections in children in daycare. Pediatr Infect Dis J 2006;25:526–532.
49. Slonim A, Steiner M, Yagupsky P. Immune response to invasive *Kingella kingae* infections, age-related incidence of disease, and levels of antibody to outer-membrane proteins. Clin Infect Dis 2003;37:521–527.
50. Yagupsky P, Slonim A. Characterization and immunogenicity of *Kingella kingae* outer-membrane proteins. FEMS Immunology and Medical Microbiology 2005;43:45–50.
51. Kehl-Fie TE, Miller SE, St Geme JW 3 rd. *Kingella kingae* expresses type IV pili that mediate adherence to respiratory epithelial and synovial cells. J Bacteriol 2008;190:7157–7163.
52. Kehl-Fie TE, Porsch EA, Miller SE, St. Geme JW 3 rd. Expression of *Kingella kingae* type IV pili is regulated by σ54, PilS, and PilR. J Bacteriol 2009;191:4976–4986.
53. Amir J, Yagupsky P. Invasive *Kingella kingae* infection associated with stomatitis in children. Pediatr Infect Dis J 1998;17:757–758.
54. Kehl-Fie TE, St Geme JW 3 rd. Identification and characterization of an RTX toxin in the emerging pathogen *Kingella kingae*. J Bacteriol 2007;189:430–436.
55. Garcia-Rodriguez JA, Fresnadillo Martinez MJ. Dynamics of nasopharyngeal colonization by potential respiratory pathogens. J Antimicrob Chemother 2002;50: Suppl. S2:59–73.
56. Yagupsky P, Bar-Ziv Y, Howard CB, Dagan R. Epidemiology, etiology, and clinical features of septic arthritis in children younger than 24 months. Arch Pediatr Adolesc Med 1995;149:537–540.
57. Fenollar F, Lévy PY, Raoult D. Usefulness of broad-range PCR for the diagnosis of osteoarticular infections. Curr Opin Rheumatol 2008;20:463–470.
58. Stahelin J, Goledenberger D, Gnehm HE, Altwegg M. Polymerase chain reaction diagnosis of *Kingella kingae* arthritis in a young child. Clin Infect Dis 1998;27:1328–1329.
59. Moumile K, Merckx J, Glorion C, Berche P, Ferroni A. Osteoarticular infections caused by *Kingella kingae* in children: contribution of polymerase chain reaction to the microbiologic diagnosis. Pediatr Infect Dis J 2003;22:837–839.
60. Rosey AL, Abachin E, Quesnes G, Cadilhac C, Pejin Z, Glorion C, Berche P, Ferroni A. Development of a broad-range 16 S rDNA real-time PCR for the diagnosis of septic arthritis in children. J Microbiol Methods 2007;68:88–93.
61. Luegmair M, Chaker M, Ploton C, Berard M. *Kingella kingae*: osteoarticular infections of the sternum in children: a report of six cases. J Child Orthop 2008; 2:443–447.
62. Verdier I, Gayet-Ageron A, Ploton C, Taylor P, Benito Y, Freydiere AM, Chotel F, Bérard J, Vanhems P, Vandenesch F. Contribution of a broad range polymerase chain reaction to the diagnosis of osteoarticular infections

caused by *Kingella kingae*: description of twenty-four recent pediatric diagnoses. Pediatr Infect Dis J 2005; 24:692–696.

63. Cherkaoui A, Ceroni D, Emonet S, Lefevre Y, Schrenzel J. Molecular diagnosis of *Kingella kingae* infections by specific real-time PCR assay. J Med Microbiol 2009;58:65–68.
64. Cherkaoui A, Ceroni D, Ferey S, Emonet S, Schrenzel J. Pediatric osteo-articular infections with negative culture results: what about *Kingella kingae*? Rev Med Suisse 2009;5:2235–2239.
65. Kugler KC, Biedenbach DJ, Jones RN. Determination of the antimicrobial activity of 29 clinical important compounds tested against fastidious HACEK group organisms. Diagn Microbiol Infect Dis 1999;34:73–76.
66. Jensen KT, Schonheyder H, Thomsen, VF. In-vitro activity of ß-lactam and other antimicrobial agents against *Kingella kingae*. J Antimicrob Chemother 1994;33:635–640.
67. Yagupsky P, Katz O, Peled N. Antibiotic susceptibility of *Kingella kingae* isolates from respiratory carriers and patients with invasive infections. J Antimicrob Chemother 2001;47:191–193.
68. Saphyakhajon P, Joshi AY, Huskins WC, Henry NK, Boyce TG. Empiric antibiotic therapy for acute osteoarticular infections with suspected methicillin-resistant *Staphylococcus aureus* or *Kingella*. Pediatr Infect Dis J 2008;27:765–767.

Influenza Pandemics

Ruth Elderfield and Wendy Barclay

1 Introduction

Influenza is a seasonal respiratory illness associated with more serious consequence and even death in the very young, old and immunocompromised. Annual epidemics are predictable and affect a relatively small percentage of the global population at any one time. Pandemics differ from epidemics in that they are a global phenomenon, affecting large numbers of people in multiple countries simultaneously. Pandemics tend to arise swiftly often out of the normal season, and affect a wider age group and spectrum of individuals than seasonal influenza. The first recorded influenza pandemic was in 1510 [1]. Since then human populations have been subjected to at least 15 pandemics, most notably in 1918 when estimates of the human deaths that resulted vary between 40 and 100 million [1, 2].

2 The Influenza Virus

The influenza virus, the etiologic agent, is a member of the family *Orthomyxoviridae*. There are three categories of influenza known as types A, B and C. Only type A causes pandemics and thus is the most widely studied. Influenza B viruses cause typical seasonal infections restricted to humans. Influenza C viruses also infect humans and have been isolated from clusters of children, but are often not recognised and may be dismissed as an untyped influenza like illness, due to the lack of diagnostic tests [3].

The virus particle is enveloped, whereby the genome is protected by a lipid bilayer derived from the host cell membrane. The appropriated membrane is studded with viral glycoproteins: the attachment spike protein haemagglutinin (HA) which binds to sialic acid (SA) receptors on the cell surface; the integral M2 protein, an ion channel involved in the uncoating of the virus inside the infected cell, and the neuraminidase protein (NA) which cleaves the cell membrane SAs that would otherwise tether the budding nascent virus particle to the infected cell.

Lining the inside of the virion membrane is the abundant matrix protein (M1), surrounding the eight genomic negative-sense RNA segments which are intertwined with nucleoprotein (NP) and each associated with one set of the three viral polymerase subunit proteins (PB1, PB2 and PA) (Fig. 1).

R. Elderfeild • W. Barclay (✉)
Barclay Influenza Group, Imperial College London, St. Mary's Campus, London, UK
e-mail: ruth.elderfield07@imperial.ac.uk; w.barclay@imperial.ac.uk

N. Curtis et al. (eds.), *Hot Topics in Infection and Immunity in Children VIII*,
Advances in Experimental Medicine and Biology 719, DOI 10.1007/978-1-4614-0204-6_8,
© Springer Science+Business Media, LLC 2011

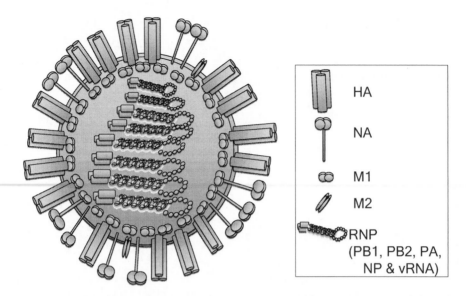

Fig. 1 Schematic cross section of the influenza A virus

The viral ribonucleoproteins (vRNPs) are the replicative units that are transported into the nucleus of the host cell after virus entry. There, the virus co-opts a number of host factors to assist the polymerase in transcribing viral mRNA, and replicating new genomes via cRNA intermediates. Finally the newly synthesized viral proteins and genomes are transported to the host cell's external membrane where the progeny virions assemble and bud.

3 The 1918 Influenza Pandemic

The 1918 pandemic dealt a devastating impact on a global population. Several features of the era, in addition to the extraordinary virulence of the virus itself that will be discussed below, contributed to the impact of the 1918 pandemic. At that time, although influenza was known as an infectious disease in terms of symptoms, the virus itself had not yet been identified. Indeed for a good proportion of the pandemic, *Hemophilius influenzae* (known as *Bacillus influenzae* at the time) was suggested as the causative agent. The influenza virus was eventually isolated from pigs by Richard Shope in 1931 [4, 5], then from humans by Andrewes, Laidlaw and Wilson Smith in 1933 [6, 7].

At the start of the influenza pandemic in 1918, the world was at war. The situation in America is absorbingly described in John M. Barry's "The Great Influenza" (Penguin, 2004) complied from military records, personal papers, oral history and newspapers of the period. There is some epidemiological and historical evidence that the outbreak may have begun in army camps either in the USA or in Europe where large numbers of young susceptible hosts were living in very crowded conditions. There is also evidence of at least two waves of disease, and indications that the second wave was more virulent than the first. For example, by the second wave historical accounts by the medical and scientific staff at the time describe symptoms such as: *cyanosis which started as mahogany spots over the cheek bones, and could expand until the patient turned black, caused by the lack of oxygen transfer in the lungs, leading to blue unoxygenated blood; extreme chills and fever, severe joint pains, vomiting and abdominal pains; earaches, headaches often localised around the eyes; and disturbing blood loss from nose, stomach, intestine and eyes.* At post mortem the lungs were often filled with the debris of destroyed cells and blood, which today would be diagnosed as Acute Respiratory Distress Syndrome (ARDS).

In Philadelphia, where the virus had been introduced from the local port, health workers and scientists were requesting widespread restrictions on gatherings and provision of information to the media. Unheeding, the governor and the senior health official sanctioned a large city wide Liberty Loan parade in order to gather funds for the war effort. Within 72 h of the parade every bed in the city's 31 hospitals was filled. The daily death rates for the city rose at an alarming rate, 3 days after the parade 117 died in one day, on day 11 more than 400 people died.

The second and third waves of the 1918 pandemic resulted in a cumulative case fatality rate (expressed as a ratio of the number of people infected to the number of people who died) of >2.5%. Later pandemics of the twentieth and twenty-first century only reached case fatality rates of less than 0.1% [8].

3.1 Reconstituting the 1918 Influenza Virus

Remarkably, although it has not been possible to isolate infectious virus directly from stored samples of that era, we do today have access to the causative agent of this pandemic following the elegant application of modern science. In 2005, Jeffrey Taubenberger and colleagues used the polymerase chain reaction to amplify small fragments of viral RNA isolated from formalin-treated post-mortem pathological slides and also from frozen lung tissue obtained from a person who died in Alaska and was buried in the permafrost, from this material the nucleotide sequence of the 1918 virus was deduced [9, 10]. Taubenberger joined forces with Terence Tumpey and colleagues at the CDC. They used the 1918 virus sequence information to generate plasmids containing the viral cDNA which when transfected into suitable mammalian cells, allowed the recovery of infectious 1918 virus [9, 11–16].

The reconstituted virus was more virulent in animal models than any other influenza virus strains studied previously [11, 16–23]. Thus although there is strong evidence that secondary bacterial infection contributed significantly to deaths from 1918 virus in humans [24, 25], the virus itself, in the absence of bacteria, is remarkably pathogenic to animals. Studies have been carried out to map the genetic determinants of this virus in the hope that this will help us to predict the virulence of future influenza strains as they emerge. The polymerase genes and the virus HA gene have been implicated in the extreme virulence of this virus, but work continues to understand the mechanisms by which this particular influenza strain is so deadly [26–29].

Phylogenetic studies suggest that the genome of the 1918 virus is most similar to viruses found in birds. However there are a number of key amino acid changes that indicate that, although it originated in an avian host, the virus underwent adaptation in order to replicate and transmit within human and swine hosts [8]. Interestingly the 1918 virus exhibits low pathogenicity in experimentally infected swine [20].

4 Twentieth Century Influenza Pandemics

The 1918 pandemic virus has been called the 'Mother of all pandemic viruses' [8] as all of the twentieth Century pandemics are derived from virus lineages descended from the 1918 virus. After 1918, viruses derived from that outbreak continued to circulate in humans causing annual epidemics of moderate or mild severity [30]. However, because of their segmented genomes, influenza viruses are particularly prone to a special form of recombination known as reassortment that occurs if one host is coinfected by two different viruses. Such mixing events allow the introduction of genetic material from viruses that usually circulate in birds with the human adapted viruses, and new viruses thereby created may be able to cause a novel outbreak. We know that the two major pandemics in the second half of the twentieth century were formed in this way: The 'Asian' pandemic of 1957 was caused by the emergence of an H2N2 sub-type virus that retained the M, NP, PB2, PA and the

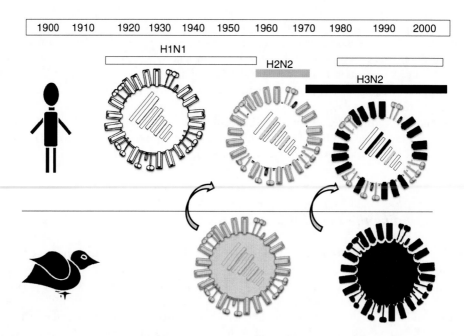

Fig. 2 A time line illustrating the emergence of the three twentieth century influenza pandemic subtypes. The H1N1 virus accepted alternative genes in recombination events with avian viruses, giving rise to the H2N2 and H3N2 subtypes

NS of viruses derived directly from the 1918 virus, but obtained a new HA, NA and PB1 from an avian H2N2 virus. Because the human population had not experienced infection with a virus of H2 antigenic type, the entire world was susceptible and the virus spread rapidly.

Nonetheless, perhaps because antibiotics were available by that time or perhaps the H2 virus itself had a milder phenotype, the death toll of this pandemic was much lower at only ~two million people. The 1968 H3N2 'Hong Kong' pandemic that followed just 11 years later resulted in between one and two million deaths. This virus was a recombinant between the circulating human H2N2 virus and an avian strain with H3 HA. The reassortant virus still retained five segments originally derived from the 1918 strain, but acquired the HA and PB1 from the avian virus. Each of the H2 and H3 pandemic events were so universal that the virus displaced the previously circulating strains [31, 32]. In 1977, the H1N1 strain re-emerged as a circulating human strain (not to be confused with the 1976 swine H1N1 outbreak at Fort Dix). The colourfully named 'Red' or 'Russian' flu appeared initially in China in May of 1977 with isolates found in Russia soon after [33]. This strain produced a relatively mild disease mainly in young children. However, subsequent genetic analysis indicated a 27 year gap in the evolutionary history of this virus. In fact it was genetically similar to virus isolated in 1950. The eventual conclusion drawn was that this virus had been deep frozen in a laboratory and its release was accidental. The H3N2 and H1N1 subtypes have continued to co-circulate and to cause human seasonal influenza outbreaks into the twenty-first century (Fig. 2) [32–36].

5 Birds Are the Natural Host for Influenza Viruses

The influenza A virus naturally circulates in aquatic birds, where it replicates in the gut [37–39]. All of the 16 HA and 9 NA subtypes have been isolated from either or both of the *Anseriformes* (an order which includes ducks and geese) and from the *Charadriiformes* (the order to which shore birds and gulls belong). No other species has been infected by all the influenza types. Different subtypes predominate in the different orders of birds; H3 and H6 for example, are found mainly in *Anseriformes*, whereas in *Charadriiformes* H4, H9, H11 and H13 are the predominant HA subtype. The virus is excreted in high titres into the water bodies that are home to domestic and migratory birds [40, 41]. The virus can then be picked up by migratory birds and spread along the migratory routes. However, the relative geographic isolation of some flocks has been proposed as one cause of genetic divergence within the HA and NA subtypes [42].

Whilst the disease caused by influenza is mainly asymptomatic in aquatic birds, some isolates are capable of developing from low pathogenic avian influenza (LPAI) into a highly pathogenic influenza (HPAI) capable of killing domestic poultry. The two subtypes that are prone to change pathogenicity are H5 and H7, as seen in Eurasia with the HPAI H5N1 viruses and the Netherlands with the H7N7 virus. Economically, the now widely-distributed H5N1 virus has been responsible for the death of over a billion head of poultry either directly through the disease or indirectly through preventative culling measures [39] Other subtypes H9, H1, H3, H4 and H14, whilst still appearing mild in aquatic birds, can be fatal in domesticated flocks [39].

Only viruses of the H1, H2 or H3 subtypes are known to have circulated among humans or pigs.

5.1 Antigenic Shift and Drift

The drastic recombination events that result in novel pandemic viruses described above are called antigenic shift.

After the introduction of the new subtype and its wide circulation in humans, the increasing prevalence of specific immunity among human hosts exerts selection pressure that drives evolutionary change in the HA protein via the accumulation of point mutations that block the antibody recognition through conformational changes or glycosylation events on the antigenic epitopes. This process is called antigenic drift.

5.2 Cyclical Nature of Seasonal Influenza Since the Hong Kong Pandemic

As the virus continues to circulate in humans in the interpandemic periods, mutations accumulate that confer antigenic drift as well as other adaptive mutations that alter the nature of the virus and may be associated with loss of virulence. This, along with increased wide vaccination campaigns for the elderly and immuno-compromised populations, could explain the gradual decrease in influenza-like illness since the Hong Kong pandemic of 1968, particularly in recent years (Fig. 3) [43, 44].

The ability of influenza A viruses to recombine so readily has worried virologists and public health planners alike, because of the risk that a seasonal strain of influenza with human adapted components might recombine with one of the highly pathogenic avian influenza (HPAI) viruses such as the notorious H5 or H7 subtypes. These two subtypes are lethal in poultry because of an extended tropism conferred by mutation in the HA gene that allows them to infect and propagate in many organs and tissues, rather than being restricted to areas where the appropriate host cell proteases exist. Consequently infection with these viruses carries a high mortality in humans of more than 60%. H5N1 virus has been responsible for over 500 cases of human infection and 300 deaths (as of 31st August 2010 according to the World Health Organisation avian influenza surveillance system) but thankfully has not yet reassorted with a human-adapted influenza virus, nor given rise to a pandemic outbreak [45].

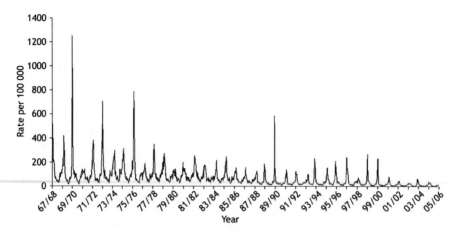

Fig. 3 Influenza-like illness incidence in England and Wales since the Hong Kong pandemic [46]

Because the research community was so focused on surveillance of and strategies to control H5 and H7 infections, the outbreak of the 2009 'swine' influenza was a surprise. We had largely overlooked the idea that the next pandemic would originate in pigs even though an outbreak in Fort Dix, New Jersey in 1976 associated with the death of a soldier from infection with an H1N1 swine flu had led to mass vaccination campaigns at that time. The Fort Dix incident did not give rise to a pandemic, the virus remained contained within the military and transmission of the virus had fizzled out by the time the now-infamous vaccination campaign began [47, 48].

5.3 Swine as a Recombining Mixing Pot

Most text books propose that swine are the mixing vessel in which influenza viruses of avian and human origin reassort. It has been evident that pigs can be infected with influenza since the early days of virus isolation. Indeed, Shope initially isolated influenza from a pig [4, 5] and Kida et al. showed that pigs could be infected by many different subtypes of avian influenza [49].

The 2009 pandemic virus illustrates just how good a mixing pot the swine host can be. The origins of its eight gene segments come from at least four different sources and three different hosts. The PB2 and PA segments appear to have originated in an avian reservoir and transmitted to the swine host around 1998. The PB1 segment derives from a human virus but was transferred to swine in 1998. The HA, NP and NS segments once again can trace their lineage back to the 1918 pandemic influenza, when the virus infected pigs and subsequently circulated through the years to become a classical swine virus. The NA and M segments are from circulating swine viruses, of the Eurasian lineage, believed to have transferred from birds in 1979 (Fig. 4) [50, 51].

Why do swine make such good mixing vessels? Ito et al. showed in 1998 that the pig respiratory tract displayed SA receptors that are bound by viruses isolated from birds as well as those used by human-adapted viruses, implying that the pig was capable of being infected by an avian and a human-adapted virus at the same time [52]. In addition, the co-expression of the avian-like and human-like receptors in swine potentially allows for the selection of avian viruses with small mutations that adapt them to bind to and replicate in mammalian cells, a process known as 'receptor switching.'

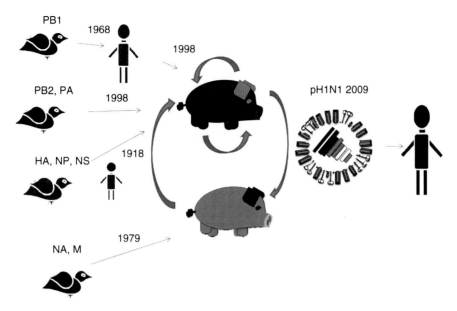

Fig. 4 The emergence of the pH1N1 2009 influenza virus. A series of avian viruses transferred into swine hosts. There multiple recombination events of classical swine influenza (*black*) resulted in a triple reassortant (TRIG), when recombined with Eurasian swine influenza (*grey*) generated the pH1N1 virus [51]

5.4 Receptor Switching

For avian influenza viruses to adapt to and transmit between humans, it is now apparent that in addition to the reassortment events that occur during antigenic shift, their HA proteins must also undergo modifications that alter their fine receptor binding specificity. The influenza HA protein binds to SA residues on the host cell surface as a prelude to cell entry. In the avian gut, these are predominantly α-2,3 linked receptors but in the human upper respiratory tract α-2,6 linked receptors predominate. Avian influenza viruses would therefore preferentially bind α-2,3 linked receptors and human-adapted influenza viruses have changed key residues at the receptor binding site allowing greater affinity for α-2,6 linked receptors (Fig. 5) [42].

In H3 HA proteins, receptor switching occurs if there is a change from glutamine (Q) at position 226 to leucine (L) (H3 numbering) and is enhanced by glycine (G) at 228 to serine (S) [53, 54]. For other subtypes the changes required for human adaptation are not exactly the same. Some H2 viruses still bind to α-2,6 human-like receptors even when the Q226 is present, a trait shared by avian H6 and H9 proteins. The H1 subtype tends to show changes at residues 225 (aspartate [D] to G) and 190 (D to glutamic acid [E]) rather than 226 and 228 but they achieve the same end [9, 10, 53–55].

Fortunately, in the case of H5 HA, none of the changes found in other subtypes have completely mediated a receptor binding switch, suggesting that the barrier to human adaptation may be particularly high for this subtype [56, 57].

One important difficulty in understanding these adaptive events is that the nature of the influenza virus receptor is not completely clear, but it is certainly more complex than a single sugar moiety [58]. The nature of carbohydrate to which influenza virus might attach has been recently studied using glycan arrays. Glycan arrays present hundreds of different carbohydrates. Different viruses or expressed HA proteins are then given the opportunity to bind to a favourite residue [59–63]. This type of experimental procedure was used recently to elucidate the receptor binding preferences of the novel pandemic H1N1 2009 virus. Interestingly, this virus along with two other swine viruses tested, was able to bind both α-2,6 and α-2,3 SA, whereas seasonal H1N1 influenza virus had a strong preference to bind carbohydrates with α-2,6 linkages and showed no binding to those with α-2,3 [63]

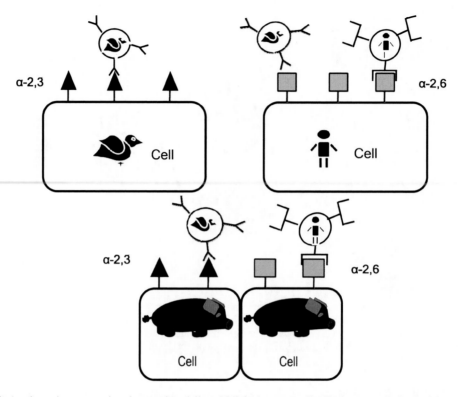

Fig. 5 A schematic representing the putative sialic acid linked receptor distribution on cells in the human upper airway, the avian gut and the swine respiratory system, with cognisant avian-like or human-like HA attachment proteins

6 The Emergence of Swine Origin Influenza (pH1N1)

The swine origin 2009 pandemic influenza virus appears to have emerged from San Luis Potosi, Mexico in late February 2009 [31], though it has been suggested that the virus was circulating at low levels in humans for some months prior to this. Indeed the most common ancestor may have emerged between August 2008 and January 2009 [31, 51, 64]. The pandemic threat of the new virus was realised as the first wave peaked in Mexico in late April 2009. The virus quickly spread across the globe. WHO moved to pandemic phase 4 after confirming human to human transmission on the 27th April, and just two days later, phase 5 was declared as the outbreak was found in two or more countries within one WHO region. Finally passage of the virus into a second WHO region triggered escalation to phase 6 on the 11th June.

6.1 How Could an H1 Virus Cause a Pandemic?

The H1N1 virus responsible for the 2009 pandemic is not the same H1N1 virus that had been circulating in humans in recent years causing seasonal H1N1 outbreaks. Both viruses have HA proteins originating from the 1918 pandemic virus. However the HA of the seasonal H1 had been under antigenic and other selective pressure as it circulated in the human population over a total of seven decades. On the other hand, since pigs are short lived, they exert little antigenic pressure to drive evolution of HA because the likelihood that a pig will be re-infected by the same influenza virus during its brief life time is very low. The rather genetically static 1918-derived pig virus became known as 'classical' swine influenza virus and it was the predominant influenza virus of swine on the North American continent throughout the twentieth century.

At the time this classical swine H1 HA recombined into what was to become the 2009 pandemic virus, it still retained 90% amino acid sequence identity to its 1918 progenitor. However, the human seasonal virus had changed so dramatically that it shared only 79% amino acids with the 1918 HA protein, and this did not allow for any antigenic cross protection for humans who had been infected with seasonal H1 in recent years against the novel 2009 pandemic strain.

6.2 The Course of the Outbreak in the UK

In the UK, the first reported case was on the 26th April 2009, brought back by those returning from holidays in Mexico. By May 29th there were 215 UK cases, of whom 52 were returning travellers, 39 were direct contacts of those travellers and 108 were people who had links to the secondary cases. Already at this early stage there were eight sporadic cases which were not linked to travel. On 21st June, the influenza-like-illness (ILI) incidence baseline was crossed and the first wave in the UK was clearly underway. This wave peaked in July and then fell well below baseline by mid-August, after schools closed for the summer holidays. Nevertheless 17% of the deaths in the UK occurred during this wave. The second wave was far shallower, but endured for a longer period from September 2009 until February 2010, and was responsible for the remaining 83% of fatal cases [65, 66].

The height of the peak of ILI cases in the second wave was much lower than had been predicted. Several factors contribute to the explanation for this: a proportion of the population (>30%) most vulnerable to seasonal influenza infection, the elderly, were already immune. In blood samples of those over 80 years old, collected in 2008 before the 2009 pandemic virus emerged, it was possible to detect antibodies that cross react with 2009 pandemic virus' HA in >30% of the samples (haemagglutination titre 1/32). In those aged 65–79 years, the seropositive frequency drops to ~20%, whereas in the 4–14 year old age bracket, the proportion with significant HAI was just ~4% [67]. These neutralising antibodies in sera collected from the elderly exist because many people in older age groups were infected in early life by closer derivatives of the 1918 H1N1 virus.

The second factor was the surprisingly mild nature of the 2009 pandemic virus in most people. Serology conducted retrospectively detected neutralising antibodies in a far larger percentage of the population than could be expected from the reports of ILI. One in three of the children in London and Birmingham were seropositive for the virus by September 2009 [67] suggesting that many people who had the infection did not report it. This could be because the symptoms were sufficiently mild that they did not feel they needed the flu service that was on offer at the time, or they did not even realise they had the infection. In this respect, the population was fortunate; this virus spread effectively but did not produce overtly pathogenic effects in most people. The case fatality rate was 0.02%, far lower than the 2.5% of the 1918 pandemic or the 0.1% of the 1957 and 1968 pandemics. Despite that, there were over 1,500 hospitalizations due to influenza like illnesses in the UK and 474 deaths [68]. Globally mortality far exceeds 18,000 deaths [69].

6.3 Who was at Risk?

In the event of a pandemic, such as in 2009, difficult decisions need to be made by health authorities to prioritise limited supplies of drugs and vaccines. Pharmaceutical companies may generate millions of doses of vaccine (300 million for the 2009 pandemic) [70] to protect the population, but this is still only enough to immunize a small proportion of the globe, so provision of vaccine to those most vulnerable to this virus needed to be prioritized.

6.3.1 Age

Usually vaccine is given primarily to those most susceptible to poor outcome from infection by seasonal influenza, namely the elderly. In the UK, from 2001 to 2009, 69% of those who died from seasonal influenza were >65 years old. However, with the 2009 pandemic virus only 15% of deaths were in the >65 demographic. True to form, those elderly individuals who did succumb to the 2009 pandemic influenza suffered a severe illness, with a case fatality rate of 0.9%, indicating the virus was able to cause serious morbidity in those who were immunologically susceptible [65].

The first infection with an influenza virus in children can often be quite severe. Johnson et al. showed that the high incidence of unexpected paediatric fatalities from the Fujian H3N2 seasonal drift variant in 2003 was linked with a higher than usual infection rate for seasonal influenza in the young in that year, possibly explained by a more drastic antigenic drift than in immediately previous years [71]. Similarly when a novel pandemic virus circulates widely, the incidence rate in the very young is particularly high and their clinical course in the face of lack of any relevant immunological experience is often severe. Indeed during the 2009 pandemic, the highest mortality rates were observed in those under 1 year of age [72]. In addition since school age children are major transmitters of influenza, there is good logic in targeting them in pandemic and seasonal immunization campaigns because overall community incidence may be curtailed in this way [73–76].

The 2009 pandemic virus was certainly able to infect and transmit well within the paediatric cohort. In the UK one in three children and in Hong Kong half of the children had been infected after the first wave [77, 78], supporting the global observations that school aged children and young adults were most likely to contract influenza [79]. Generally seasonal influenza causes two paediatric deaths per million people, while pH1N1, by mid-2010, had been responsible for 5–6 deaths per million in the Netherlands and the UK respectively and 11 per million in Argentina in the paediatric cohort [72, 80–82].

In their 2010 study of paediatric mortality from pandemic influenza in the UK, Sachedina and Donaldson identified 70 deaths in the 0–18 year old age group directly attributable to infection with the virus. As in similar studies across the world, they described common symptoms including fever, cough and shortness of breath [72, 79, 83–86]. The UK study observed that a combination of neurological, gastrointestinal and respiratory disease was present in more than half the deaths, an observation again echoed in other countries. Indeed pre-existing neurological disorders have frequently been listed as a co-morbidity [65, 66, 83–85, 87–89]. Half of those of school age who died in the UK attended schools for those with special needs [72]. As with adults, bacterial coinfections were often observed; in the US 43% of paediatric deaths were associated with secondary bacterial infections and in the UK 20% were associated with laboratory confirmed cases [72, 73].

6.3.2 Pregnancy

Pregnant women appear to be especially at risk from complications of influenza of either a seasonal or pandemic nature, especially during the second and third trimesters [90]. This increased susceptibility has been ascribed to mechanical changes within the body, which act to increase the pressure on the cardiovascular system, including an increased heart rate, stroke volume and oxygen consumption set against a decreased lung capacity [91, 92]. This may also account for increased risk in obesity.

In addition, hormonal changes in pregnancy cause what might be broadly termed a swing away from cell mediated immunity and a bias towards the humoral system that may affect the ability to clear the virus [92, 93]. On the other hand humoral immunity is also not complete in pregnancy and depletion in the levels of IgG2 were observed in pregnant women who died in the 2009 pH1N1 pandemic [94]. There is evidence from previous pandemics that the mortality rate was high amongst pregnant women: In the 1918 pandemic between 27% and 47% of those pregnant who contracted influenza died. In 1957 the percentage of deaths was lower, but still considerable at 20% [92]. There is also evidence of complications; of those pregnant women who developed pneumonia but survived during the 1918 pandemic, more than 50% did not carry the foetus to full term [92].

6.4 Other Co-morbidities

In the 2009 pandemic it was very evident that other co-morbidities increased susceptibility to severe influenza infection. These included obesity, asthma and chronic obstructive pulmonary disease (COPD), diabetes, immunosuppression, heart conditions and neurological complications. Whilst only a third of those who were admitted to intensive care had co-morbidities, over three quarters of those who died did [73]. In the UK, the FLUCIN database collected data from all those admitted to hospital with influenza during the pandemic first wave. Co-morbidities were described for around 50% cases, leaving a significant number of young healthy adults that suffered severe disease and even death despite no obvious prior predisposition for bad outcome [95].

6.5 Bacterial Coinfections

In the US, a study of the first 100 fatal cases caused by pH1N1 found that 25% of them had bacterial coinfections, with *Staphylococcus aureus* and *Streptococcus pneumoniae* being the most common pathogens [96]. Other fatal case studies have put the incidence of bacterial coinfections between 28% and 36% and also included *Streptococcus pyogenes* in the list of common bacterial coinfections [89, 97, 98]. In Argentina, coinfection with streptococci increased the likelihood of severe outcome with an odds ratio of 17 [99]. However in the UK, high incidence of bacterial coinfection was less evident [95]. The effect of bacterial superinfection on the outcome of infection with pandemic H1N1 2009 is likely to have been affected by differences in the bacterial strains circulating in communities around the world at the time and this may, in part, account for the widely different case fatality rates seen in different areas.

6.6 Influenza Vaccination for the Pandemic

There is a well-established system in place for the generation of seasonal influenza vaccines; the dominant circulating strains are carefully monitored and predictions are made annually about which viruses are likely to predominate in the forth-coming 'flu season. The chosen seasonal viruses are recombined with the internal segments of the high growth A/Puerto Rico/08/34 (PR8) vaccine backbone strain, to create viruses with HA and NA antigens from seasonal strains that can be readily amplified in eggs. Similarly for the 2009 pandemic vaccine, a reassortant virus bearing the H1 HA and N1 NA genes of the pandemic strain A/California/07/2009 on a high growth body was used to generate high yield virus in eggs. However, vaccine production problems became apparent early during production phase when manufacturers realized that the growth of the Cal/07/09 reassortant was only 30–50% of that seen for the seasonal strains. Eventually a higher growth variant was obtained. In all about 30 versions of H1N1/2009 pandemic vaccine were generated in multiple countries by different manufacturers, with either wild type or reassortant viruses grown traditionally in eggs or in cell based systems and vaccines produced as spilt (just the HA and NA genes) preparations, whole inactivated virion preparations or, in one case, a live attenuated virus. Using a strategy based on development of H5N1 pandemic vaccines, the most widely used pandemic vaccine in the UK was an inactivated vaccine generated by GSK that was administered combined with AS03 adjuvant (composed of α-tocopherol, squalene and polysorbate 80 emulsion) a chemical mix added to enhance and prolong the immunogenic response and reduce the amount of HA protein required per dose to achieve immunity (antigen sparing). The immunogenicity of the pandemic vaccine was in fact much higher than expected based on experience of clinical trials with the H5N1 equivalent. In the end, a single dose of adjuvanted vaccine was sufficient to achieve seroconversion in adults. Although it was expected that two doses would be needed in children, who usually require a prime boost regimen for effective levels of antibody to be achieved, a single dose was eventually used as it turned out to be adequately immunogenic and significantly less reactogenic after the first dose than the second [100]. This vaccine was recorded as having a 72% effectiveness despite the relatively small doses of 3.75 µg of HA protein (unadjuvanted vaccines typically contain 15–30 µg). Other adjuvants were also trialled globally including alum (aluminium hydroxide) in China and Russia and another oil-in-water adjuvant broadly similar to ASO3 called MF59 in Korea and Italy. Clinical trials were run using the proprietary Sanofi-Pasteur AF03 adjuvant in the USA, Europe and Asia [70].

Live attenuated vaccine (LAIV) for pandemic 2009 was widely administered in the US. Other vaccine strategies that were not yet licensed have been researched using the 2009 H1N1 virus as antigen. These include: the use of virosomes (lipid vesicles) that have the HA and NA proteins scattered through the bilayer, live recombinant adenovirus vaccines that express the HA protein, virus like particles in which the HA, NA and M are expressed in insect cells which are infected with a recombinant baculovirus, these are purified and self assembled into immunogenic particles, finally plants infected with a transformed Agrobacterum vector that generate HA proteins [70].

6.7 The Virulence of H1N1 2009 Virus in Animal Models

Although the outcome of infection in most people infected with pandemic 2009 virus was mild, in animal models this virus causes more severe disease than recent seasonal H1N1 viruses. Itoh et al., compared seasonal H1N1 (A/Kawasaki/UTK-4/09) and pandemic H1N1 (A/California/04/2009) viruses in a number of animal models. Interestingly, infected mini-pigs remained relatively asymptomatic. In contrast, pH1N1 virus caused severe lung pathology in mice, ferrets and macaques including lung lesions and damage caused by the infiltration of inflammatory mediators to a greater extent than was observed with infection with the seasonal viruses [17, 101, 102]. Van de Brand et al., infected ferrets intratracheally with very high doses of seasonal H1N1, pH1N1 or HPAI H5N1 virus and found that infection with pH1N1 caused pathology intermediate between seasonal influenza and H5N1 and could lead to severe pneumonia and death in this model [17]. Infection of alveolar pneumocytes, not observed with the seasonal virus [17, 102] may correlate with the more profound binding of the pH1N1 HA to α-2,3 linked SA [63] which tends to be located deeper into the lung [103]. The difference between the animal models and the epidemiology in humans suggests that a basal level of existing immunity in the human population has protected against the moderately severe disease this virus can cause in immunologically naive experimental animals.

6.8 Oseltamivir Resistance of pH1N1 Virus

There are two antiviral drug classes currently available to treat influenza. Adamantanes (amantadine and rimantadine) are directed against the ion channel M2 protein and prevent the uncoating of the virus genome early in infection. The second class of drugs, the neuraminidase inhibitors (oseltamivir -Tamiflu® and zanamivir- Relenza®) were rationally designed to block the active site of the neuraminidase of the influenza virus. NA acts to cleave the SA receptors on the surface of the host cell, allowing the release of newly formed virions which can then infect uninfected cells.

Unfortunately, influenza viruses readily developed resistance to the adamantanes through point mutations at residues 26, 27, 30, 31 or 34 in the M2 protein, with no compromise in viral fitness [104, 105]. 90% of the seasonal H3N2 isolated in the US and Asia contain a resistant phenotype and the S31N mutation was already present in the 2009 pH1N1 at the time it crossed into humans [106, 107].

The NAI drug class has therefore become a favourite for stockpiling anti-influenza therapies. The first of these drugs to reach the clinic was zanamivir (Relenza) but use of this drug is hindered by the necessity to inhale it because it is not orally bioavailable. The second NAI oseltamivir (Tamiflu®) benefits from a convenient oral formulation, good bioavailability and is suitable for use in paediatric and adult populations. Data have not yet emerged fully for its effects in those > 65 years of age or the immunocompromised [108].

According to the Cochrane review of efficacy in 2005, if zanamivir or oseltamivir are used to treat an infection 48 h after onset of symptoms, there is a 'modest' reduction in influenza symptoms within 0.78 days in adults and 1 day in children for zanamivir and 0.86 days in adults and 0.87 days in children for oseltamivir [108, 109]. As prophylactics, both drugs fare well, in control groups taking the medications during seasonal influenza, there was a reduction in incidence of 69% and 74% for zanamivir and oseltamivir respectively. For post-exposure prophylaxis there were 81% and 90% relative reductions in infection after zanamivir and oseltamivir administration respectively [108, 109].

Resistance to oseltamivir can emerge, but early experiments indicated that mutations such as H275Y (N1 numbering, H274Y in N2), would confer a fitness cost to the virus by reducing NA affinity for the SA substrate [110, 111]. Thus, it was assumed that the resistance mutation would not persist in the community. However, by 2008, ~99% of seasonal H1N1 viruses had acquired the H275Y mutation that was associated with the resistant phenotype without fitness cost [108, 112]. The lack of fitness cost can be ascribed to compensating mutations in the NA protein [113, 114].

The obvious fear was that the pH1N1 virus would develop oseltamivir resistance given the widespread use of the drug in the early waves of the pandemic. pH1N1 viruses with the H275Y mutation have been isolated, however these were predominantly in those undergoing prolonged treatment regimens (often in immunocompromised patients) or through the use of low dose prophylaxis [115, 116]. At the time of writing, there had not been widespread transmission spread of the resistant strain. Indeed the fitness cost to this strain of virus is still under debate, with different labs publishing conflicting results [117–119].

6.9 Other Mutations in pH1N1 Virus that May Impact the Course of the Pandemic

6.9.1 D225G

There is historical evidence that implies that during the 1918 pandemic, the second and third waves were more severe than the first. It has been suggested that the virus acquired mutations as it circulated in its new human host and these 'hotted up' its virulence. Thus it is important to identify any mutations that may similarly increase virulence of the 2009 pH1N1 strain. In particular it was possible that critically ill or deceased patients had been infected with a virus variant that had more pathogenic potential than the viruses that predominated in the community. The sequence of viruses from such cases has been analysed in a number of studies [120–125]. One of the interesting mutations observed is a D225G (H3 numbering, D222G in H1) mutation in the HA protein. 18% of critical cases in Norway and 12.5% of the critically ill in Hong Kong had this mutation [123]. In Scotland this mutation was only found in patients who were critically ill (4.1%) [120].

However it is not clear that this mutation alone is responsible for poor outcome: The D225G mutation was found in the virus from a nasopharyngeal swab and tracheal aspirate from a 25 year old man admitted to intensive care with pneumonia and ARDS. However the same virus transmitted to a contact case, but did not lead to severe illness despite the latter individual having two hallmark co-morbidities, namely obesity and diabetes [121]. Although D225G was detected predominately in viruses from critical cases in Greece, it was also isolated in two mild cases of the disease [122]. It has been suggested that some of the reported isolates with this change are the results of egg adaptation during the culture period [126]. In addition, the prevalence of this mutation in the critically ill has been ascribed to factors such as sampling bias, the critically ill being more likely to be genotyped than the mild cases.

The presence of this mutation enhances binding of H1 HA to $\alpha2,3$ linked SA receptors [124]. The proposed mechanism by which such a mutation may enhance virulence is that the ability to bind more efficiently to $\alpha2,3$SA receptors extends the lung tropism of the virus to bind ciliated cells that may then be unable to clear virus efficiently via the mucociliary escalator [124]. Additionally, increased binding to type II pneumocytes and macrophages in the alveoli and to submucosal glands in the trachea and bronchi may enhance lung damage [127]. However using reverse genetics to engineer this point mutation into an otherwise isogenic background, it was shown that the D225G change was not associated with an increase in virulence in the ferret or guinea pig models and remains easily transmitted between guinea pigs [127]. The mutation did result in a lower infectious dose for infection of mice who predominately express the $\alpha2,3$, linked form of SA receptor [128].

6.10 Presence or Absence of Other Virulence Factors in pH1N1Virus

6.10.1 PB1-F2

The PB1 segment has a second reading frame (+1) which encodes a small protein (87-90 amino acids), PB1-F2 [129]. This protein has been assigned two functions, induction of apoptosis through its mitochondrial targeting C terminal domain and a role in lung inflammation [129–133]. A proposed third function, relating to polymerase function and reflected by the retention of the PB1 protein in the nucleus appears to be strain specific [134, 135].

Viruses with intact PB1-F2 genes cause increased pathology in the mouse model [131] and also predispose the host to secondary bacterial infections and subsequent pneumonia. Mice infected with a PR8 virus containing the full length PB1-F2 suffered greater weight loss and increased mortality when subjected to a secondary bacterial infection than mice infected with a PR8 with a truncated form of PB1-F2 [133].

However in natural isolates, particularly those from swine, the PB1-F2 gene is not always full length. Zell et al. analysed the influenza A sequences available in Genbank in 2007 [136]. They found that 96% avian strains possessed the full length PB1-F2, but in contrast only 75% of swine viruses and 81% of human viruses had the full length gene. Classical swine influenza strains have truncated forms of PB1-F2 with premature stop codons after 11, 25 and 34 residues [136].

Truncation of the PB1-F2 gene to just 57 amino acids also occurred in seasonal H1N1 viruses in the 1950s. Loss of the C terminus of PB1-F2 removes the mitochondrial targeting sequence of the protein, abrogating its interaction with host proteins ANT3 and VDAC1 and reducing its ability to trigger apoptosis in immune cells [130, 137]. In addition this region of the protein appears to harbour pro-inflammatory properties. Indeed, peptides generated to contain amino acid sequence from the C-terminal region of PB1-F2 generated an inflammatory response when administered to mouse lung. Two days post exposure mice lost up to 15% body weight. Interestingly the same peptide derived from recent H3N2 seasonal virus contains 5 amino acid differences from early H3N2 homologues that appear to abrogate the pro-inflammatory function. Inflammation triggered by PB1-F2 peptides from highly virulent strains such as 1918 influenza may predispose to secondary bacterial pneumonia [138]. Indeed, some highly virulent viruses such as H5N1 and 1918 H1N1 viruses possess a point mutation in this region of PB1-F2, N66S, that is partly responsible for their enhanced morbidity and mortality in mice [139].

The 2009 pH1N1 influenza virus has a PB1-F2 gene truncated to just 11 amino acids in length which is inactive. Using reverse genetics to engineer viruses in which full length protein was restored, Hai et al. noted no increase in virulence in mice or ferrets [140]. Even the introduction of the notorious 1918 like N66S point mutation did not affect the outcome of infection in these models. Thus acquisition of virulence by restoration of this gene to the pandemic virus seems unlikely [140].

6.10.2 Cleavage of Influenza HA Leads to Extended Tropism

The tropism of influenza virus is not only determined by its receptor use. SA is a widely distributed cell surface sugar but influenza in humans is largely restricted to the respiratory tract. The reliance on host cell proteases to cleave and thus activate the fusogenic properties of the HA protein determines the organs in which the virus can undergo productive infection. In humans the abundance of Clara Tryptase in respiratory secretions allows the virus efficient replication in the lung [150]. In highly pathogenic avian influenza viruses such as H5 and H7 strains, the insertion of a polybasic motif allows the HA to be cleaved by ubiquitous proteases such as furin, facilitating systemic infection [151]. Despite the high mortality rates of those afflicted with the 1918 virus, it does not contain the polybasic cleavage site found in the highly pathogenic H5N1 viruses. The acquisition of this virulence motif in H5 and H7 subtypes of HA occurs during amplification in poultry [152, 153]. This motif has not been seen in any pH1N1 isolates in 2009 or 2010, and pH1N1 viruses remain dependent on the addition of trypsin to growth media for their propagation in cell culture.

6.10.3 The NS1 Protein

The influenza virus counteracts the otherwise suppressive effect of the interferon response using a nonstructural protein NS1, reviewed in detail by Hale et al., [141]. NS1 works in at least two ways to prevent induction of interferon. Firstly in the cytoplasm NS1 binds dsRNA and other RNAs that are the likely triggers of innate immunity as well as forming a complex with the host cell pattern recognition receptor RIG-I and its controlling protein TRIM25. Secondly in the nucleus some NS1 proteins can bind to the host cell factor CPSF30 and in doing so they suppress the processing of newly synthesized mRNAs and prevent their export to the cytoplasm. This latter function is strain specific. It has been suggested that viruses that have enhanced ability to perform both these functions may induce a more severe disease because they can evade the innate immune response more efficiently. Indeed introduction of CPSF30 binding ability to the lab adapted PR8 vaccine strain that usually lack this function enhanced its virulence in mice [142]. The pH1N1 virus lacks CPSF binding capacity. However Hale et al. have shown that reintroduction of this phenotype did not affect virulence of pH1N1 in ferrets or mice [143]. Despite lacking CPSF30 binding capacity, the pH1N1 virus induces very low levels of interferon in infected cells [144].

7 The Future

The 1968 and 1957 pandemic viruses both displaced the previously circulating subtypes. In contrast the re-emergence of H1N1 in humans in 1977 was not associated with subtype displacement likely because a large cohort of the older population was not susceptible to the virus and therefore remained viable hosts for the contemporary H3N2 viruses. Similarly due to residual immunity in the elderly, pH1N1 2009 has not displaced the H3N2 subtype. Initially it was believed that the seasonal H1N1 subtype may have gone extinct after pH1N1 emerged, as there was a period of many months where this virus was not isolated, however recently seasonal H1N1 isolates have been detected in Texas [145]. The trivalent vaccine administered in 2010 contains pH1N1, H3N2 and influenza B virus antigens but no seasonal H1N1 component.

Influenza is a seasonal disease. Infections peak once a year in the cold, dry season in the Northern or Southern hemispheres, although in the tropics the seasons are less clearly separated and it may be that virus continually circulates [146]. The Royal College of General Practitioners (RCGP) scheme in the UK has monitored the incidence of ILI since the emergence of the H3N2 subtype in the 1968 pandemic (Fig. 3), the re-emergence of related H3N2 and seasonal H1N1 strains has been observed year on year and is due to the capacity of the virus to accumulate small point mutations in HA and NA antigens, the process called antigenic drift. These mutations occur at antigenic sites and allow the circulating virus to evade immune suppression by throwing off antibody binding through confirmation changes or glycosylation events. Accumulated drift mutations may ultimately change the phenotype of the virus. The virus may alter its affinity or specificity for the receptors on the host cell surface in its efforts to avoid the immune response [147]. Indeed it is clear that, as it has evolved over four decades in humans, the H3N2 virus has changed its receptor binding affinities with phenotypic consequence [43, 148].

As herd immunity increases against the newly emerged pH1N1 virus, it is not in doubt that antigenic drift will occur. However, the resulting phenotypic changes are unknown and currently unpredictable. Moreover since the virus has re-infected swine, a species in which frequent reassortments occur [149], the evolution of this 2009 H1N1 virus and the consequence of reassortment events in animals for human disease remain to be observed in the coming years.

References

1. Morens DM, Taubenberger JK, Harvey HA, Memoli MJ. The 1918 influenza pandemic: lessons for 2009 and the future. Critical Care Medicine. 2010;38(4 Suppl):e10–20.
2. Johnson NPAS, Mueller J. Updating the accounts: global mortality of the 1918–1920 "Spanish" influenza pandemic. Bulletin of the History of Medicine. 2002 Jan;76(1):105–15
3. Gouarin S, Vabret A, Dina J, Petitjean J, Brouard J, Freymuth F. Study of Influenza C Virus Infection in France. IBIS. 2008;1446(April):1441–1446.
4. Shope RE. Swine Influenza : I. Experimental Transmission and Pathology. The Journal of Experimental Medicine. 1931 Jul;54(3):349–59.
5. Shope RE. The Etiology of Swine Influenza. Science (New York, N.Y.). 1931 Feb;73(1886):214–5.
6. Laidlaw PP, Smith W, with C. H. The susceptibility of mice to the viruses of human and swine influenza. Lancet, The. 1934;2(859):
7. Andrewes CH, Laidlaw PP, with W. Virus obtained from influenza patients. Lancet, The. 1933;19(66):

8. Taubenberger JK, Morens DM. 1918 Influenza: the mother of all pandemics. Emerging Infectious Diseases. 2006 Jan;12(1):15–22.

9. Reid AH, Fanning TG, Hultin JV, Taubenberger JK. Origin and evolution of the 1918 "Spanish" influenza virus hemagglutinin gene. Proceedings of the National Academy of Sciences of the United States of America. 1999 Feb;96(4):1651–6.

10. Taubenberger JK. Initial Genetic Characterization of the 1918 "Spanish" Influenza Virus Science. 1997 Mar;275(5307):1793–1796.

11. Tumpey TM, Basler CF, Aguilar PV, Zeng H, Solórzano A, Swayne DE, et al. Characterization of the reconstructed 1918 Spanish influenza pandemic virus. Science (New York, N.Y.). 2005 Oct;310(5745):77–80.

12. Reid AH, Fanning TG, Janczewski TA, Taubenberger JK. Characterization of the 1918 "Spanish" influenza virus neuraminidase gene. Proceedings of the National Academy of Sciences of the United States of America. 2000 Jun;97(12):6785–90.

13. Basler CF, Reid AH, Dybing JK, Janczewski TA, Fanning TG, Zheng H, et al. Sequence of the 1918 pandemic influenza virus nonstructural gene (NS) segment and characterization of recombinant viruses bearing the 1918 NS genes. Proceedings of the National Academy of Sciences of the United States of America. 2001 Feb;98(5):2746–51.

14. Reid AH, Fanning TG, Janczewski TA, McCall S, Taubenberger JK. Characterization of the 1918 "Spanish" influenza virus matrix gene segment. Journal of Virology. 2002 Nov;76(21):10717–23.

15. Reid AH, Fanning TG, Janczewski TA, Lourens RM, Taubenberger JK. Novel origin of the 1918 pandemic influenza virus nucleoprotein gene. Journal of Virology. 2004 Nov;78(22):12462–70.

16. Taubenberger JK, Reid AH, Lourens RM, Wang R, Jin G, Fanning TG. Characterization of the 1918 influenza virus polymerase genes. Nature. 2005 Oct;437(7060):889–93.

17. Brand JMA van den, Stittelaar KJ, Amerongen G van, Rimmelzwaan GF, Simon J, Wit E de, et al. Severity of pneumonia due to new H1N1 influenza virus in ferrets is intermediate between that due to seasonal H1N1 virus and highly pathogenic avian influenza H5N1 virus. The Journal of Infectious Diseases. 2010 Apr;201(7):993–9.

18. Memoli MJ, Tumpey TM, Jagger BW, Dugan VG, Sheng Z-M, Qi L, et al. An early "classical" swine H1N1 influenza virus shows similar pathogenicity to the 1918 pandemic virus in ferrets and mice. Virology. 2009;393(2):338–345.

19. Tumpey TM, García-Sastre A, Taubenberger JK, Palese P, Swayne DE, Basler CF. Pathogenicity and immunogenicity of influenza viruses with genes from the 1918 pandemic virus. Proceedings of the National Academy of Sciences of the United States of America. 2004;101(9):3166–3171.

20. Weingartl HM, Albrecht RA, Lager KM, Babiuk S, Marszal P, Neufeld J, et al. Experimental infection of pigs with the human 1918 pandemic influenza virus. Journal of Virology. 2009;83(9):4287–4296.

21. Meunier I, Pillet S, Simonsen JN, Messling V von. Influenza pathogenesis: lessons learned from animal studies with H5N1, H1N1 Spanish, and pandemic H1N1 2009 influenza. Critical Care Medicine. 2010 Apr;38 (4 Suppl):e21-9.

22. Watanabe T, Watanabe S, Shinya K, Kim JH, Hatta M, Kawaoka Y. Viral RNA polymerase complex promotes optimal growth of 1918 virus in the lower respiratory tract of ferrets. Proceedings of the National Academy of Sciences of the United States of America. 2009 Jan;106(2):588–92.

23. Belser J a, Wadford DA, Pappas C, Gustin KM, Maines TR, Pearce MB, et al. Pathogenesis of pandemic influenza A (H1N1) and triple-reassortant swine influenza A (H1) viruses in mice. Journal of Virology. 2010 May;84(9):4194–203.

24. Klugman KP, Mills Astley C, Lipsitch M. Time from Illness Onset to Death, 1918 Infl uenza and Pneumococcal Pneumonia Clinical infectious diseases : an official publication of the Infectious Diseases Society of America. 2009 Apr;15(2):346–347.

25. Morens DM, Taubenberger JK, Fauci AS. Predominant role of bacterial pneumonia as a cause of death in pandemic influenza: implications for pandemic influenza preparedness. The Journal of Infectious Diseases. 2008 Oct;198(7):962–70.

26. Brown JN, Palermo RE, Baskin CR, Gritsenko M, Sabourin PJ, Long JP, et al. Macaque Proteome Response to Highly Pathogenic Avian Influenza and 1918 Reassortant Influenza Virus Infections. Journal of Virology. 2010 Sep;

27. Billharz R, Zeng H, Proll SC, Korth MJ, Lederer S, Albrecht R, et al. The NS1 protein of the 1918 pandemic influenza virus blocks host interferon and lipid metabolism pathways. Journal of Virology. 2009 Oct;83(20):10557–70.

28. Kash JC, Basler CF, García-Sastre A, Carter V, Billharz R, Swayne DE, et al. Global host immune response: pathogenesis and transcriptional profiling of type A influenza viruses expressing the hemagglutinin and neuraminidase genes from the 1918 pandemic virus. Journal of Virology. 2004 Sep;78(17):9499–511.

29. Kobasa D, Jones SM, Shinya K, Kash JC, Copps J, Ebihara H, et al. Aberrant innate immune response in lethal infection of macaques with the 1918 influenza virus. Nature. 2007 Jan;445(7125):319–23.

30. Nelson MI, Viboud C, Simonsen L, Bennett RT, Griesemer SB, St George K, et al. Multiple reassortment events in the evolutionary history of H1N1 influenza A virus since 1918. PLoS Pathogens. 2008 Feb;4(2):e1000012.

31. Lagacé-Wiens PRS, Rubinstein E, Gumel A. Influenza epidemiology–past, present, and future. Critical Care Medicine. 2010 Apr;38(4 Suppl):e1-9.

32. Cox NJ, Subbarao K. Global epidemiology of influenza: past and present. Annual Review of Medicine. 2000 Jan;51407–21.

33. Wertheim JO. The re-emergence of H1N1 influenza virus in 1977: a cautionary tale for estimating divergence times using biologically unrealistic sampling dates. PloS One. 2010 Jan;5(6):e11184.

34. Zimmer SM, Burke DS. Historical perspective–Emergence of influenza A (H1N1) viruses. The New England Journal of Medicine. 2009 Jul;361(3):279–85.

35. Scholtissek C, Hoyningen V von, Rott R. Genetic relatedness between the new 1977 epidemic strains (H1N1) of influenza and human influenza strains isolated between 1947 and 1957 (H1N1). Virology. 1978 Sep;89(2): 613–7.

36. Nakajima K, Desselberger U, Palese P. Recent human influenza A (H1N1) viruses are closely related genetically to strains isolated in 1950. Nature. 1978 Jul;274(5669):334–9.

37. Kishida N, Sakoda Y, Isoda N, Matsuda K, Eto M, Sunaga Y, et al. Pathogenicity of H5 influenza viruses for ducks. Archives of Virology. 2005 Jul;150(7):1383–92.

38. Homme PJ, Easterday BC. Avian influenza virus infections. IV. Response of pheasants, ducks, and geese to influenza A-turkey-Wisconsin-1966 virus. Avian Diseases. 1970 May;14(2):285–90.

39. Krauss S, Webster R. Avian Influenza Virus Surveillance and Wild Birds: Past and Present. Avian Diseases. 2010;(54):394–398.

40. Stallknecht DE, Kearney MT, Shane SM, Zwank PJ. Effects of pH, temperature, and salinity on persistence of avian influenza viruses in water. Avian Diseases. 1990;34(2):412–8.

41. Stallknecht DE, Shane SM, Kearney MT, Zwank PJ. Persistence of avian influenza viruses in water. Avian Diseases. 1990;34(2):406–11.

42. Donis RO, Bean WJ, Kawaoka Y, Webster RG. Distinct lineages of influenza virus H4 hemagglutinin genes in different regions of the world. Virology. 1989 Apr;169(2):408–17.

43. Thompson CI, Barclay WS, Zambon MC, Pickles RJ. Infection of human airway epithelium by human and avian strains of influenza a virus. Journal of Virology. 2006 Aug;80(16):8060–8.

44. Thompson CI, Barclay WS, Zambon MC. Changes in in vitro susceptibility of influenza A H3N2 viruses to a neuraminidase inhibitor drug during evolution in the human host. The Journal of Antimicrobial Chemotherapy. 2004 May;53(5):759–65.

45. WHO. WHO | Cumulative Number of Confirmed Human Cases of Avian Influenza A/(H5N1) Reported to WHO 2010.

46. Elliot AJ, et al. Surveillance of influenza-like illness in England and Wales during 1966–2006. Euro Surveill. 2006;11(10):pii=651. Available from: http://www.eurosurveillance.org/ViewArticle.aspx?ArticleId=651.

47. Lessler J, Cummings DAT, Fishman S, Vora A, Burke DS. Transmissibility of swine flu at Fort Dix, 1976. Journal of the Royal Society, Interface/The Royal Society. 2007 Aug;4(15):755–62.

48. Gaydos JC, Top FH, Hodder RA, Russell PK. Swine Influenza A Outbreak, Fort Dix, New Jersey, 1976. Emerging Infectious Diseases. 2006;12(1):23–28.

49. Kida H, Ito T, Yasuda J, Shimizu Y, Itakura C, Shortridge KF, et al. Potential for transmission of avian influenza viruses to pigs. Journal of General Virology. 1994 Sep;75(9):2183–2188.

50. Garten RJ, Davis CT, Russell C a, Shu B, Lindstrom S, Balish A, et al. Antigenic and genetic characteristics of swine-origin 2009 A(H1N1) influenza viruses circulating in humans. Science (New York, N.Y.). 2009 Jul;325(5937):197–201.

51. Smith GJD, Vijaykrishna D, Bahl J, Lycett SJ, Worobey M, Pybus OG, et al. Origins and evolutionary genomics of the 2009 swine-origin H1N1 influenza A epidemic. Nature. 2009 Jun;459(7250):1122–5.

52. Ito T, Couceiro JNSS, Kelm S, Baum LG, Krauss S, Castrucci MR, et al. Molecular Basis for the Generation in Pigs of Influenza A Viruses with Pandemic Potential J. Virol. 1998;72(9):7367–7373.

53. Naeve CW, Hinshaw VS, Webster RG. Mutations in the hemagglutinin receptor-binding site can change the biological properties of an influenza virus. Journal of Virology. 1984 Aug;51(2):567–9.

54. Matrosovich M, Tuzikov A, Bovin N, Gambaryan A, Klimov A, Castrucci MR, et al. Early alterations of the receptor-binding properties of H1, H2, and H3 avian influenza virus hemagglutinins after their introduction into mammals. Journal of Virology. 2000 Sep;74(18):8502–12.

55. Pappas C, Viswanathan K, Chandrasekaran A, Raman R, Katz JM, Sasisekharan R, et al. Receptor Specificity and Transmission of H2N2 Subtype Viruses Isolated from the Pandemic of 1957. PloS One. 2010 Jan;5(6):e11158.

56. Ayora-Talavera G, Shelton H, Scull M a, Ren J, Jones IM, Pickles RJ, et al. Mutations in H5N1 influenza virus hemagglutinin that confer binding to human tracheal airway epithelium. PloS One. 2009 Jan;4(11):e7836.

57. Chutinimitkul S, Riel D van, Munster VJ, Brand JMA van den, Rimmelzwaan GF, Kuiken T, et al. In vitro assessment of attachment pattern and replication efficiency of H5N1 influenza A viruses with altered receptor specificity. Journal of Virology. 2010 Jul;84(13):6825–33.

58. Nicholls JM, Chan RWY, Russell RJ, Air GM, Peiris JSM. Evolving complexities of influenza virus and its receptors. Trends in Microbiology. 2008 Apr;16(4):149–57.

59. Stevens J, Blixt O, Glaser L, Taubenberger JK, Palese P, Paulson JC, et al. Glycan microarray analysis of the hemagglutinins from modern and pandemic influenza viruses reveals different receptor specificities. Journal of Molecular Biology. 2006 Feb;355(5):1143–55.

60. Stevens J, Blixt O, Chen L-M, Donis RO, Paulson JC, Wilson I a. Recent avian H5N1 viruses exhibit increased propensity for acquiring human receptor specificity. Journal of Molecular Biology. 2008 Sep;381(5):1382–94.

61. Blixt O, Head S, Mondala T, Scanlan C, Huflejt ME, Alvarez R, et al. Printed covalent glycan array for ligand profiling of diverse glycan binding proteins. Proceedings of the National Academy of Sciences of the United States of America. 2004 Dec;101(49):17033–8.

62. Liao H-Y, Hsu C-H, Wang S-C, Liang C-H, Yen H-Y, Su C-Y, et al. Differential Receptor Binding Affinities of Influenza Hemagglutinins on Glycan Arrays. Journal of the American Chemical Society. 2010 Sep;286–291.

63. Childs RA, Palma AS, Wharton S, Matrosovich T, Liu Y, Chai W, et al. Receptor-binding specificity of pandemic influenza A (H1N1) 2009 virus determined by carbohydrate microarray. Nature Biotechnology. 2010 Feb;28(2):178–178.

64. Trifonov V, Khiabanian H, Rabadan R. Influenza A (H1N1) Virus. Emerging Infectious Diseases. 2009; 115–119.

65. Pebody RG, McLean E, Zhao H, Cleary P, Bracebridge S, Foster K, et al. Pandemic Influenza A (H1N1) 2009 and mortality in the United Kingdom: risk factors for death, April 2009 to March 2010. Euro Surveillance : Bulletin Européen Sur Les Maladies Transmissibles=European Communicable Disease Bulletin. 2010 Jan;15(20):1–11.

66. HPA. The role of the Health Protection Agency in the "containment" phase during the first wave of pandemic influenza in England in 2009. 2010.

67. Miller E, Hoschler K, Hardelid P, Stanford E, Andrews N, Zambon M. Incidence of 2009 pandemic influenza A H1N1 infection in England: a cross-sectional serological study. The Lancet. 2010 Apr;375(9720): 1100–1108.

68. HPA. Weekly epidemiological update 2010;36[cited 2010 Sep 13] Available from: http://www.hpa.org.uk/web/HPAweb&Page&HPAwebAutoListName/Page/1243928258560.

69. WHO. WHO | Pandemic (H1N1) 2009 - update 112 World Health Organisation, Global Alert Response. 2010;

70. Girard MP, Katz J, Pervikov Y, Palkonyay L, Kieny M-P. Report of the 6th meeting on the evaluation of pandemic influenza vaccines in clinical trials World Health Organization, Geneva, Switzerland, 17–18 February 2010. Vaccine. 2010 Oct;28(42):6811–20.

71. Johnson BF, Wilson LE, Ellis J, Elliot AJ, Barclay WS, Pebody RG, et al. Fatal cases of influenza a in childhood. PloS One. 2009 Jan;4(10):e7671.

72. Sachedina N, Donaldson LJ. Paediatric mortality related to pandemic influenza A H1N1 infection in England: an observational population-based study. The Lancet. 2010 Oct;6736(10):1–7.

73. Rothberg MB, Haessler SD. Complications of seasonal and pandemic influenza. Critical Care Medicine. 2010;38(4 Suppl):e91-7.

74. Ferguson NM, Cummings DAT, Fraser C, Cajka JC, Cooley PC, Burke DS. Strategies for mitigating an influenza pandemic. Nature. 2006;442(7101):448–452.

75. Kawaguchi R. Influenza (H1N1) 2009 Outbreak and School Closure, Osaka Prefecture, Japan. Emerging Infectious Diseases. 2009 Oct;15(10):2009-2009.

76. Wu JT. School Closure and Mitigation of Pandemic (H1N1) 2009, Hong Kong. Emerging Infectious Diseases. 2010 Mar;16(3):10–13.

77. Miller E, Hoschler K, Hardelid P, Stanford E, Andrews N, Zambon M. Incidence of 2009 pandemic influenza A H1N1 infection in England: a cross-sectional serological study. The Lancet. 2010;375(9720):1100–1108.

78. Wu JT, Ma ESK, Lee CK, Chu DKW, Ho P, Shen AL, et al. The Infection Attack Rate and Severity of 2009 Pandemic H1N1 Influenza in Hong Kong. Clinical Infectious Diseases. 2010 Nov;51(10):1184–1191

79. Reyes L, Arvelo W, Estevez A, Gray J, Moir JC, Gordillo B, et al. Population-based surveillance for 2009 pandemic influenza A (H1N1) virus in Guatemala, 2009. Influenza and Other Respiratory Viruses. 2010 May;4(3):129–40.

80. Libster R, Bugna J, Coviello S, Hijano DR, Dunaiewsky M, Reynoso N, et al. Pediatric hospitalizations associated with 2009 pandemic influenza A (H1N1) in Argentina. The New England Journal of Medicine. 2010;362(1):45–55.

81. 't Klooster T van, Wielders C, Donker T, Isken L, Meijer A, Den Wijngaard C van, et al. Surveillance of Hospitalisations for 2009 Pandemic Influenza A(H1N1) in the Netherlands, 5 June – 31 December 2009 Eurosurveillance. 2010;15(2):4.

82. Pitman RJ, Melegaro A, Gelb D, Siddiqui MR, Gay NJ, Edmunds WJ. Assessing the burden of influenza and other respiratory infections in England and Wales. The Journal of Infection. 2007 Jun;54(6):530–8.

83. Zhao C, Gan Y, Sun J. Radiographic study of severe Influenza-A (H1N1) disease in children. European Journal of Radiology. 2010 Oct;1–5.

84. Zheng Y, He Y, Deng J, Lu Z, Wei J, Yang W, et al. Hospitalized children with 2009 influenza a (H1N1) infection in Shenzhen, China, november-december 2009. Pediatric Pulmonology. 2010 Oct;(July):1–7.

85. Halasa NB. Update on the 2009 pandemic influenza A H1N1 in children. Current Opinion in Pediatrics. 2010 Feb;22(1):83–7.

86. Feiterna-Sperling C, Edelmann A, Nickel R, Magdorf K, Bergmann F, Rautenberg P, et al. Pandemic Influenza A (H1N1) Outbreak among 15 School-Aged HIV-1-Infected Children. Clinical Infectious Diseases: An Official Publication of the Infectious Diseases Society of America. 2010 Nov;518–12.

87. Falagas ME, Koletsi PK, Baskouta E, Rafailidis PI, Dimopoulos G, Karageorgopoulos DE. Pandemic A(H1N1) 2009 influenza: review of the Southern Hemisphere experience. Epidemiology and Infection. 2010 Oct;5:1–14.

88. Sasbón JS, Centeno M a, García MD, Boada NB, Lattini BE, Motto E a, et al. Influenza A (pH1N1) infection in children admitted to a pediatric intensive care unit: Differences with other respiratory viruses. Pediatric Critical Care Medicine: A Journal of the Society of Critical Care Medicine and the World Federation of Pediatric Intensive and Critical Care Societies. 2010 Apr;12(1):1–5.

89. Bautista E, Chotpitayasunondh T, Gao Z, Harper S a, Shaw M, Uyeki TM, et al. Clinical aspects of pandemic 2009 influenza A (H1N1) virus infection. The New England Journal of Medicine. 2010 May;362(18):1708–19.

90. Siston AM, Rasmussen S a, Honein M a, Fry AM, Seib K, Callaghan WM, et al. Pandemic 2009 influenza A(H1N1) virus illness among pregnant women in the United States. JAMA : The Journal of the American Medical Association. 2010 Apr;303(15):1517–25.

91. Goodnight WH, Soper DE. Pneumonia in pregnancy. Critical Care Medicine. 2005 Oct;33(Supplement): S390-S397.

92. Rasmussen S a, Jamieson DJ, Bresee JS. Pandemic influenza and pregnant women. Emerging Infectious Diseases. 2008 Jan;14(1):95–100.

93. Jamieson DJ, Theiler RN, Rasmussen S a. Emerging infections and pregnancy. Emerging Infectious Diseases. 2006 Nov;12(11):1638–43.

94. Gordon CL, Johnson PDR, Permezel M, Holmes NE, Gutteridge G, McDonald CF, et al. Association between severe pandemic 2009 influenza A (H1N1) virus infection and immunoglobulin G(2) subclass deficiency. Clinical Infectious Diseases: An Official Publication of the Infectious Diseases Society of America. 2010 Mar;50(5):672–8.

95. Nguyen-Van-Tam JS, Openshaw PJM, Hashim a, Gadd EM, Lim WS, Semple MG, et al. Risk factors for hospitalisation and poor outcome with pandemic A/H1N1 influenza: United Kingdom first wave (May-September 2009). Thorax. 2010 Jul;65(7):645–51.

96. Shieh W-J, Blau DM, Denison AM, Deleon-Carnes M, Adem P, Bhatnagar J, et al. 2009 pandemic influenza A (H1N1): pathology and pathogenesis of 100 fatal cases in the United States. The American Journal of Pathology. 2010 Jul;177(1):166–75.

97. Gill JR, Sheng Z-M, Ely SF, Guinee DG, Beasley MB, Suh J, et al. Pulmonary pathologic findings of fatal 2009 pandemic influenza A/H1N1 viral infections. Archives of Pathology & Laboratory Medicine. 2010 Feb;134(2):235–43.

98. Mauad T, Hajjar LA, Callegari GD, Silva LFF da, Schout D, Galas FRBG, et al. Lung pathology in fatal novel human influenza A (H1N1) infection. American Journal of Respiratory and Critical Care Medicine. 2010 Jan;181(1):72–9.

99. Palacios G, Hornig M, Cisterna D, Savji N, Bussetti AV, Kapoor V, et al. Streptococcus pneumoniae coinfection is correlated with the severity of H1N1 pandemic influenza. PloS One. 2009 Jan;4(12):e8540.

100. Waddington CS, Walker WT, Oeser C, Reiner A, John T, Wilkins S, et al. Safety and immunogenicity of AS03B adjuvanted split virion versus non-adjuvanted whole virion H1N1 influenza vaccine in UK children aged 6 months-12 years: open label, randomised, parallel group, multicentre study. BMJ (Clinical research ed.). 2010 Jan;340c2649.

101. Munster VJ, Wit E de, Brand JM a van den, Herfst S, Schrauwen EJ a, Bestebroer TM, et al. Pathogenesis and transmission of swine-origin 2009 A(H1N1) influenza virus in ferrets. Science (New York, N.Y.). 2009 Jul;325(5939):481–3.

102. Itoh Y, Shinya K, Kiso M, Watanabe T, Sakoda Y, Hatta M, et al. In vitro and in vivo characterization of new swine-origin H1N1 influenza viruses. Nature. 2009;460(7258):1021–1025.

103. Shinya K, Ebina M, Yamada S, Ono M, Kasai N, Kawaoka Y. Avian flu: influenza virus receptors in the human airway. Nature. 2006 Mar;440(7083):435–6.

104. Bright RA, Medina M-jo, Xu X, Perez-Oronoz G, Wallis TR, Davis XM, et al. Incidence of adamantane resistance among influenza A (H3N2) viruses isolated worldwide from 1994 to 2005: a cause for concern. The Lancet. 2005 Oct;366(9492):1175–1181.

105. Sweet C, Hayden FG, Jakeman KJ, Grambas S, Hay a J. Virulence of rimantadine-resistant human influenza A (H3N2) viruses in ferrets. The Journal of Infectious Diseases. 1991 Nov;164(5):969–72.

106. Nelson MI, Simonsen L, Viboud C, Miller M a, Holmes EC. The origin and global emergence of adamantane resistant A/H3N2 influenza viruses. Virology. 2009 Jun;388(2):270–8.

107. Rungrotmongkol T, Intharathep P, Malaisree M, Nunthaboot N, Kaiyawet N, Sompornpisut P, et al. Susceptibility of antiviral drugs against 2009 influenza A (H1N1) virus. Biochemical and Biophysical Research Communications. 2009 Jul;385(3):390–4.

108. Moss RB, Davey RT, Steigbigel RT, Fang F. Targeting pandemic influenza: a primer on influenza antivirals and drug resistance. The Journal of Antimicrobial Chemotherapy. 2010 Jun;65(6):1086–93.

109. Jefferson T, Jones M, Doshi P, Del Mar C. Neuraminidase inhibitors for preventing and treating influenza in healthy adults: systematic review and meta-analysis BMJ. 2009 Dec;339(dec07 2):b5106-b5106.

110. Carr J, Ives J, Kelly L, Lambkin R, Oxford J, Mendel D, et al. Influenza virus carrying neuraminidase with reduced sensitivity to oseltamivir carboxylate has altered properties in vitro and is compromised for infectivity and replicative ability in vivo. Antiviral Research. 2002 May;54(2):79–88.

111. Ives J a L, Carr J a, Mendel DB, Tai CY, Lambkin R, Kelly L, et al. The H274Y mutation in the influenza A/ H1N1 neuraminidase active site following oseltamivir phosphate treatment leave virus severely compromised both in vitro and in vivo. Antiviral Research. 2002 Aug;55(2):307–17.

112. Meijer A. Oseltamivir-Resistant Influenza Virus A (H1N1), Europe, 2007–08 Season. Emerging Infectious Diseases. 2009 Apr;15(4):552–560.

113. Collins PJ, Haire LF, Lin YP, Liu J, Russell RJ, Walker P a, et al. Structural basis for oseltamivir resistance of influenza viruses. Vaccine. 2009 Oct;27(45):6317–23.

114. Rameix-Welti M-A, Enouf V, Cuvelier F, Jeannin P, Werf S van der. Enzymatic properties of the neuraminidase of seasonal H1N1 influenza viruses provide insights for the emergence of natural resistance to oseltamivir. PLoS Pathogens. 2008 Jul;4(7):e1000103.

115. Chen H. Oseltamivir-Resistant Influenza A Pandemic (H1N1) 2009 Virus, Hong Kong, China Emerging Infectious Diseases. 2009 Dec;15(12):1970–1972.

116. Baz M, Abed Y, Papenburg J, Bouhy X, Hamelin M-E, Boivin G. Emergence of oseltamivir-resistant pandemic H1N1 virus during prophylaxis. The New England Journal of Medicine. 2009 Dec;361(23):2296–7.

117. Duan S, Boltz DA, Seiler P, Li J, Bragstad K, Nielsen LP, et al. Oseltamivir-resistant pandemic H1N1/2009 influenza virus possesses lower transmissibility and fitness in ferrets. PLoS Pathogens. 2010 Jan;6(7):e1001022.

118. Seibert CW, Kaminski M, Philipp J, Rubbenstroth D, Albrecht RA, Schwalm F, et al. Oseltamivir-resistant variants of the 2009 pandemic H1N1 influenza A virus are not attenuated in the guinea pig and ferret transmission models. Journal of Virology. 2010 Aug;

119. Hamelin M-È, Baz M, Abed Y, Couture C, Joubert P, Beaulieu É, et al. Oseltamivir-Resistant Pandemic A/ H1N1 Virus Is as Virulent as Its Wild-Type Counterpart in Mice and Ferrets. PLoS Pathogens. 2010 Jul;6(7):e1001015.

120. Miller RR, MacLean a R, Gunson RN, Carman WF. Occurrence of haemagglutinin mutation D222G in pandemic influenza A(H1N1) infected patients in the West of Scotland, United Kingdom, 2009–10. Euro Surveillance: Bulletin Européen Sur Les Maladies Transmissibles=European Communicable Disease Bulletin. 2010 Jan;15(16):19534–19534.

121. Puzelli S. Transmission of Hemagglutinin D222G Mutant Strain of Pandemic (H1N1) 2009 Virus. Emerging Infectious Diseases. 2010 May;16(5):2009–2011.

122. Melidou A, Gioula G, Exindari M, Chatzidimitriou D, Diza E, Malisiovas N. Molecular and phylogenetic analysis of the haemagglutinin gene of pandemic influenza H1N1 2009 viruses associated with severe and fatal infections. Virus Research. 2010 Aug;151(2):192–9.

123. Chen H, Wen X, To KKW, Wang P, Tse H, Chan JFW, et al. Quasispecies of the D225G substitution in the hemagglutinin of pandemic influenza A(H1N1) 2009 virus from patients with severe disease in Hong Kong, China. The Journal of Infectious Diseases. 2010 May;201(10):1517–21.

124. Liu Y, Childs RA, Matrosovich T, Wharton S, Palma AS, Chai W, et al. Altered receptor specificity and cell tropism of D222G haemagglutinin mutants from fatal cases of pandemic A(H1N1) 2009 influenza. Journal of Virology. 2010 Sep

125. Antón A, Marcos MA, Martínez MJ, Ramón S, Martínez A, Cardeñosa N, et al. D225G mutation in the hemagglutinin protein found in 3 severe cases of 2009 pandemic influenza A (H1N1) in Spain. Diagnostic Microbiology and Infectious Disease. 2010 Jun;67(2):207–8.

126. WHO. Preliminary review of D222G amino acid substitution in the haemagglutinin of pandemic influenza A (H1N1) 2009 viruses. Relevé épidémiologique hebdomadaire/Section d'hygiène du Secrétariat de la Société des Nations = Weekly epidemiological record/Health Section of the Secretariat of the League of Nations. 2010 Jan;85(4):21–2.

127. Chutinimitkul S, Herfst S, Steel J, Lowen AC, Ye J, Riel D van, et al. Virulence-associated substitution D222G in hemagglutinin of 2009 pandemic influenza A(H1N1) virus affects receptor binding. Journal of Virology. 2010 Sep

128. Ning Z-Y, Luo M-Y, Qi W-B, Yu B, Jiao P-R, Liao M. Detection of expression of influenza virus receptors in tissues of BALB/c mice by histochemistry. Veterinary Research Communications. 2009 Aug;895–903.

129. Chen W, Calvo PA, Malide D, Gibbs J, Schubert U, Bacik I, et al. A novel influenza A virus mitochondrial protein that induces cell death. Nature Medicine. 2001;7(12):1306–1312.

130. Zamarin D, García-Sastre A, Xiao X, Wang R, Palese P. Influenza virus PB1-F2 protein induces cell death through mitochondrial ANT3 and VDAC1. PLoS Pathogens. 2005 Sep;1(1):e4.

131. Zamarin D, Ortigoza MB, Palese P. Influenza A virus PB1-F2 protein contributes to viral pathogenesis in mice. Journal of Virology. 2006 Aug;80(16):7976–83.

132. Gibbs JS, Malide D, Hornung F, Bennink JR, Yewdell JW. The Influenza A Virus PB1-F2 Protein Targets the Inner Mitochondrial Membrane via a Predicted Basic Amphipathic Helix That Disrupts Mitochondrial Function. Journal of Virology. 2003 Jul;77(13):7214–7224.

133. McAuley JL, Hornung F, Boyd KL, Smith AM, McKeon R, Bennink J, et al. Expression of the 1918 influenza A virus PB1-F2 enhances the pathogenesis of viral and secondary bacterial pneumonia. Cell Host & Microbe. 2007 Oct;2(4):240–9.

134. McAuley JL, Zhang K, McCullers JA. The effects of influenza A virus PB1-F2 protein on polymerase activity are strain specific and do not impact pathogenesis. Journal of Virology. 2010 Jan;84(1):558–64.

135. Mazur I, Anhlan D, Mitzner D, Wixler L, Schubert U, Ludwig S. The proapoptotic influenza A virus protein PB1-F2 regulates viral polymerase activity by interaction with the PB1 protein. Cellular Microbiology. 2008 May;10(5):1140–52.

136. Zell R, Krumbholz A, Eitner A, Krieg R, Halbhuber K-J, Wutzler P. Prevalence of PB1-F2 of influenza A viruses. The Journal of General Virology. 2007 Feb;88(Pt 2):536–46.

137. Henkel M, Mitzner D, Henklein P, Meyer-Almes F-J, Moroni A, Difrancesco ML, et al. The Proapoptotic Influenza A Virus Protein PB1-F2 Forms a Nonselective Ion Channel. PloS One. 2010 Jan;5(6):e11112.

138. McCullers JA, English BK. Improving therapeutic strategies for secondary bacterial pneumonia following influenza. Future Microbiology. 2008 Aug;3397–404.

139. Conenello GM, Zamarin D, Perrone LA, Tumpey T, Palese P. A single mutation in the PB1-F2 of H5N1 (HK/97) and 1918 influenza A viruses contributes to increased virulence. PLoS Pathogens. 2007 Oct;3(10):1414–21.

140. Hai R, Schmolke M, Varga ZT, Manicassamy B, Wang TT, Belser JA, et al. PB1-F2 expression by the 2009 pandemic H1N1 influenza virus has minimal impact on virulence in animal models. Journal of Virology. 2010 May;84(9):4442–50.

141. Hale BG, Randall RE, Ortín J, Jackson D. The multifunctional NS1 protein of influenza A viruses. The Journal of General Virology. 2008 Oct;89(Pt 10):2359–76.

142. Steidle S, Martínez-Sobrido L, Mordstein M, Lienenklaus S, García-Sastre A, Stäheli P, et al. Glycine 184 in the non-structural protein NS1 determines virulence of influenza A virus strain PR8 without affecting the host interferon response. Journal of Virology. 2010 Oct

143. Hale BG, Steel J, Medina RA, Manicassamy B, Ye J, Hickman D, et al. Inefficient control of host gene expression by the 2009 pandemic H1N1 influenza A virus NS1 protein. Journal of Virology. 2010 Jul;84(14):6909–22.

144. Osterlund P, Pirhonen J, Ikonen N, Rönkkö E, Strengell M, Mäkelä SM, et al. Pandemic H1N1 2009 influenza A virus induces weak cytokine responses in human macrophages and dendritic cells and is highly sensitive to the antiviral actions of interferons. Journal of Virology. 2010 Feb;84(3):1414–22.

145. Texas State Health services. Texas Influenza surveillance Report 2010–2011 Season MMWR Week 42 Texas Influenza surveillance Report 2010–2011 Season MMWR Week 42. 2010

146. Russell C a, Jones TC, Barr IG, Cox NJ, Garten RJ, Gregory V, et al. The global circulation of seasonal influenza A (H3N2) viruses. Science (New York, N.Y.). 2008 Apr;320(5874):340–6.

147. Hensley SE, Das SR, Bailey AL, Schmidt LM, Hickman HD, Jayaraman A, et al. Hemagglutinin receptor binding avidity drives influenza A virus antigenic drift. Science (New York, N.Y.). 2009 Oct;326(5953):734–6.

148. Thompson CI, Barclay WS, Zambon MC. Changes in in vitro susceptibility of influenza A H3N2 viruses to a neuraminidase inhibitor drug during evolution in the human host. The Journal of Antimicrobial Chemotherapy. 2004 May;53(5):759–65.

149. Vijaykrishna D, Poon LLM, Zhu HC, Ma SK, Li OTW, Cheung CL, et al. Reassortment of pandemic H1N1/2009 influenza A virus in swine. Science (New York, N.Y.). 2010 Jun;328(5985):1529.

150. Kido H, Yokogoshi Y, Sakai K, Tashiro M, Kishino Y, Fukutomi a, et al. Isolation and characterization of a novel trypsin-like protease found in rat bronchiolar epithelial Clara cells. A possible activator of the viral fusion glycoprotein. The Journal of Biological Chemistry. 1992 Jul 5;267(19):13573–9.
151. Stieneke-Grober A, Vey M, Angliker H, et al. Influenza virus hemagglutinin with multibasic cleavage site is activated by furin, a subtilisin-like endoprotease. EMBO J 1992;11:2407–2414.
152. Kawaoka Y, Nestorowicz A, Alexander DJ, Webster RG. Molecular analyses of the hemagglutinin genes of H5 influenza viruses: origin of a virulent turkey strain. Virology 1987 May;158(1):218–27.
153. Bosch FX, Von Hoyningen-Huene V, Scholtissek C, Rott R. The overall evolution of the H7 influenza virus haemagglutinins is different from the evolution of the proteolytic cleavage site. The Journal of General Virology. 1982 Jul;61 (Pt l):101–4.

Management of Shunt Related Infections

Mona Al-Dabbagh and Simon Dobson

1 Introduction

Cerebrospinal fluid (CSF) shunts are used to reduce the increased pressure inside the brain's ventricles and divert that fluid to other body sites for absorption. The proximal portion of the shunt is usually placed in one of the cerebral ventricles, or sometimes inside a brain cyst, subarachnoidal, or lumbar spaces and the distal portion can be internalized or externalized. Internalized devices are usually placed in the peritoneal cavity (Ventriculo-peritoneal 'VP' shunt) or less likely in the heart atrium (Ventriculo-atrial 'VA' shunt) and rarely in the pleural cavity (Ventriculo-pleural shunt). Sometimes shunts can be externalized temporarily for CSF therapeutic diversion or intracranial pressure monitoring (External Ventricular Drain 'EVD') and sometimes they are externalized for administration of cancer chemotherapy into a brain tumor or administration of antibiotics (Ommaya reservoir).

Shunt infection is a fairly common complication of CSF shunt devices that is related to a substantial morbidity and mortality. Infection of CSF shunt devices occurs at an incidence rate of 5–15% [1–3]; and this rate decreases as a function of time, where the odds of infection reduces by fourfold per year of shunt function [4]. The reported infection rate per procedure ranges from 7.8–12.7% [3, 5, 6]. Increased rate of shunt infections have been reported with the following factors: prematurity [4, 7], the initial first month after shunt insertion [3], patients requiring serial shunt revisions [3, 4], extremes of age at shunt insertion (< 5 years and > 60 years) [3, 4, 8], shunt revision after treatment for an infected shunt [6], limited surgeon experience with CSF shunts and surgeons' case volume [3, 5, 6], etiology of Intraventricular Haemorrhage (IVH) [6], or meningomyelocele, especially if associated with late shunt placement or CSF leak after meningomyelocele closure [9].

Infections of shunt catheters are mainly caused by pathogens colonizing the skin of patients, and it is believed that bacterial colonization of the catheters starts during the surgical procedure of insertion. Coagulase negative staphylococci (CoNS), mostly *S. epidermidis*, are the most commonly isolated pathogens accounting for between 36% and 54% of infections [3, 8–10]; the next most frequent pathogen is *Staphylococcus aureus* which accounts for around 22% of the cases [3].

M. Al-Dabbagh (✉) • S. Dobson
Division of Infectious and Immunological Diseases, Department of Pediatrics,
BC Children's Hospital, Vancouver, Canada
e-mail: maldabbagh@cw.bc.ca; sdobson@cw.bc.ca

N. Curtis et al. (eds.), *Hot Topics in Infection and Immunity in Children VIII*,
Advances in Experimental Medicine and Biology 719, DOI 10.1007/978-1-4614-0204-6_9,
© Springer Science+Business Media, LLC 2011

VP shunt infection with Gram negative pathogens (including nosocomial Enterobacteriaceae and *Pseudomonas aeruginosa*) is reported in around 17% of cases [8], and is usually associated with intraperitoneal infection/ inflammation or with the haematogenous spread of the infection from other body sites. Other rarer infectious organisms include Propionibacterium spp., enterococci, β-hemolytic streptococci and Candida spp. [11–13]. Infection with Candida spp. occurs mostly in premature newborns, immunosuppressed patients and in patients that require prolonged use of broad spectrum antibiotics or in those who require bladder and intravenous catheters. It is associated with a mortality rate of 5.8% [12, 14].

The classic manifestations of shunt infection include fever, headache, irritability, convulsions, nuchal rigidity, vomiting and occasionally hyperemia of the shunt track. Abdominal pain or peritonitis may occur if the infection involves the distal portion of the shunt device, and is usually caused by Gram negative bacilli and occasionally by Propionibacterium spp. [15, 16]. Leukocytosis in addition to CSF pleiocytosis usually accompanies the infection.

Staphylococci, enterococci, Propionibacterium spp and *Pseudomonas aeruginosa* have the ability to colonize the device, forming a thick biofilm attached to the shunt surfaces, which protects them from host defense system cells. In addition, bacteria living in a biofilm usually exhibit high MIC's resulting in antimicrobial tolerance and reduced susceptibility to conventional antibiotics, making eradication of the infection more complicated and difficult unless the device is removed [17, 18].

Recurrence of shunt infection after treatment is reported to occur at a rate approaching 26% [19]. Around two thirds are caused by the same organism and one third by a different organism, where *Staphylococcus epidermidis* accounts for 29% of the cases [19]. Infection within the preceding 6 months is the most significant risk factor for recurrence of the infection [19].

2 Treatment of Shunt Infection

Defining optimal treatment measures for CSF shunt infections is rendered difficult by the absence of controlled clinical trials; all recommendations are based on observational and uncontrolled studies, and include the following:

2.1 Removal of Shunt Device

This is usually followed by insertion of temporary EVD device until clearance of CSF infection. This, in combination with appropriate antimicrobial therapy, is the most effective treatment, and has success rates that are higher than when parenteral antibiotics are given with the shunt kept in situ [20–22].

2.2 Parenteral Antibiotic Therapy

Empirical antibiotics should target the most likely pathogens. In children, vancomycin alone or in combination with an agent that covers Gram negative organisms (e.g. cefotaxime) should be initiated if CSF Gram stain reveals the presence of Gram-negative bacilli, whereas in adults, vancomycin plus an agent that covers nosocomial Gram negative pathogens (cefipime, ceftazidime or meropenem) should be considered [22]. Once culture and susceptibility results are provided, antibiotics should be tailored accordingly (Table 1). If CSF shunt infection is caused by staphylococci and the shunt cannot be removed or the infection was refractory despite appropriate antibiotic therapy, addition of rifampin to vancomycin is recommended [22, 23].

Table 1 Recommended antimicrobial therapy for shunt infections based on isolated pathogen and drug susceptibility

Microorganism [references]	Recommended therapy	Alternative therapy	Treatment duration[a]	Comments
CoNS [22, 26, 27]	Vancomycin	Linezolid	– CSF normal after shunt removal 7 days – CSF abnormal after shunt removal 10 days	Consider the addition of Rifampin
Staphylococcus aureus [22, 26, 27]				
– Methicillin sensitive	Naficillin or Oxacillin	Vancomycin, Meropenem	10 days	
– Methicillin resistant	Vancomycin	Linezolid, TMP-SXT[b]	10 days	Consider the addition of Rifampin
Enterococcus species [22, 24, 26, 27, 56, 57]				
– Ampicillin sensitive	Ampicillin (+/– Sulbactam) + Gentamicin	Vancomycin, Linizolid	14–21 days	In severe multi-resistant infections, addition of Rifampin (if susceptible) may be considered
– Ampicillin resistant/ Vancomycin sensitive	Vancomycin + Gentamicin	Quinupristin-dalfopristin[c]		
– Vancomycin resistant VRE	Linizolid	Quinupristin-dalfopristin[c]		
Escherichia coli and Enterobacteriacea [22]	Third generation Cefalosporin	Floeroquinolone, Aztreonam or Meropenem	10–14 days	Longer duration of therapy may be indicated depending on clinical response
Pseudomonas aeruginosa [22]	Cefipime or Ceftazidime	Pipracillin/Tazobactam, Ciprofloxacin, Aztreonam or Meropenem	10–14 days	– Consider the addition of an Aminoglycoside – Longer duration of therapy may be indicated depending on clinical response
Propionibacterium spp. [58, 59]	Penicillin G	Cefalosporins , Clindamycin, Vancomycin, Linizolid	7–14 days	– 14 days of treatment might be necessary to avoid relapse – Shunt removal is advised – Consider the addition of rifampin – 7–10 days of anaerobic incubation are required for better growth

(continued)

Table 1 (continued)

Microorganism [references]	Recommended therapy	Alternative therapy	Treatment duration[a]	Comments
Candida [14, 32, 60]	LFAmB[b] with or without 5-FC[b, d] Followed by oral Fluconazole	Fluconazole	Treat with Intravenous medications for several weeks until clinical improvement is observed; then switch to Fluconazole until CSF is normal, and resolution of clinical and radiological abnormalities is demonstrated	– Immediate removal of the device is recommended – Most of these recommendations are based on adult data

[a]These recommendations depend on the use of appropriate antibiotics according to antimicrobial susceptibility testing and after confirming SCF normalization and sterility

[b]*TMP-SXT* trimethoprim sulfamethoxazole, *LFAmB* lpid formula amphotericin B, *5-FC* 5 flourouracil

[c]Consider intravenous and intraventricular administration, although experience with intraventricular Quinupristin-dalfopristin is limited

[d]5-FC is not routinely recommended in neonatal CNS Candida meningitis

Linezolid had been used for treatment of CSF shunt infections in a number of case reports and case series with good success [24–26] and it may be an option for the treatment of infections that are not responding to vancomycin therapy. It can also be used in cases where shunt removal is not an option, or in the treatment of meningitis and ventriculitis caused by Gram positive multi-drug resistant strains (VRE and MRSA) [26, 27]. However Linezolid needs to be used with caution, because of its potential drug interactions and toxic effects.

2.3 Intraventricular Antibiotics

Intrathecal or intraventricular antibiotics should be reserved for cases failing conventional therapy. Their use can be considered in cases whose CSF is failing to sterilize on parenteral therapy, or whose device cannot be removed or when appropriate CSF penetration cannot be achieved (e.g. patients with an Ommaya reservoir in a brain tumor, or infection caused by a resistant organism that is susceptible to an antibiotic with poor CSF penetration) [22]. The best experience is with both vancomycin and gentamicin, for which intraventricular administration can be done for patients who do not respond well to parenteral therapy and serial CSF drug concentration monitoring can be provided [28]. In addition, the routine use of intraventricular antibiotics in combination with systemic antibiotic therapy after externalization of a shunt device was reported in a case series by Arneli et al. [29]. They demonstrated a fast CSF sterilization rate, with low morbidity and mortality rates.

A few case reports on the use of intraventricular amphotericin B for treatment of candida shunt infection, in conjunction to parenteral therapy and shunt removal, have shown good response to treatment with clinical cure [30, 31].

2.4 Duration of Antibiotic Therapy and Timing of Shunt Reinsertion

This depends mainly on the infective pathogens and on the repeated CSF Gram stain, chemistry and culture results. Most recommendations are based on observational studies and not randomized controlled trials (Table 1).

- Infection caused by CoNS:

    - If the CSF chemistry and Gram stain were normal, and CSF culture was confirmed to be negative after removal of the hardware, the shunt can be replaced 3 days after removal [22].
    - If CSF chemistry was abnormal (CSF protein > 200 mg/dl), but normalized after hardware removal with confirmed negative cultures afterwards, intravenous antibiotics should be continued for 7 days prior to re-shunting [22].
    - If CSF chemistry was abnormal (CSF protein > 200 mg/dl), and continued to be abnormal after removal of the shunt and/ or there were persistently positive cultures, then 10 consecutive days of therapeutic antibiotics after CSF sterilization are required before reinsertion of the shunt [22].

- Infection caused by *Staphylococcus aureus* requires 10 days of therapeutic antibiotics after negative cultures prior to shunt re-insertion [22].
- Infection caused by Gram negative bacilli require a minimum of 10–14 days of antibiotic therapy after CSF sterilization before shunt replacement; longer duration of antibiotic treatment may be required if response to therapy was delayed [22].
- Infection caused by Candida species: adult studies recommend the use of liposomal amphotericin B with or without 5-FC for several weeks until clinical improvement is observed; oral fluconazole can be used afterwards as step down therapy until all clinical, radiological and CSF abnormalities have resolved [32]; removal of the shunt device is highly recommended in such circumstances.

- Infections caused by multiple organisms and compartmentalized infections with multiple hardware devices: Hector et al. [33], reported success in treating these complicated infections by the use of 3 week courses of intravenous antibiotic therapy (based on culture/susceptibility results) in combination with 2 weeks of daily intraventricular antibiotics. Intraventricular antibiotics were either administered via an EVD after shunt removal or via an externalized shunt device.

3 Prevention of Shunt Infection

Measures for the prevention of shunt infection include perioperative antibiotic prophylaxis, use of antibiotic impregnated shunt catheters and careful adherence to sterile aseptic techniques [34]. Despite the abundance of reports on different techniques used for prevention of CSF shunt infections, there is a lack of level-1 evidence from well designed randomized control trials (RCTs) on most of them in the literature. The best evidence is with the use of prophylactic perioperative antibiotics during insertion of internal shunts, while all other techniques require further evaluation in larger prospective randomized clinical trials. Choksey et al. reported that rigid adherence to a protocol of sterile shunt placement technique in neurosurgical practice (use of peri- and post-operative antibiotics, aseptic surgical technique, liberal application of topical antiseptics and avoidance of hematoma formation) is associated with shunt infection rate of 0.57%, which is much lower than historically reported infection rates [34].

3.1 Prophylactic Perioperative Antibiotics

In a meta-analysis of 15 RCTs (1684 patients) comparing the use of prophylactic systemic perioperative antibiotics versus placebo (or no antibiotic therapy) in patients undergoing shunt placement, prophylactic antibiotic therapy was associated with a significantly lower rate of infections in patients undergoing internal shunt placement compared to the control group (OR 0.51, 95% CI 0.36–0.73; Number Needed to Treat (NNT) 12, 95% CI 7–30). No additional benefit was found when prophylaxis was given for prolonged (>24 h) periods of time, and no conclusion could be reached regarding prophylactic antibiotic administration for EVDs [35].

Few trials evaluated the effectiveness of different regimens of systemic antibiotics compared to placebo. For internal shunts, Nejat et al. demonstrated that perioperative ceftriaxone and trimethoprim-sulphamethoxazole (TMP/SXT) had similar efficacies in prevention of shunt infection [36] and Tacconelli et al. compared the periprocedural use of prophylactic vancomycin with cefazolin in a university hospital with high rate of nosocomial MRSA infection. Shunt infection rate was significantly lower in patients who received vancomycin than cefazolin (4% vs. 14%; p=0.03) [37].

Wong et al. compared the use of continuous prophylactic cefipime vs. ampicillin/sulbactam and aztreaonam in neurosurgical patients with external ventricular drains (EVDs) in situ. Results showed no difference between CSF infection, wound infection or extracranial infection rates in the two groups, suggesting that either of these broad spectrum prophylactic agents used singly is equivalent in prevention of infections in these patients [38].

Current recommendations on prophylactic antibiotic regimens used for prevention of shunt infections are based on available information on host risk factors, epidemiology of infectious pathogens and environmental risk factors. Data on which antibiotic regimen is most effective for prevention of internal shunt and EVD infection are sparse, thus it will be necessary to evaluate the effectiveness of different systemic antibiotic regimens for prevention of shunt infections in future large clinical trials.

3.2 Use of Antibiotic Impregnated Catheters (AIC)

This strategy is intended to prevent bacterial colonization of and biofilm formation on shunt catheters and subsequent CNS infections associated with shunt devices. There are two types of antibiotic impregnated catheters: antibiotic impregnated shunts (AIS), which are internalized shunts, and antibiotic impregnated EVDs (AI-EVDs). The currently available AIS are impregnated with both clindamycin and rifampin [39], while AI-EVDs are impregnated either with clindamycin and rifampin or with minocyclin and rifampin [40, 41].

Data on the use of AICs are controversial with some overlap between results that were only focusing on AIS or AI-EVDs and those that were studying AICs in general. Here, the available evidence related to each of the three different entities will be discussed separately.

3.2.1 Antibiotic Impregnated Catheters (AIC) in General

Richards et al. evaluated AICs in a large retrospective cohort study. Shunt revision due to infection was demonstrated in 30/994 procedures in which AICs were used, while shut revision was needed in 47/994 procedures in the matched control group (p=0.048) [42]. A recent retrospective cohort study also demonstrated a significant decrease in overall infection rate specifically related to staphylococci when using any AICs. The number needed to treat (NNT) with AICs to prevent any infection was 8, and NNT to prevent staphylococcal infections was 14. However, subgroup analysis showed that the reduction in infection rate was more significant among patients who had internalized antibiotic impregnated shunt devices [43].

3.2.2 Antibiotic Impregnated Shunts (AIS)

Studies on the use of AIS are heterogeneous. Most are observational studies and many used historical controls for comparison. The majority of published trials show evidence favouring their use or trends toward reduction in infection rates [39, 42–47].

A randomized control trial comparing the use of AIS to non-AIS showed no statistically significant difference in shunt infection rates between groups [46], although a significant reduction in staphylococcal infections was observed in the treatment group. In an Australian prospective study comparing patients who required AIS insertion with historical controls who had had non-AIS, significant reduction in the rate of infections (from 6.5% to 1.2%) was shown for AIS devices [39]. Significant reduction in the rate of shunt infections was also reported by Parker et al., where infection rates were 11.2% and 3.2% in children with non-AIS and AIS respectively. This significant reduction in the rate of infections was documented in a high risk population, including premature neonates, patients with prolonged hospital stay and patients who had AIS insertion following acute meningitis or following conversion of EVDs [45]. Sciubba et al. performed a retrospective review of 353 pediatric patients. Comparison was done between 208 non-AIS procedures and 145 AIS procedures. Within a 6 month follow-up period, infection of the shunt device was reported in 12% of the non-AIS group compared with 2.4% of the AIS group (p<0.001) [44]. Multivariate analysis showed that the use of AIS was associated with a 2.4-fold decreased likelihood of shunt infection [44].

3.2.3 Antibiotic Impregnated External Ventricular Drains (AI-EVDs)

The use of AIC has been shown to be as effective as prophylactic broad spectrum systemic antibiotics in the prevention of ventriculostomy infections [48]. Zabramski et al. compared using AI-EVDs to non AI-EVDs in a randomized controlled trial of 288 adult subjects. Significant reduction in the rate of shunt related infections was identified among the AI-EVD group [40]. Muttaiyah et al. compared 60 cases that required insertion of AI-EVD with 60 historical controls that had non AI-EVD. Results showed trends toward reduction of ventriculitis rate among the AI-EVD group (p=0.0627) with significant delay in time to EVD infection (p=0.0091) [41]. On the other hand, in a retrospective cohort analysis among children requiring shunt implantation, the rate of shunt infection was compared in the periods before and after the introduction of AIS. Overall shunt infection rate was similar in the two groups, but subgroup analysis revealed a trend towards reduction of infection rate among neonates, while an increased rate of infection among patients who had prior AI-EVDs was noticed [49]. Stevens et al. demonstrated that the risk of false negative culture results may be increased when CSF samples are taken for culture from an AI-EVD, which may lead to inappropriately short antibiotic therapy and thus increased likelihood of shunt re-infection [50]. This may explain the rising rate of shunt re-infections described [49].

3.3 Double Gloving

The technique of double gloving was primarily implemented to decrease rates of glove perforation and associated risks of infection transmission to operating personnel. It is also theoretically possible that use of double gloves might reduce the risk of surgical wound contamination by the operating staff's hands, although there is no evidence to confirm that this technique reduces the risk of shunt infection [51]. However, based on the idea that it is the patient's skin flora that are most responsible for shunt related infections in the first post operative month, Sorensen et al. studied the possibility that surgeons' gloves transmit infection, and found that they were contaminated by Propionibacterium acnes in 10/10 surgical procedures, and by CoNS in 8/10. Atiqu-urRahman et al. performed a retrospective cohort study that evaluated the rate of shunt infection after removing the outer pair of surgeons' gloves (after initial double gloving) and before handling the shunt catheter. Results showed a statistically significant reduction in infection rate in this group compared to the group that continued with standard double gloving technique [52]. This technique was also studied as a part of a suggested strict evidence based protocol (including intra-operative and post-operative care) for the insertion and management of EVDs by Dasic et al. Results demonstrated significant reduction in the rate of EVD infection compared to historical controls [53].

3.4 Antimicrobial Suture Wound Closure (AMS)

AMS is a new product that, to our knowledge, has not been widely evaluated to assess its effect in shunt infection reduction. These sutures are triclosan-coated, which has anti-MSSA, anti-MRSA and anti-CoNS activity. Rozelle et al. evaluated the role of AMS in prevention of shunt related infections. He performed a double blinded randomized controlled study of 61 patients with 84 CSF shunt procedures (46 cases and 38 controls). Reported shunt infection rates were 4.3% in cases versus 21% in controls (p=0.038), and no suture related adverse events were reported [54]. A follow-up analysis of this study suggested that AMS use is cost effective [55].

4 Conclusion

- Skin flora account for the majority of shunt related infections, most of which are capable of colonizing the shunts with biofilm formation which makes eradication of the infection more complicated and often necessitates shunt removal.

- The suggested treatment regimens of infected shunts include the use of parenteral antibiotic therapy together with shunt removal and establishing an external device.
- The timing of shunt re-implantation is dependent upon the isolated pathogen, CSF sterilization and on CSF chemistry.
- Prophylactic peri-operative antibiotics are warranted in the first 24 h after internal device implantation. Further controlled clinical trials are needed to study the best prophylactic antibiotic regimens, but prophylactic vancomycin administration may be advisable over cefazolin given the predominance of CoNS infection in such circumstances and in settings with high MRSA infection rates.
- Antibiotic Impregnated Catheters appear to have a significant effect in reducing infection rates among neurosurgical patients requiring shunt insertion with increasing evidence that support using AIS. However, there are data that demonstrate increased risk of shunt infection relapse among patients with impregnated external ventricular drains secondary to false negative culture results when CSF samples are obtained from them, and the question of potential development of resistant organisms secondary to antibiotic impregnation also needs to be studied in detail.
- Prevention of shunt infections requires multiple intra-operative and post-operative preventative measures. This lends itself to continuous quality improvement "bundling" strategies. Some simple preventative measures such as removing or changing the outer pair of gloves during surgery and before handling the shunt catheters should be considered and studied in larger prospective trials.

References

1. Borgbjerg BM, Gjerris F, Albeck MJ, Borgesen SE. Risk of infection after cerebrospinal fluid shunt: an analysis of 884 first-time shunts. Acta Neurochir (Wien) 1995;36(1–2):1–7.
2. Caldarelli M, Rocco C, Marca FL. Shunt complications in the first postoperative year in children with myelomeningocele. Childs Nerv Syst 1996;12(12):748–54.
3. George R, Leibrock L, Epstein M Long-term analysis of cerebrospinal fluid shunt infections. A 25 year experience. J Neurosurg. 1979;51(6):804–11.
4. McGirt MJ, Leveque J-C, Wellons III JC, Villavicencio AT, Hopkins JS, Fuchs HE, et al. Cerebrospinal Fluid Shunt Survival and Etiology of Failures: A Seven-Year Institutional Experience. Pediatric Neurosurgery 2002;36(5):248–55.
5. Cochrane DD, Kestle J. Ventricular shunting for hydrocephalus in children: patients, procedures, surgeons and institutions in English Canada, 1989–2001. Eur J Pediatr Surg 2002;12(S11):S6–11.
6. Simon TD, Hall M, Riva-Cambrin J, Albert JE, Jeffries HE, Lafleur B, Dean JM, Kestle JR ; Hydrocephalus Clinical Research Network. Infection rates following initial cerebrospinal fluid shunt placement across pediatric hospitals in the United States. Neurosurg Pediatr 2009;4(2):156–65.
7. Kulkarni AV, Drake JM, Lamberti-Pasculli M. Cerebrospinal fluid shunt infection: a prospective study of risk factors. J Neurosurg. 2001;94(2):195–201.
8. Langley JM, Gravel D, Moore D, Matlow A, Embree J, MacKinnon-Cameron D, Conly D. Study of Cerebrospinal Fluid Shunt- Associated Infections in the First Year Following Placement, by the Canadian Nosocomial Infection Surveillance Program. infection control and hospital epidemiology 2009;30(3):285–88.
9. Clemmensen D, Rasmussen MM, Mosdal C. A retrospective study of infections after primary VP shunt placement in the newborn with myelomeningocele without prophylactic antibiotics. Childs Nerv Syst. 2010;26(11): 1517–21.
10. Sarguna P, Lakshmi V. Ventriculoperitoneal shunt infections. Indian Journal of Medical Microbiology 2006;24(1):52–4.
11. Brook I. Meningitis and shunt infection caused by anaerobic bacteria in children. Pediatr Neurol. 2002;26(2): 99–105.
12. Montero A, Romero J, Vargas JA, Regueiro CA, Sánchez-Aloz G, De Prados F, De la Torre A, and Aragon G. Candida Infection of Cerebrospinal Fluid Shunt Devices: Report of Two Cases and Review of the Literature. Acta Neurochir (Wien) 2000;142(1):67–74.
13. Benca J, Ondrusova A, Huttova M >, Rudinsky B, Kisac P, Bauer F. Neuroinfections due to Enterococcus faecalis in children. Neuro Endocrinol Lett. 2007;28(S2):S32–3.

14. Sanchez-Portocarrero J, Martin-Rabadan P, Saldana CJ, Perez- Cecilia E. Candida cerebrospinal Fluid shunt infections. Report of two new cases and review of the literature. Diagn Microbiol Infect Dis 1994;20(1): 33–40.
15. Turgut M, Alabaz D, Erbey F, Kocabas E, Erman T, Alhan E, Aksaray N. Cerebrospinal fluid shunt infections in children.. Pediatr Neurosurg. 2005; 41(3):131–6.
16. Montano N, Sturiale C, Paternoster G, Lauretti L, Fernandez E, Pallini R. Massive ascites as unique sign of shunt infection by Propionibacterium acnes. Br J Neurosurg. 2010;24(2):221–3.
17. Fux CA, Quigley M, Worel AM , Post C , Zimmerli S , Ehrlich G , Veeh RH.. Biofilm-related infections of cerebrospinal fluid shunts. Clin Microbiol Infect. 2006;12(4):331–7.
18. Braxton EE Jr, Ehrlich GD, Hall-Stoodley L, Stoodley P, Veeh R, Fux C, Hu FZ, Quigley M, Post JC. Role of biofilms in neurosurgical device-related infections. Neurosurg Rev. 2005;28(4):249–55.
19. Kestle JRW, Garton HJL, Whitehead WE, Drake JM, Kulkarni AV, Cochrane DD, Muszynski C, and Walker ML.. Management of shunt infections: a multicenter pilot study. J Neurosurgery 2006;105(S3):S177-81.
20. Schreffler RT, Schreffler A, Wittler RR. Treatment of cerebrospinal fluid shunt infections: a decision analysis. Pediatr Infect Dis J 2002;21(7):632–6.
21. James HE, Walsh JW, Wilson HD, Connor JD, Bean JR, Tibbs PA. The management of cerebrospinal fluid shunt infections: a clinical experience. Acta Neurochir (Wien) 1981;59(3–4):157–66.
22. Tunkel AR, Hartman BJ, Kaplan SL, Kaufman BA, Roos KL, Scheld WM, et al. Practice guidelines for the management of bacterial meningitis. Clin Infect Dis 2004;39(9):1267–84.
23. Gombert ME, Landesman SH, Corrado ML, Stein SC, Melvin ET, Cummings M. Vancomycin and rifampin therapy for Staphylococcus epidermidis meningitis associated with CSF shunts: report of three cases. J Neurosurg. 1981;55(4):633–6.
24. Maranich AM, Rajnik M. Successful treatment of vancomycin-resistant enterococcal ventriculitis in a pediatric patient with linezolid. Mil Med. 2008;173(9):927–9.
25. Castro P, Soriano A, Escrich C , Villalba G , Sarasa M , Mensa J. Linezolid treatment of ventriculoperitoneal shunt infection without implant removal. Eur J Clin Microbiol Infect Dis. 2005;24(9):603–6.
26. Ntziora F, Falagas ME. Linezolid for the treatment of patients with central nervous system infection. Ann Pharmacother. 2007;41(2):296–308.
27. Yilmaz A, Dalgic N, Müslüman M , Sancar M , Colak I , Aydin Y. Linezolid treatment of shunt-related cerebro-spinal fluid infections in children. J Neurosurg Pediatr. 2010;5(5):443–8.
28. Wen DY, Bottini AJ, Hall WA, Haines SJ. The intraventricular use of antibiotics. Neurosurg Clin North Am 1992;3(2):343–54.
29. Arnell K, Enblad P, Wester T, Sjölin J. Treatment of cerebrospinal fluid shunt infections in children using sys-temic and intraventricular antibiotic therapy in combination with externalization of the ventricular catheter: efficacy in 34 consecutively treated infections. J Neurosurg. 2007;107(S 3): S 213–9.
30. Chiou CC, Wong TT, Lin HH, Hwang B, Tang RB, Wu KG , Lee BH. Fungal infection of ventriculoperitoneal shunts in children. Clin Infect Dis 1994;19(6):1049–53.
31. Shapiro S, Javed T, Mealey J,. Candida albicans shunt infection. Pediatr Neurosci 1989; 15:125–30.
32. Pappas PG, Kauffman C, Andes D, Benjamin DK Jr, Calandra TF, Edwards JE Jr, Filler SG, Fisher JF, Kullberg BJ, Ostrosky-Zeichner L, Reboli AC, Rex JH, Walsh TJ, Sobel JD; Infectious Diseases Society of America. Clinical practice guidelines for the management of candidiasis: 2009 update by the Infectious Diseases Society of America. Clin Infect Dis. 2009;48(5):503–35.
33. James HE, Bradley JS. Management of complicated shunt infections: a clinical report. J Neurosurg Pediatr. 2008;1(3):223–8.
34. Choksey MS, Malik IA. Zero tolerance to shunt infections: can it be achieved? J Neurol Neurosurg Psychiatry 2004;75(1):87–91.
35. Ratilal B, Costa J, Sampaio C. Antibiotic prophylaxis for surgical introduction of intracranial ventricular shunts: a systematic review. J Neurosurg Pediatr. 2008;1(1):48–56.
36. Nejat F, Tajik P, El Khashab M, Kazmi SS, Khotaei GT, Salahesh S. A randomized trial of ceftriaxone versus trimethoprim-sulfamethoxazole to prevent ventriculoperitoneal shunt infection. J Microbiol Immunol Infect. 2008;41(2):112–7.
37. Tacconelli E, Cataldo MA, Albanese A, Tumbarello M, Arduini E, Spanu T, et al. Vancomycin versus cefazolin prophylaxis for cerebrospinal shunt placement in a hospital with a high prevalence of meticillin-resistant Staphylococcus aureus. J. Hosp Infect 2008;69(4):337–44.
38. Wong GK, Poon WS, Lyon D, Wai S. Cefepime vs. Ampicillin/Sulbactam and Aztreonam as antibiotic prophy-laxis in neurosurgical patients with external ventricular drain: result of a prospective randomized controlled clinical trial. J Clin Pharm Ther. 2006;31(3):231–5.
39. Pattavilakom A, Xenos C, Bradfield O, Danks RA. Reduction in shunt infection using antibiotic impregnated CSF shunt catheters: an Australian prospective study. J Clin Neurosci. 2007;14(6):526–31.

40. Zabramski JM, Whiting D, Darouiche RO, Horner TG, Olson J, Robertson C, et al. Efficacy of antimicrobial-impregnated external ventricular drain catheters: a prospective, randomized, controlled trial. J Neurosurg. 2003;98(4):725–30.

41. Muttaiyah S, Ritchie S, John S, Mee E, Roberts S. Efficacy of antibiotic-impregnated external ventricular drain catheters. J Clin Neurosci. 2010;17(3):296–8.

42. Richards HK, Seeley HM, Pickard JD. Efficacy of antibiotic-impregnated shunt catheters in reducing shunt infection: data from the United Kingdom Shunt Registry. J Neurosurg Pediatr. 2009;4(4):389–93.

43. Gutiérrez-González R, Boto GR, Fernández-Pérez C, Del Prado N. Protective effect of rifampicin and clindamycin impregnated devices against Staphylococcus spp. infection after cerebrospinal fluid diversion procedures. BMC Neurol. 2010;10:93.

44. Sciubba DM, Lin LM, Woodworth GF, McGirt MJ, Carson B, Jallo GI. Factors contributing to the medical costs of cerebrospinal fluid shunt infection treatment in pediatric patients with standard shunt components compared with those in patients with antibiotic impregnated components. Neurosurg Focus. 2007;22(4):E9.

45. Parker SL, Attenello FJ, Sciubba DM, Garces-Ambrossi GL, Ahn E, Weingart J, et al. Comparison of shunt infection incidence in high-risk subgroups receiving antibiotic-impregnated versus standard shunts. Childs Nerv Syst. 2009;25(1):77–83.

46. Govender ST, Nathoo N, van Dellen JR. Evaluation of an antibiotic-impregnated shunt system for the treatment of hydrocephalus. J Neurosurg Pediatr. 2003;99(5):831–9.

47. Albanese A, De Bonis P, Sabatino G, Capone G, Marchese E, Vignati A, et al. Antibiotic-impregnated ventriculo-peritoneal shunts in patients at high risk of infection. Acta Neurochir (Wien) 2009;151(10):1259–63.

48. Wong GK, Poon WS, Ng SC, Ip M. The impact of ventricular catheter impregnated with antimicrobial agents on infections in patients with ventricular catheter: interim report.. Acta Neurochir Suppl. 2008;102: 53–5.

49. Hayhurst C, Williams D, Kandasamy J, O'Brien DF, Mallucci CL. The impact of antibiotic-impregnated catheters on shunt infection in children and neonates. Childs Nerv Syst. 2008;24(5):557–62.

50. Stevens EA, Palavecino E, Sherertz RJ, Shihabi Z, Couture DE. Effects of antibiotic-impregnated external ventricular drains on bacterial culture results: an in vitro analysis. J. Neurosurg. 2010;113(1):86–92.

51. Gruber TJ, Reimer S, Rozelle C. Pediatric Neurosurgical Practice Patterns Designated to prevent Cerebrospinal Fluid shunt infection. Pediatric Neurosurgery 2009;45(6):456–60.

52. Rehman AU, Rehman TU, Bashir HH, Gupta V. A simple method to reduce infection of ventriculoperitoneal shunts. J Neurosurg Pediatr. 2010;5(6):569–72.

53. Dasic D, Hanna SJ, Bojanic S, Kerr RS. External ventricular drain infection: the effect of a strict protocol on infection rates and a review of the literature. Br J Neurosurg. 2006;20(5):296–300.

54. Rozzelle CJ, Leonardo J, Li V. Antimicrobial suture wound closure for cerebrospinal fluid shunt surgery: a prospective, double-blinded, randomized controlled trial. J Neurosurg Pediatr. 2008;2(2):111–7.

55. Stone J, Gruber TJ, Rozzelle CJ. Healthcare savings associated with reduced infection rates using antimicrobial suture wound closure for cerebrospinal fluid shunt procedures. Pediatr Neurosurg. 2010;46(1):19–24.

56. Shaikh ZH, Peloquin CA, Ericsson CD Successful treatment of vancomycin-resistant Enterococcus faecium meningitis with linezolid: case report and literature review. Scand J Infect Dis. 2001;33(5):375–9.

57. Williamson JC, Glazier SS, Peacock JEJ. Successful treatment of ventriculostomy-related meningitis caused by vancomycin-resistant Enterococcus with intravenous and intraventricular quinupristin/dalfopristin Clin Neurol Neurosurg 2002;104(1):54–6.

58. Arnell K, Cesarini K, Lagerqvist-Widh A, Wester T, Sjölin J. Cerebrospinal fluid shunt infections in children over a 13-year period: anaerobic cultures and comparison of clinical signs of infection with Propionibacterium acnes and with other bacteria. J Neurosurg Pediatr. 2008;1(5):366–72.

59. Bayston R, Nuradeen B, Ashraf W, Freeman BJ. Antibiotics for the eradication of Propionibacterium acnes biofilms in surgical infection. J Antimicrob Chemother. 2007;60(6):1298–301.

60. Murphy K, Bradley J, James HE. The treatment of Candida albicans shunt infections. Childs Nerv Syst. 2000;16(1):4–7.

Nontuberculous Lymphadenopathy in Children: Using the Evidence to Plan Optimal Management

Julia E. Clark

1 Introduction

Nontuberculous mycobacterial (NTM) lymphadenopathy in children is managed and seen by surgeons and paediatricians. Despite a wealth of literature (but only one randomised controlled trial [1]) publications continue to express uncertainty about the optimal management of these lesions [2]. This review explores the literature to provide a practical, evidence-based approach to their management.

2 Clinical Presentation

Nontuberculous mycobacterial (NTM) lymphadenopathy in children is well recognised and well described. The literature incorporates many case series over the last 40 years outlining the recognised clinical spectrum [3–9]. These typically describe a child aged between 1 and 10 years, who has chronic lymphadenopathy (i.e. > 3 weeks) often with sudden onset, but who remains well, is apyrexial and has had no response to broad spectrum antibiotics. The lymphadenopathy is usually unilateral and either cervical or submandibular. After a period of weeks or months the skin overlying the lymph node becomes involved, becoming discoloured, darkish red to purple. This then thins and the lesion may spontaneously discharge. Investigations are often unremarkable and include a normal full blood count, inflammatory markers and negative serology. A chest x-ray is normal and there is usually no history of contact with tuberculosis. A Mantoux test is usually negative [4, 10], but may be positive [11]. Interferon gamma release assays (IGRA) are negative in the majority of cases [12].

2.1 Microbiology

It is important when discussing and evaluating treatment options and natural history that microbiologal aspects are included, as there have been evolving changes in the mycobacteria detected over time and there are marked geographical differences. For instance, publications in the late 1970s and

J.E. Clark (✉)
Great North Children's Hospital, Newcastle, UK
e-mail: Julia.Clark@nuth.nhs.uk

N. Curtis et al. (eds.), *Hot Topics in Infection and Immunity in Children VIII*,
Advances in Experimental Medicine and Biology 719, DOI 10.1007/978-1-4614-0204-6_10,
© Springer Science+Business Media, LLC 2011

1980s describe MAIS complex (*M. avium intracellulare*, and *M. scrofulaceum*) [3, 13] as being the predominant species detected. However since the 1990s *M. avium intracellulare* has been predominantly noted [5, 14, 15]. Both *Mycobacterium malmoense* and *M. haemophilum* are also reported in some countries but significantly in the UK (*M. malmoense*) [14–19] and Israel and Holland (*M. haemophilum*) [20, 21]. These geographical variations are important as anti-mycobacterial susceptibility differs between species. *M. avium* is almost universally resistant to all antibiotic therapies except for the macrolides, whereas *M. malmoense* is frequently sensitive to quinolones, rifamycins and ethambutol, as well as macrolides. Frequently there is poor correlation between *in vitro* antibiotic susceptibility and clinical response.

3 Optimal Management

Surprisingly, there continues to be discussion about management. Four different approaches can be taken: surgical excision, medical treatment alone, combination of surgical treatment and medical treatment and conservative (i.e. no treatment or intervention).

3.1 Surgical Treatment

There has been plenty of literature over the last 40 years on surgical intervention. This includes one randomised controlled trial and numerous case series. The randomised controlled trial compared 50 children with excision of their nontuberculous mycobacterial lymphadenopathy and 50 children who had medical treatment with a combination of clarithromycin and rifabutin for 12 weeks [1]. The primary end point was cure at 6 months, and the excision group achieved 96% cure against a 66% cure in the medical treatment group. This excellent 6 months cure rate for excision is mirrored in all the other cases series over the last 40 years most of which compare different surgical treatments [2–4, 7, 13, 14, 24–26]. In these, most achieve an over 90% cure rate with excision compared with zero to 63% cure rate with incision and drainage. It is also clear that incision and drainage produces a discharging, delayed healing scar and a poorer cosmetic outcome [27] than excision and is therefore not the optimal surgical approach. Other surgical modalities explored in the literature include curettage, generally with a success rate similar to incision and drainage (33–70%) [7, 14, 28].

There is therefore good evidence for the role of excision in NTM lymphadenopathy in children. It is usually curative, produces an improved aesthetic outcome if complete excision is performed within 1 month after the onset of lymph node swelling [5, 29], and disfiguring scars are less common. Excision is also superior to antibiotic treatment as a primary therapy [1].

However, excision may not always be appropriate or possible. The lymph node may be extremely large, maybe in a difficult position running across the route of the facial nerve, there may be a significant amount of skin involved and the final cosmetic result achieved with a large excision maybe therefore less acceptable. Incision and drainage or curettage may be useful in this situation, recognising that the wound may then take some months to heal.

3.2 Medical Treatment

With the development of potentially effective anti mycobacterial treatment there has been a great interest in using combinations originally developed for disseminated *M. avium* infections in HIV-infected patients to medically treat NTM lymphadenopathy in children. However, combinations of clarithromycin and rifabutin are not without side effects, including uveitis and teeth staining, but despite this case reports appeared from the mid 1990s suggesting there was some possible role for these antibiotics in achieving cure [7, 30–34]. As described, a randomised control trial compared medical treatment with excision and found that 66% of children resolved on rifabutin and clarithromycin at 6 months [1]. It is highly likely however, that most cases of nontuberculous mycobacterial lymphadenopathy will resolve spontaneously, albeit gradually, and so without a control arm it is difficult to assess whether the observed proportion resolving with antibiotics is significant [35].

3.3 Conservative Treatment

A study in Israeli from 2008 described 92 children who had no medical or excision treatment; 71% resolved completely by 6 months [21]. All, however, had fine needle aspirates as part of their diagnostic work up. It is possible that fine needle aspiration promoted healing, confounding the true natural history. Interestingly, it is extremely hard to find any description of cases in the literature where observation alone without any surgical or medical intervention has occurred. In the few cases described, healing does occur, taking between 6 months and 5 years [25].

4 Summary

There is no evidence that drug treatment improves healing more rapidly or is associated with an improved cosmetic outcome compared to spontaneous resolution, and no studies have related therapy and outcome to mycobacterial species and susceptibility. It is interesting that widespread and accepted use of drug treatment has developed with no good evidence that drugs facilitate healing [36]. It is therefore essential, given spontaneous healing will occur, that any future studies compare drug treatment with spontaneous resolution.

In conclusion there is good evidence that excision of nontuberculous mycobacterial lymphadenopathy is usually curative and should be performed where possible. Where lesions are too large or too difficult to surgically excise, alternatives could include de-bulking with incision and drainage or curettage, recognising that treated this way lesions will be slow to heal. Until there is evidence about the efficacy of antimycobacterial drug treatment it should not be used routinely, though it may be considered in extensive, complex disease. Also, there is no evidence to suggest that antimycobacterial drugs confer an additional benefit when the lesion is excised.

References

1. Lindeboom JA, Kuijper EJ, Bruijnesteijn van Coppenraet ES, Lindeboom R, Prins JM. Surgical excision versus antibiotic treatment for nontuberculous mycobacterial cervicofacial lymphadenitis in children: a multicente, randomized, controlled trial. Clin Infect Dis 2007;44(8):1057–64.
2. Harris RL, Modayil P, Adam J, Sharland M, Heath P, Planche T, et al. Cervicofacial nontuberculous mycobacterium lymphadenitis in children: is surgery always necessary? Int J Pediatr Otorhinolaryngol 2009;73(9):1297–301.
3. Schaad UB, Votteler TP, McCracken GH, Jr., Nelson JD. Management of atypical mycobacterial lymphadenitis in childhood: a review based on 380 cases. J Pediatr 1979;95(3):356–60.
4. Margileth AM, Chandra R, Altman RP. Chronic lymphadenopathy due to mycobacterial infection: Clinical features, diagnosis, histopathology, and management. Am J Dis Child 1984;138(10):917–22.

5. Maltezou HC, Spyridis P, Kafetzis DA. Nontuberculous mycobacterial lymphadenitis in children. Pediatr Infect Dis J 1999;18(11):968–70.
6. Danielides V, Patrikakos G, Moerman M, Bonte K, Dhooge C, Vermeersch H. Diagnosis, management and surgical treatment of non-tuberculous mycobacterial head and neck infection in children. ORL J Otorhinolaryngol Relat Spec 2002;64(4):284–9.
7. Panesar J, Higgins K, Daya H, Forte V, Allen U. Nontuberculous mycobacterial cervical adenitis: a ten-year retrospective review. Laryngoscope 2003;113(1):149–54.
8. Spyridis P, Maltezou HC, Hantzakos A, Scondras C, Kafetzis DA. Mycobacterial cervical lymphadenitis in children: clinical and laboratory factors of importance for differential diagnosis. Scand J Infect Dis 2001;33(5):362–6.
9. MacKellar A. Diagnosis and management of atypical mycobacterial lymphadenitis in children. J Pediatr Surg 1976;11(1):85–9.
10. Daley AJ, Isaacs D. Differential avian and human tuberculin skin testing in non-tuberculous mycobacterial infection. Arch Dis Child 1999;80(4):377–9.
11. Haimi-Cohen Y, Zeharia A, Mimouni M, Soukhman M, Amir J. Skin indurations in response to tuberculin testing in patients with nontuberculous mycobacterial lymphadenitis. Clin Infect Dis 2001;33(10):1786–8.
12. Detjen AK, Keil T, Roll S, Hauer B, Mauch H, Wahn U, et al. Interferon-gamma release assays improve the diagnosis of tuberculosis and nontuberculous mycobacterial disease in children in a country with a low incidence of tuberculosis. Clin Infect Dis 2007;45(3):322–8.
13. Taha AM, Davidson PT, Bailey WC. Surgical treatment of atypical mycobacterial lymphadenitis in children. Pediatr Infect Dis 1985;4(6):664–7.
14. Tunkel DE. Surgery for cervicofacial nontuberculous mycobacterial adenitis in children: an update. Arch Otolaryngol Head Neck Surg 1999;125(10):1109–13.
15. Wolinsky E. Mycobacterial lymphadenitis in children: a prospective study of 105 nontuberculous cases with long-term follow-up. Clin Infect Dis 1995;20(4):954–63.
16. Clark JE, Magee JG, Cant AJ. Non-tuberculous mycobacterial lymphadenopathy. Arch Dis Child 1995;72(2):165–6.
17. White MP, Bangash H, Goel KM, Jenkins PA. Non-tuberculous mycobacterial lymphadenitis. Arch Dis Child 1986;61(4):368–71.
18. Colville A. Retrospective review of culture-positive mycobacterial lymphadenitis cases in children in Nottingham, 1979–1990. Eur J Clin Microbiol Infect Dis 1993;12(3):192–5.
19. Evans MJ, Smith NM, Thornton CM, Youngson GG, Gray ES. Atypical mycobacterial lymphadenitis in childhood--a clinicopathological study of 17 cases. J Clin Pathol 1998;51(12):925–7.
20. Lindeboom JA, Prins JM, Bruijnesteijn van Coppenraet ES, Lindeboom R, Kuijper EJ. Cervicofacial lymphadenitis in children caused by Mycobacterium haemophilum. Clin Infect Dis 2005;41(11):1569–75.
21. Zeharia A, Eidlitz-Markus T, Haimi-Cohen Y, Samra Z, Kaufman L, Amir J. Management of nontuberculous mycobacteria-induced cervical lymphadenitis with observation alone. Pediatr Infect Dis J 2008;27(10):920–2.
22. Wright JE. Non-tuberculous mycobacterial lymphadenitis. Aust N Z J Surg 1996;66(4):225–8.
23. Rahal A, Abela A, Arcand PH, Quintal MC, Lebel MH, Tapiero BF. Nontuberculous mycobacterial adenitis of the head and neck in children: experience from a tertiary care pediatric center. Laryngoscope 2001;111(10):1791–6.
24. Mushtaq I, Martin HC. Atypical mycobacterial disease in children: a personal series. Pediatr Surg Int 2002;18(8):707–11.
25. Mandell DL, Wald ER, Michaels MG, Dohar JE. Management of nontuberculous mycobacterial cervical lymphadenitis. Arch Otolaryngol Head Neck Surg 2003;129(3):341–4.
26. Wei JL, Bond J, Sykes KJ, Selvarangan R, Jackson MA. Treatment outcomes for nontuberculous mycobacterial cervicofacial lymphadenitis in children based on the type of surgical intervention. Otolaryngol Head Neck Surg 2008;138(5):566–71.
27. Thegerstrom J, Friman V, Nylen O, Romanus V, Olsen B. Clinical features and incidence of Mycobacterium avium infections in children. Scand J Infect Dis 2008;40(6–7):481–6.
28. Fergusson JA, Simpson E. Surgical treatment of atypical mycobacterial cervicofacial adenitis in children. Aust N Z J Surg 1999;69(6):426–9.
29. Lindeboom JA, Lindeboom R, Bruijnesteijn van Coppenraet ES, Kuijper EJ, Tuk J, Prins JM. Esthetic outcome of surgical excision versus antibiotic therapy for nontuberculous mycobacterial cervicofacial lymphadenitis in children. Pediatr Infect Dis J 2009;28(11):1028–30.
30. Berger C, Pfyffer GE, Nadal D. Treatment of nontuberculous mycobacterial lymphadenitis with clarithromycin plus rifabutin. J Pediatr 1996;128(3):383–6.
31. Losurdo G, Castagnola E, Cristina E, Tasso L, Toma P, Buffa P, et al. Cervical lymphadenitis caused by nontuberculous mycobacteria in immunocompetent children: clinical and therapeutic experience. Head Neck 1998;20(3):245–9.

32. Hazra R, Robson CD, Perez-Atayde AR, Husson RN. Lymphadenitis due to nontuberculous mycobacteria in children: presentation and response to therapy. Clin Infect Dis 1999;28(1):123–9.
33. Luong A, McClay JE, Jafri HS, Brown O. Antibiotic therapy for nontuberculous mycobacterial cervicofacial lymphadenitis. Laryngoscope 2005;115(10):1746–51.
34. Coulter JB, Lloyd DA, Jones M, Cooper JC, McCormick MS, Clarke RW, et al. Nontuberculous mycobacterial adenitis: effectiveness of chemotherapy following incomplete excision. Acta Paediatr 2006;95(2):182–8.
35. Starke JR. Commentary: The natural history of nontuberculous mycobacterial cervical adenitis. Pediatr Infect Dis J 2008;27(10):923–4.
36. Pilkington EF, MacArthur CJ, Beekmann SE, Polgreen PM, Winthrop KL. Treatment patterns of pediatric nontuberculous mycobacterial (NTM) cervical lymphadenitis as reported by nationwide surveys of pediatric otolaryngology and infectious disease societies. Int J Pediatr Otorhinolaryngol 2010;74(4):343–6.

Pediatric Brucellosis: An (Almost) Forgotten Disease

Pablo Yagupsky

1 Introduction

Brucellosis is the most common bacterial zoonotic infection causing human infections worldwide. Although the disease in domestic animals has been controlled in most developed countries, human brucellosis remains an endemic public health threat in several parts of the world, affecting especially the pediatric population. The aim of this chapter is to summarize the epidemiology, pathogenesis, clinical features, diagnosis, and treatment of childhood brucellosis, and discuss the public health agenda for eradicating the disease.

2 The Organism

Brucellae are small Gram-negative coccobacilli (Fig. 1) and, along with other important human pathogens of zoonotic origin such as bartonellae and rickettsiae, are members of the α2 subdivision of the proteobacteria [1]. To date, seven terrestrial *Brucella* species have been recognized of which four, namely *B. melitensis, B. abortus, B. suis, and B. canis,* respectively maintained in nature by sheep and goats, bovines, swine, and dogs, can infect humans. Recently, two novel species, *B. ceti* and *B. pinnipedalis*, endemic among sea mammals and capable of causing human disease, have been described [2].

The genetic homology between *Brucella* species is greater than 87% and, therefore, they should be considered mere subtypes of a single species that probably diverged from a common ancestor closely similar to *B. suis* biotype 3 [3]. However, the traditional division of the *Brucella* genus into species, largely based on oxidative metabolism patterns, phage susceptibility, and preferred hosts, has been kept for epidemiological and practical considerations.

P. Yagupsky (✉)
Clinical Microbiology Laboratory, Soroka University Medical Center,
Ben-Gurion University of the Negev, Beer-Sheva, Israel
e-mail: yagupsky@bgu.ac.il

N. Curtis et al. (eds.), *Hot Topics in Infection and Immunity in Children VIII*,
Advances in Experimental Medicine and Biology 719, DOI 10.1007/978-1-4614-0204-6_11,
© Springer Science+Business Media, LLC 2011

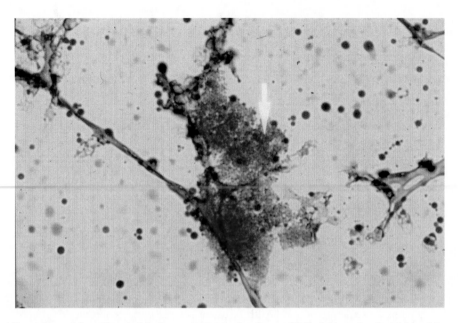

Fig. 1 Microcolony of *B. melitensis* (*black arrow*) growing in a Bactec 9240 vial inoculated with blood from a child with arthritis (Note clustering of small, Gram-negative coccobacilli)

3 Epidemiology

Brucellosis appears to have emerged as a human disease as the result of domestication of bovines, camels, sheep, and goats, and 17% of the skeletons uncovered in the city of Herculaneum dating to 79 A.D. already show typical signs of brucellar spondylitis [4].

Brucellosis is highly transmissible to man because of its low infecting dose and multiple routes of transmission, including the gastrointestinal and respiratory tracts, conjunctivae, abraded skin, and even the venereal one [5, 6]. Neonates can be infected by the transplacental route in the course of a bacteremic maternal episode, by exposure to blood, urine, or genital secretions during delivery, by breast feeding, and through blood and exchange transfusions [7]. Person-to-person transmission of the disease, usually resulting from massive exposure to patient's blood or exudates or by sexual contact, is exceptional and, therefore, isolation of infected patients is not recommended [8].

The epidemiology of human brucellosis shows marked differences between developing and developed countries (Table 1). In many Mediterranean countries, the Middle East, South America and central Asia, brucellosis is still endemic [9], and humans usually acquire the disease through consumption of unpasteurized dairy products, especially raw milk, soft cheeses, and yogurt derived from infected animals, and the children's population is the most affected. In a recent surveillance

Table 1 Epidemiology of brucellosis in developing and developed countries

	Developing world	Developed world
Character	Endemic	Sporadic
Source/exposure	Contaminated food	Mostly occupational
Portal of entry	Gastrointestinal	Respiratory, conjunctival, skin
Population affected	Children (~50%)	Adults (>90%)

study conducted in an area endemic for *B. melitensis* in southern Israel, 382 of 828 (46%) patients diagnosed in the 8-year period 2002–2010 were younger than 15 years, of whom 90 were less than 5 years of age (unpublished data). In countries endemic for brucellosis, the majority of cases of human disease is detected in the spring and summer months, following the onset of the animal parturition season [10, 11] (Fig. 2).

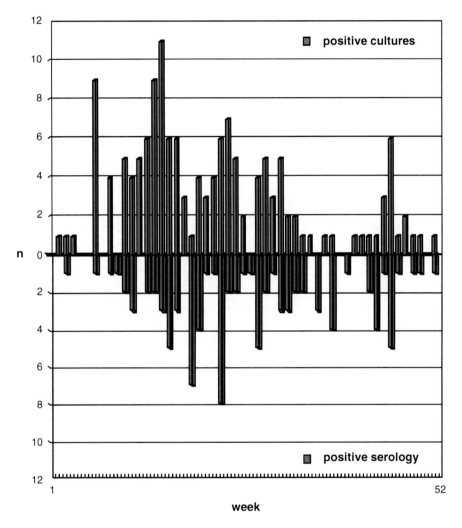

Fig. 2 Weekly distribution of positive cultures and serology in an endemic area for *B. melitensis* in southern Israel [10]

In most developed-world countries, brucelosis has been controlled by animal vaccination and food safety measures, and the rare occurrence of human disease is usually related to occupational skin, conjunctival, or respiratory exposure of veterinarians, abattoir workers, and laboratory personnel, travel to endemic areas, and illegal import of dairy products [6]. It should be pointed-out that, even in developed countries, brucellosis remains endemic among wildlife, representing a potential reservoir of infection to domestic animals and humans [12]. The possibility that brucellae might be used in the future as bioterrorism agents is also feared [13].

4 Pathogenesis

Following ingestion of contaminated food, *Brucella* organisms penetrate through the intestinal mucosa, enter the lymphatic system, and reach the bloodstream causing bacteremia. The organism disseminates to many body systems and organs, invading particularly tissues rich in reticuloen-dothelial cells. After binding to macrophages, brucellae are internalized inside vesicles that fuse with endosomes resulting in the killing of ingested bacteria [5]. However, a small bacterial subpopu-lation survives, replicates in the endoplasmic reticulum, and is released with the help of hemolysins inducing tissue necrosis.

Brucella species show a remarkable ability to subvert the host's innate immune response to establish a persistent infection. By adopting a facultative intracellular lifestyle, the organism remains remote from antibodies, complement, and antibiotics, and the disease may progress from the acute phase to a chronic form that may last for years. A pumping system that selectively trans-ports macromolecules across membranes (type IV secretion system VirB) appears to be important for the intracellular survival and trafficking of infecting brucellae, and strains harboring mutations in the VirB operon show clear virulence attenuation [6]. The transported effector of type IV secre-tion systems of other bacteria is frequently a toxin that is injected into the host's cells or secreted directly to the bloodstream. The molecule (s) transported by the VirB system of *Brucella* species remain (s) unknown, and analysis of the complete genome of the organism has consistently failed to identify any of the classic virulence factors found in other pathogenic bacteria [6].

All smooth (virulent) *Brucella* organisms bear two non-endotoxic lipopolysaccharide antigens –A– and M– in a variable ratio that enables further classification of the different species in serovars. The lipopolysaccharide layer of *Brucella* organisms helps to block the innate and specific immune response and, thus, is involved in the pathogenesis of the disease. This bacterial component appears to have multiple additional functions, regulating the entry and immune evasion in the infected mac-rophage, and inhibiting apoptosis ensuring, thus, persistent infection. The O-side chain of the lipopolisaccharide plays a crucial role in virulence, and *Brucella* strains lacking the O-side chain lose their ability to evade the host's immune response, are less virulent and, thus, have been used as animal vaccines [6].

Tumor necrosis factor-α (TNF-α) plays a crucial role in fighting brucellosis during the early non-specific phases of the immune response, by directly enhancing the antimicrobial activity of phagocytes and, indirectly, by activating natural killer cells to produce interferon-γ (IFN-γ)– a potent activator of macrophage bactericidal activity– and exert cytotoxic action against *Brucella*-infected macrophages [5].

A component of the *Brucella* wall (outer-membrane protein 25) induces inhibition of TNF-α production, disrupting the bactericidal effect of natural killer cells and macrophages and blocking the normal Th1 specific immunity, which is crucial for the bacterial eradication [14].

5 Immunity to *Brucella* Organisms

Based on observations made with volunteers inoculated with the attenuated Rev 1 animal vaccine strain of *B. melitensis*, the dynamics of the human immune response to the organism can be pre-cisely determined [5]. Shortly after *Brucella* infection has occurred, IgM antibodies against the bacterial lipopolisaccharide can be detected in blood, followed by appearance of IgG immunoglobu-lins after the second week. Both antibody types reach a peak concentration during the fourth week of infection and decline thereafter. Immunoglobulin M levels are usually higher and persist longer than IgG class antibodies. The appearance of IgA coupled with persistent IgG antibodies is consistent with chronic disease [5]. Because brucellae become sequestrated inside reticuloendothe-lial cells, the humoral immune response has only a limited role in eliminating the infection.

Interferon-γ plays a central and complex role in the pathogenesis of human brucellosis by activating macrophages, producing oxygen-reactive intermediates and nitric oxide killing in host phagocytes, enhancingcell differentiation and cytokine production, inducing apoptosis, and by increasing the expression of antigen-presenting molecules [5].

6 Clinical Presentation of Brucellosis in Children

The protean and non-specific manifestations of human brucellosis may mimic other infectious and non-infectious conditions, and the true nature of the disease may be easily missed. Childhood brucellosis is characterized by a remittent febrile course and a variety of symptoms and signs related to the ability of the organism to infect all organs and body systems, and particularly macrophage-rich tissues such as bone marrow, lymph nodes, liver, and spleen (Table 1). Fever, sweats, and arthralgia or arthritis, usually involving a single large joint, are present in the majority of patients, and hepatomegaly, and enlargement of spleen and lymph nodes are also frequently found [15, 16]. Aspiration of affected joints of children with culture-proven arthritis frequently yields leukocyte counts less than 50,000 WBC/mL, which is the laboratory cut-off value recommended to exclude pediatric infectious arthritis [17]. Elevated liver enzymes and hematological manifestations, consistent with bone marrow invasion, such as anemia, leucopenia and thrombocytopenia are common laboratory abnormalities [15–21] (Table 2).

Decreased appetite, respiratory symptoms, lymphadenopathy, skin rashes, and splenomegaly appear to be significantly more common in children than in adult patients with brucellosis, whereas urogenital involvement and spondylitis, which are not unusual in adults, are uncommon in children [20–22]. Severe or life-threatening clinical manifestations of human brucellosis, such as meningo-encephalitis and endocarditis, usually involving the aortic valve, are rare in children, and mortality is exceptional [16, 19, 20, 23].

7 Diagnosis of Human Brucellosis

Because of the non-specific clinical presentation of children with brucellosis and the need for prolonged and potentially toxic antimicrobial regimens, the diagnosis should be confirmed before instituting specific antibiotic therapy. The laboratory diagnosis of brucellosis relies on three approaches: culture, serology, and nucleic acid amplification methods.

Table 2 Common clinical and laboratory features of childhood brucellosis

Clinical finding	(%)
Fever	90
Weakness	30
Arthritis/arthralgia	70
Hepatomegaly	70
Splenomegaly	80
Laboratory finding	
Leucopenia	20
Thrombocytopenia	5
Elevated liver enzymes	40

7.1 Culture

Isolation of the *Brucella* organisms remains the only irrefutable proof of the disease. Specimens suitable for culture are blood, bone marrow, synovial fluid, semen, milk, tissues, exudates, and cerebrospinal fluid. In the past, bone marrow was recommended as the preferable specimen because of the high concentration of brucellae in the reticuloendothelial tissue. However, this has been disputed in two studies in which improved detection of the organism in peripheral blood cultures, compared to bone marrow cultures, was demonstrated [24]. The possibility of obtaining larger specimens and drawing multiple samples also make peripheral blood cultures more practical. In addition, even in areas endemic for the disease, brucellosis is frequently diagnosed by the unexpected isolation of the organism in a febrile patient, whereas obtaining bone marrow for culturing brucellae requires a high index of clinical suspicion beforehand [24].

Isolation of *Brucella* species is made difficult by the slow growth of the organism and limited by the low sensitivity of cultures in chronic cases or antibiotic-treated patients. In addition, living brucellae pose a serious threat to laboratory personnel and are considered the most common cause of laboratory-acquired disease. Manipulation of suspicious cultures inside a biological safety II level biological cabinet, and avoidance of performing superfluous or risky microbiological procedures is strongly advised [11, 13]. To improve detection, use of biphasic media, prolonged incubation of blood culture vials (for up to 35 days), and periodic performance of blind subcultures of negative vials has been traditionally recommended [24]. Recent experience with the use of automated blood culture technology, such as the BACTEC 9240, BacT/Alert, and Vital systems, has shown that this modern methods enable detection of more than 95% of cultures positive for brucellae within the routine 1-week incubation period instituted in most clinical microbiology laboratories, and performance of subcultures of negative blood culture vials is no longer necessary [24]. Automated blood culture systems are also faster and more sensitive for the detection of circulating brucellae than the lysis-centrifugation (Isolator) blood culture method [24].

7.2 Serology

Serological testing for brucellae has the advantages of being cheap and simple to perform, and it can be used to prove cure. However, its use is limited by low sensitivity in chronic cases and difficulties in interpreting borderline antibody titers in endemic areas where a substantial fraction of the population may have suffered from the disease in the past. Specificity may be also a problem because the immunodominant lipopolysaccharide antigen shares epitopes with a variety of bacterial species such as *Yersinia enterocolitica* O:9, *Pseudomonas maltophilia*, and some *Salmonella* serotypes [25]. In addition, blocking antibodies or a "prozone phenomenon", consisting of inhibition of agglutination by an excess of antibodies, may give false-negative results [5]. In the first case, use of the indirect Coombs (antihuman globulin) test improves sensitivity, whereas dilution of serum samples up to 1:1280 may overcome the latter problem [26].

The veterinarian Rose Bengal slide agglutination test is used to screen sera for *Brucella* antibodies and shows remarkable sensitivity and specificity [27]. A positive result is, then, confirmed by the standard agglutination test, which is based on the presence of both, IgG and IgM antibodies. A reciprocal titer ≥160, in the presence of a compatible clinical picture, is considered diagnostic. Because IgM antibodies tend to persist for prolonged periods, even in successfully treated patients, the test is repeated after destroying the IgM with 2-mercaptoethanol or dithiothreitol [6]. The new titer, based solely on IgG antibodies is used to follow-up patients. A declining IgG titer represents successful eradication of the organism, while persisting or increasing titers may be observed in patients with recrudescence of the disease.

A newly developed indirect enzyme-linked immunosorbent assay (ELISA) Brucellacapt (Vircell SL, Granada, Spain) has shown improved sensitivity compared to the traditional agglutination methods. However, evaluation of the test indicated that the Brucellacapt performance is highly dependent population chosen. In a published evaluation that included patients with culture-proven brucellosis and healthy individuals as the control group, the sensitivity was 98% and the specificity 96% [25]. The specificity rate, however, dropped to 63% when patients with conditions other than brucellosis were tested [25]. As expected, employing a higher cut-off value to improve specificity resulted in decreased sensitivity.

7.3 Nucleic Acid Amplification Techniques

Because of the difficult recovery of fastidious brucellae in culture, the organism appears as a natural candidate for detection by nucleic acid amplification methods. This culture-independent approach has the advantage of being rapid as well as highly sensitive and specific, and can be applied to any body tissue and fluid. Of the two major sequences used as targets, the 16 S rRNA shows an improved performance compared to that of the *BCSP31* gene, which encodes an immunogenic outer-membrane protein of *B. abortus* [28]. The recent incorporation of a robust DNA extraction procedure such as the diatom-guanidinium isothiocyanate method that removes PCR reaction inhibitors present in many clinical specimens, and the development of real-time PCR assays have substantially improved the performance of nucleic acid amplification tests [6]. However, these assays are expensive, lack between-laboratories standardization, require high technical expertise and special laboratory facilities and equipment and, therefore, are not usually available in developing countries where brucellosis is endemic. In addition, the significance of persistently positive PCR tests in adequately treated and apparently cured patients is unclear, although it may suggest that brucellae cannot be completely eradicated from the intracellular sanctuary, accounting for the high relapse rate of the human infection [29].

8 Treatment

Therapy of brucellosis requires penetration of antibiotics into macrophages where organisms hide, rendering β-lactam and other drugs clinically ineffective. Acquired antibiotic resistance is rare in brucellae and organisms recovered from patients with bacteriological relapse exhibit identical antibiotic susceptibility patterns compared to the original pre-treatment isolates [30]. This unusual observation in an era of increasing antimicrobial resistance affecting many bacterial species has been attributed to the solitude of *Brucella* organisms in the intracellular compartment, and the fact that the genus is devoid of plasmids [6]. The current laboratory methods used to test for *in-vitro* susceptibility of brucellae do not properly represent the intracellular milieu and, therefore, routine testing of isolates for antibiotic resistance is not recommended [31].

Successful eradication of brucellosis is frequently hampered by the requirement of prolonged and continuous antibiotic administration, need to use combination therapy to achieve bactericidal effect, and administration of parenteral drugs such as aminoglycosides, resulting in poor compliance and the risk for relapse. The impaired activity of macrolides in the acidic intracellular environment, the poor ability of aminoglycosides to cross cellular membranes, and drug interactions (i.e. down-regulation of serum doxycycline levels by rifampin) pose additional therapeutic difficulties [5, 32].

The treatment of brucellosis in children is further complicated by pediatric-specific adverse reactions to tetracyclines (teeth staining) and fluoroquinolones (potential damage to cartilages), and in case of neonates, for the contraindication to use sulfa drugs. The optimal therapy for children younger than 8 years has not been definitively determined. The WHO recommends a regimen of oral trimethoprim-sulfamethoxazole (8/40 mg/kg/day twice daily) for 6 weeks in combination with either intramuscular streptomycin (30 mg/kg/day once daily) for 3 weeks or parenteral (intravenous or intramuscular) gentamicin (5 mg/kg/day, once daily) for 7–10 days [32]. Alternatively, an oral combination of trimethoprim-sulfamethoxazole and rifampin (15 mg/kg/day) each for 6 weeks, or rifampin plus an aminoglycoside may be used [32]. Children aged 8 years or more should be treated with a combination of oral doxycycline twice a day for 6 weeks and an aminoglycoside for the first 2–3 weeks [32]. Although a combination of doxycycline and rifampin is considered a suitable alternative regimen [32], meta-analysis of randomized controlled studies (that included mostly adult patients) showed that the relative risk of relapse with this combination is significantly higher than that observed with the traditional tetracycline-aminoglycoside regimen, whereas the incidence of side effects is comparable [33]. Systematic review of the literature also demonstrated that antibiotic therapy should be administered for 6 weeks or longer to reduce the risk of relapse, and the authors concluded that a dual or triple antimicrobial regimen comprising an aminoglycoside (either streptomycin or gentamicin) for the first 2–3 weeks is preferable [33]. A 3-drug regimen that includes an aminoglycoside is advised for patients with endocarditis or meningitis [33].

Even when patients are adequately treated, relapses of the disease, usually milder than the initial episode, may occur at some time during the following year. In a study by Solera et al. comprising 200 adult patients and children older than 6 years, relapses were associated with initial fever ≥38.3°C, positive blood cultures, and duration of symptoms before administration of therapy for less than 10 days [34]. Because relapses are not usually related to acquire drug resistance, these events should be treated using the same antimicrobial drugs recommended in patients with newly diagnosed disease. Pediatricians involved in the management of children with brucellosis should assure compliance with the prescribed antibiotic regimen through education of patients and their families, and assess treatment results through rigorous long-term follow-up.

Physicians also have a duty to search for additional cases of brucellosis among family members to detect unrecognized cases, leading to early diagnosis prompt administration of specific antibiotic therapy to prevent clinical complications and chronicity. In a study conducted in southern Israel, active search for the disease among relatives of four index patients with brucellosis, resulted in the detection of 12 undiagnosed cases [35]. A similar experience was reported in Saudi Arabia. Screening of 404 family members of 55 diagnosed cases revealed additional 53 (13%) serologically positive individuals of whom nine were also bacteremic [36].

9 A Brucellosis Agenda for the New Millennium

Despite being long recognized and controllable, brucellosis remains a problem in many countries worldwide, affecting humans, domestic and wild animals. In endemic areas, the disease still causes substantial morbidity, affecting especially the young population, as well as substantial economic losses to animal husbandry.

Although brucellosis poses serious challenges in the areas of prevention, diagnosis and therapy, ultimately control of the disease depends on public health policies. Because of the complex economic, social, and health implications of brucellosis, a comprehensive agenda is needed to eliminate the disease threat (Table 3).

Improving reporting of human and animal cases is of paramount importance in assessing the severity and geographical distribution of the disease and setting interventional priorities. Vigorous, sustainable, and costly efforts aimed at eradicating of the organism from the food chain by animal vaccination, and educating the public to avoid exposure to contaminated food remain key issues for preventing disease in endemic areas. The renewed interest in brucellosis fuelled by the potential use

Table 3 The public health agenda on brucellosis for the new millennium

- Improvement of case detection and reporting
- Eradication by animal vaccination and disease surveillance
- Development of an effective, non-reactogenic human vaccine
- Education to avoid exposure
- Rapid detection by nucleic acid amplification methods
- Search for short/oral antimicrobial regimens
- Improved antibiotic delivery systems
- Manipulation of the intracellular milieu to eliminate persistent infection

of the organism as a biological weapon, could translate in the development of a vaccine for use in humans in the future, and finding ways to modify the intracellular environment to eradicate persisting organisms. Technical refinements of nucleic acid amplification assays aimed to making them affordable in poor-resources countries may contribute to an early and more precise diagnosis of the disease. Search for shorter antimicrobial regimens based on discovery of novel and more potent oral antibiotics, and development of improved drug delivery systems to gain clinical efficacy should be prioritized.

References

1. Moreno E, Stackebrandt E, Dorsch M, Wolters J, Busch M, Mayer H. *Brucella abortus* 16 S rRNA and lipid A reveal a phylogenetic relationship with members of the alpha-2 subdivision of the class proteobacteria. J Bacteriol 1990;172:3569–3576.
2. Maquart M, Zygmunt MS, Cloekaert A. Marine mammal *Brucella* isolates with different genomic characteristics display differential response with infecting human macrophages in culture. Microbes Infect 2009;11:361–366.
3. Moreno E, Cloeckaert A, Moriyon I. *Brucella* evolution and taxonomy. Vet Microbiol 2002;90:209–227.
4. D'Anastasio R, Staniscia T, Milia ML, Manzoli L, Capasso L. Origin, evolution and epidemiology of brucellosis. Epidemiol Infect 2010, in press.
5. Pappas G, Akritidis N, Bosilkovski M, Tsianos E. Brucellosis. N Eng J Med 2005;352:2325–2336.
6. Franco MP, Mulder M, Gilman RH, Smits HL. Human brucellosis. Lancet Infect Dis 2007;7:775–786.
7. Yagupsky P. Neonatal brucellosis: rare and preventable. Ann Trop Paediatr 2010;30:177–179.
8. Mesner O, Riesenberg K, Biliar N, Borstein E, Bouhnik L, Peled N, Yagupsky P. The many faces of human-to-human transmission of brucellosis: congenital infection and outbreak of nosocomial disease related to an unrecognized clinical case. Clin Infect Dis 2007;45:e135–140.
9. Pappas G, Papadimitriou P, Akritidis N, Christou L, Tsianos EV. The new global map of human brucellosis. Lancet Infect Dis 2006;6:91–99.
10. Khuri-Bulos NA, Daoud AH, Azab SM. Treatment of childhood brucellosis: results of a prospective trial on 113 children. Pediatr Infect Dis 1990;12:377–381.
11. Yagupsky P, Peled N, Riesenberg K, Banai M. Exposure of hospital personnel to *Brucella melitensis* and occurrence of laboratory-acquired disease in an endemic area. Scand J of Infect Dis 2000;32:31–35.
12. Her M, Cho DH, Kang SI, Lim JS, Kim HJ, Cho YS, Hwqng IY, Lee T, Jung SC, Yoo HS. Outbreak of brucellosis in domestic elk in Korea. Zoonoses Public Health 2010;57:155–161.
13. Yagupsky P, Baron EJ. Laboratory-exposures to brucellae and implications for bioterrorism. Emerg Infect Dis 2005;11:1180–1185.
14. Dornand J, Gross A, Lafont V, Liautard J, Liautard P. The innate immune response against *Brucella* in humans. Veterinary Microbiol. 2002;90:383–394.
15. Street L, Grant WW, Alva JD. Brucellosis in childhood. Pediatrics 1975;55:416–421.
16. Tanir G, Tufecki SB, Tuygun N. Presentation, complications and treatment outcome of brucellosis in Turkish children. Pediatr International 2009;51:114–119.
17. Press J, Buskila D, Yagupsky P. Leukocyte count in the synovial fluid of children with culture-proven brucellar arthritis. Clin Rheumatol 2002;21:191–193.

18. Tsolia M, Drakonaki S, Messaritaki A, Farmakakis T, Kostaki M, Tsapra H, Karpathios T. Clinical features, complications and treatment outcome of childhood brucellosis in central Greece. J Infect 2001;44:257–262.
19. Gedalia A, Watenberg N, Rotschild M. Childhood brucellosis in the Negev. Harefuah 1990;119:313–315.
20. Al-Eissa YA, Kambal AM, Al-Nasser MN, Al-Habib SA, Al-Fawaz IM, Al-Zamil FA. Childhood brucellosis; a study of 102 cases. Pediatr Infect Dis 1990;9:74–79.
21. Gur A, Geyik MF, Dikici B, Nas K, Cevik R, Sarac J, Hosoglu S. Complications of brucellosis in different age groups: a study of 283 cases in Southeastern Anatolia in Turkey. Yonsei Med J 2003;44:33–44.
22. Yinnon AM, Morali GA, Goren A, Rudensky B, Isacsohn M, Michel J, Hershko C. Effect of age and duration of disease on the clinical manifestations of brucellosis: A study of 73 consecutive patients in Israel. Israel J Med Sci 1993;29:11–16.
23. Al-Eissa, Kambal AM, Alrabeeah AA, Abdullah AMA, Al-Jurayyan NA, Al-Jishi NM. Osteoarticular brucellosis in children. Ann Rheum Dis 1990; 49:896–900.
24. Yagupsky P. Detection of brucellae in blood cultures. J Clin Microbiol 1999;37:3437–3442.
25. Orduña A, Almaraz A, Prado A, Gutierrez P, GarciaPascual A, Dueñas A, Cuervo M, Abad R, Hernandez B, Lorenzo B, Bratos MA, RodriguezTorres A. Evaluation of an inmunocapture-agglutination test (Brucellacapt) for serodiagnosis of human brucellosis. J Clin Microbiol 2000;38:4000–4005.
26. Araj GF. Update on laboratory diagnosis of human brucellosis. Int J Antimicrob Agents 2010; (in press).
27. Ruiz-Mesa JD, Sanchez-Gonzalez J, Reguera JM, Martin L, Lopez-Palmero S, Colmenero JD. Rose Bengal test: diagnostic yield and use for the rapid diagnosis of human brucellosis in emergency departments in endemic areas. Clin Microbiol Infect 2005;11:221–225.
28. Navarro E, Escribano J, Fernandeez J, Solera J. Comparison of three different PCR methods for detectiomn of *Brucella* spp in human blood samples. FEMS Immunol Med Microbiol 2002;34:147–151.
29. Vrioni G, Pappas G, Privali E, Gartzonika C, Levidiotou S. An eternal microbe: *Brucella* DNA load persists for years after clinical cure. Clin Infect Dis 2008;46:e131–136.
30. Ariza J, Bosch Gudiol F, Liñares J, Viladrich PF, Martin R. Relevance of *in vitro* antimicrobial susceptibility of *Brucella melitensis* to relapse rate in human brucellosis. Antimicrob Agents Chemother 1986;30:958–960.
31. WrightValderas M, Barrow WW. Establishment of a method for evaluating intracellular antibiotic efficacy in *Brucella abortus*-infected Mono Mac 6 monocytes. J Antimicrob Chemother 2008;61:28–34.
32. Treatment of brucellosis in humans. *In:* Brucellosis in humans and animals. Food and Agriculture Organization of the United Nations, World Organization for Animal Health, World Health Organization. 2006; pp. 36–41.
33. Skalsky K, Yahav D, Bishara J, Pitlik S, Leibovici L, Paul M. Treatment of human brucellosis: systemic review and meta-analysis of randomized controlled trials. Br Med J 2008;336:701–704.
34. Solera J, Martinez-Alfaro E, Espinosa A, Castillejos ML, Geijo P, Rodriguez-Zapata M. Multivariate model for predicting relapse in human brucellosis. J Infect 1998;36:85–92.
35. Abramson O, Rosenvasser Z, Block C, Dagan R. Detection and treatment of brucellosis by screening a population at risk. Pediatr Infect Dis 1991;10:434–438.
36. Almuneef MA, Memish ZA, Balkhy HH, Alotaibi B, Algoda S, Abbas M, Alsubaie S. Importance of screening household members of acute brucellosis cases in endemic areas. Epidemiol Infect 2004;132:533–540.

Q Fever: Still More Queries than Answers

Corine E. Delsing, Adilia Warris, and Chantal P. Bleeker-Rovers

1 Introduction

Q fever is an ubiquitous zoonosis that is caused by *Coxiella burnetii*. Q stands for "query," dating from the time when the causative pathogen was still unknown. During the last decade, understanding of the disease has improved, although many questions including some concerning optimal treatment, still remain. There are few reports of Q fever in children which suggests they are relatively resistant to development of the disease.

2 History

Q fever was first described in 1935 by Edward Derrick, who was in charge of investigating a febrile illness among 20 abattoir employees in Brisbane, Australia [1]. He inoculated guinea pigs with urine and serum from patients, which in turn developed fever (Fig. 1). In collaboration with Frank MacFarlane Burnet, he isolated a microorganism and named it *Rickettsia burnetii*, because he believed it to be a rickettsial disease, despite the fact that patients did not present with skin manifestations. Comparison with the "Nine mile" strain that was simultaneously isolated from ticks from the area around Nine Mile creek, by Herald Cox in the Rocky Mountain Laboratory, Montana, USA, showed that it was indeed the same micro-organism. In honour of this contribution, the bacterium was later renamed *Coxiella burnetii* [2]. Since then, the disease has primarily presented as an occupational illness of farm animal handlers, or as a result of close proximity of infected herds and urban areas, although small outbreaks due to exposure to *Coxiella* from parturient pets (such as cats and dogs) have also been reported [3, 4].

C.E. Delsing (✉) • C.P. Bleeker-Rovers
Department of Internal Medicine and Radboud Expertise Centre for Q fever, Nijmegen Institute for Infection, Inflammation and Immunity; Radboud University Nijmegen Medical Centre, Nijmegen, The Netherlands
e-mail: C.Delsing@AIG.umcn.nl; C.Bleeker-Rovers@aig.umcn.nl

A. Warris
Department of Pediatric Infectious Diseases, Nijmegen Institute for Infection, Inflammation and Immunity, Radboud University Nijmegen Medical Centre, Nijmegen, The Netherlands
e-mail: A.Warris@cukz.umcn.nl

N. Curtis et al. (eds.), *Hot Topics in Infection and Immunity in Children VIII*,
Advances in Experimental Medicine and Biology 719, DOI 10.1007/978-1-4614-0204-6_12,
© Springer Science+Business Media, LLC 2011

Fig. 1 Edward Derrick measuring the temperature of Q fever infected guinea pigs (Courtesy of Professor Robin Cooke)

3 Microbiology

Coxiella burnetii is a small Gram-negative obligate intracellular bacterium that is phylogenetically related to the Legionnellales. The bacterium displays a unique antigenic variation of its surface lipopolysaccharide (LPS), which can be used to distinguish between acute infection and chronic Q fever. In acute Q fever, patients become infected with the virulent phase I form. During multiplication, chromosomal deletions occur in the genes encoding for this LPS, resulting in a loss of length of the polysaccharide structure [5] and a less virulent form develops (phase II). Phase I LPS, with its extended carbohydrate structure, sterically blocks access of antibody to surface proteins making it more difficult to elicit a humoral immune response. For this reason acute infection is marked by the development of IgG mainly directed against phase II antigens. When infection is not cleared by the immune system and virulent phase I bacteria persist (chronic Q fever), with time high titres of antibodies directed at phase I antigens develop.

Another important property of *C. burnetii* is its ability to form a spore like structure, the small dense variant form [6]. This morphological variant is extremely resistant to heat, desiccation and chemical agents, including disinfectants, making it possible for the bacterium to survive in soil for many months. The organism can be killed by pasteurization.

4 Transmission

The main reservoir of *Coxiella* is domestic ruminants such as dairy goats, sheep and cows, but pets have also been a source of urban outbreaks. In ruminants *Coxiella burnetii* causes abortion and stillbirth, but otherwise most animals remain asymptomatic, often resulting in delayed diagnosis of Q fever infection of dairy herds.

The animals shed the bacterium in milk, urine and faeces, and in very high concentrations in birth by-products (up to 10^9 micro-organisms per gram of placental tissue). *C. burnetii* is extremely infectious as was illustrated by an experiment demonstrating that inhalation of a single bacterium could cause seroconversion in humans [7]. Because of this high infectious potential it has been classified as a category B bioterrorism agent.

The most important mode of transmission of Q fever is through inhalation of infected aerosols (Fig. 2). The ingestion of infected milk products is associated with seroconversion, although it is uncertain whether this can also lead to clinical illness [8]. Human-to-human transmission does not usually occur, although it has been described following contact with infected parturient women [9], after blood transfusion [10] and through sexual contact [11].

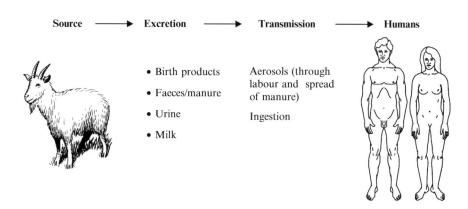

Fig. 2 Transmission of Q fever

5 Epidemiology

The epidemiology varies from country to country. Most outbreaks involve a limited number of patients, although a few larger epidemics have been described. In 1956 there were 1358 clinically suspected cases among workers of meat-processing plants, of which 814 were serologically confirmed [12]. In 1983 in Bagnes, a Swiss alpine valley, 415 cases of acute Q fever were reported after inhabitants of the villages located along the road were exposed to aerosols from a flock of 900 sheep that descended from the mountain pastures [13]. And in 2003 a lambing ewe, infected 299 vendors and visitors at a farmers' market in Germany [14].

Starting in 2007, there was a large outbreak of Q fever in The Netherlands with almost 4,000 confirmed cases up to September 2010. Before then, 10–20 cases of Q fever were reported annually and a seroprevalence study showed that, before 2007, the infection rate was low (2.4%) [15]. Already in 2007, an association with intense goat farming in the region was suggested [16]. A strong seasonal variation was observed, probably related to lambing season and dry weather facilitating airborne spread of *Coxiella* (Fig. 3). In 2008, a large human cluster of Q fever in an urban area was clearly linked to a dairy goat farm with a Q fever related to an abortion episode which occurred a few weeks before the first human cases presented [17]. The high relative risk (31.1 [95% CI 16.4–59.1]) to contract Q fever when living within a 2 km radius of a dairy goat farm compared to persons living more than 5 km away, supported this hypothesis.

Besides proximity to an infected herd, smoking and male sex have been shown to be important risk factors for development of symptomatic disease [18, 19].

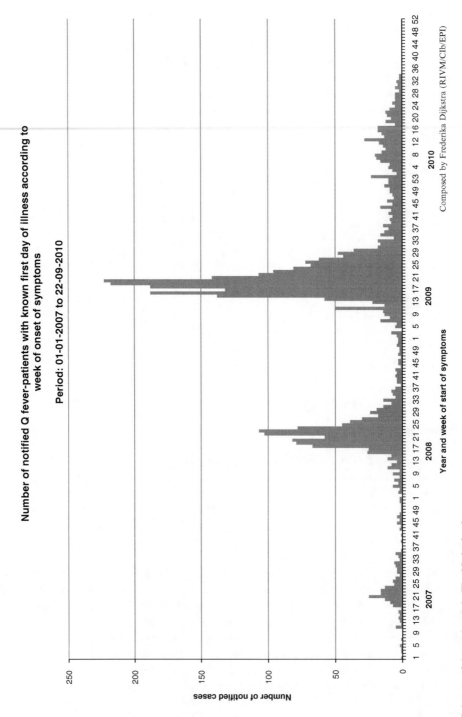

Fig. 3 Epicurve of the outbreak in The Netherlands

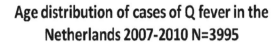

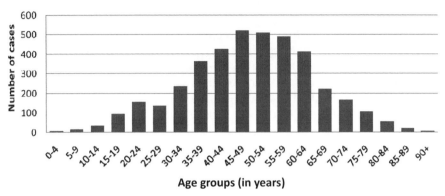

Fig. 4 Age distribution of Q fever cases in The Netherlands (Source RIVM/CIb/Epi)

Q fever in children is rarely reported and few scientific data are available. A prospective study in Greece among children hospitalized in a tertiary hospital, showed an age-specific increase of the incidence rate of acute Q fever from 0.15% in patients younger than 5 years of age, to 2.6% of patients 11–14 years of age [20]. During the epidemic in The Netherlands, only 120 of 3,995 reported cases were below 18 years of age. The mean age of patients was 51 years (Fig. 4). A study performed in the USA showed comparable age distribution [21]. Possible explanations for the low incidence of Q fever among children are differences in exposure to *C. burnetii*, a different clinical expression with less symptomatic cases in children, or a difference in consideration of and diagnostic evaluation for Q fever in this age group [22]. Although exposure in some outbreaks was clearly limited to adults (such as the meat workers in Brisbane), other outbreaks (such as the one in The Netherlands starting in 2007 and the one in 1983 in Bagnes, Switzerland) presumably led to similar environmental exposure of children. However, a survey among the inhabitants of the affected area in Bagnes, showed that only 10 out of 80 (12.5%) children, who had evidence of recent Q fever infection, had symptomatic disease versus 181 out of 335(54%) adults [13].

Children with Q fever more often present with gastro-intestinal symptoms [20, 23] and since such symptoms are common in children, usually caused by self-limiting viral infections, this may be a reason why specific investigations are undertaken less frequently than in adults with acute Q fever, who usually present with pneumonia.

6 Clinical Manifestations

6.1 Acute Q Fever

Acute Q fever occurs 2–6 weeks after exposure depending on the infective dose. The infection remains asymptomatic in up to 60% of adult cases. Patients with symptomatic disease usually present with mild flu-like symptoms. Fever is almost always present. Children present with gastro-intestinal symptoms in 50–80% of cases, whereas adults mostly present with pulmonary complaints such as dyspnoea and cough [20, 23, 24]. Rash is uncommon in adults, occurring only in about 11% of patients [25] compared to up to 50% of children [20]. It is usually a maculopapular rash on the trunk [26]. The percentage of patients presenting with elevated liver enzymes or even frank hepatitis, differs from country to country. It is not yet clear whether this is related to variations in the micro-organism, the host or the dominant route of infection. Complications of acute Q fever include myocarditis, pericarditis [27–30], meningitis, cerebellitis and encephalitis [20, 31–35] develop in 1% of acute Q fever cases. Other rare manifestations described in children include haemolytic uraemic syndrome [36], lymphadenopathy [37], haemophagocytosis [38] and rhab-domyolysis [28].

It is not clear whether Q fever leads to more severe disease in immunocompromised patients. Two children with Q fever infection and acute leukemia have been described; both were treated with antibiotics and survived [39, 40]. An 11-year-old patient with an undiagnosed Q-fever pneumonia who had chronic granulomatous disease died, the diagnosis being made at autopsy [41]. Other fatal cases include a 15-year-old boy with myocarditis [27] and a 2-year-old boy with fulminant hepatic failure [42].

6.2 Chronic Q Fever

Chronic Q fever develops in 1–2% of patients with acute Q fever although some patients with chronic Q fever do not recall having had an acute infection at all [43]. It usually develops insidiously, months or even years after acute infection and patients often present with non-specific symptoms such as low grade fever, night sweats and weight loss. In a large retrospective study from France, endocarditis was found to be the predominant manifestation of chronic Q fever in adults, constituting 73% of chronic Q fever cases. Other manifestations were vascular infection (8%), chronic infection in pregnancy (6%), and chronic hepatitis (3%) [25]. However, in The Netherlands, a substantially higher percentage of chronic Q fever cases are patients with infected aneurysms and vascular prostheses: in a recent report, 12 of 22 (55%) chronic Q fever patients had vascular infection [44].

Patients who are most at risk of developing chronic disease after acute infection are those with pre-existing valvulopathy and vascular prostheses and patients who are immunocompromised. A study from France suggests that endocarditis may develop in up to 40% of patients with acute Q fever and pre-existing valvular defects [43]. Diagnosis of Q fever endocarditis is difficult since vegetations are often absent on echocardiography and conventional blood cultures are negative. Revised Duke criteria have been developed for the diagnosis of Q fever which included a positive *Coxiella* PCR or a high antibody titer to phase I antigens among the major criteria [45].

For early detection of chronic Q fever, it has been recommended that serological follow-up be performed for at least 1 year after acute infection [43].

Chronic Q fever in children is rare and only 11 cases have been published. The two described manifestations of chronic Q fever in children are endocarditis and osteomyelitis. Five paediatric cases of endocarditis have been published [46–49], of whom four had pre-existing valvulopathy. Chronic (often relapsing and multifocal) osteomyelitis has been described in six previously healthy children [50, 51]. Chronic osteomyelitis with negative conventional cultures and histology showing granulomas, should prompt investigation for Q fever [51].

6.3 Q Fever During Pregnancy

Chronic infection with *Coxiella burnetii* seems to have a tropism for vascular structures, such as aneurysms and cardiac valves but also for placental tissue. In dairy cattle this is illustrated by the fact that, although the animal seems to be asymptomatic, placentitis leads to stillbirth and excretion of high concentrations of microorganisms. In a series of 23 women with chronic Q fever during pregnancy, a relationship was found between a positive PCR on placental tissue and fetal death, suggesting that placentitis also plays an important role in chronic Q fever during pregnancy on humans [52].

Although two large seroprevalence studies found no significant association between seropositivity for *Coxiella* and adverse pregnancy outcome [53, 54], a case series of 53 women who had evidence of Q fever during pregnancy, described obstetric complications such as spontaneous abortion, intrauterine growth retardation, fetal and neonatal death and premature delivery [52] in over 80% of women who did not receive long-term cotrimoxazole treatment. Women who did receive long-term corimoxazole treatment had significantly fewer complications (43.8%). The authors therefore recommended the treatment of women who contract Q fever during pregnancy with cotrimoxazole throughout the remainder of their pregnancy.

7 Diagnosis of Q Fever

Diagnosis of acute Q fever is based on serology. There are different techniques available but the reference method is the immunofluorescence assay (IFA). A seroconversion or a fourfold rise in antibody titer is diagnostic for acute Q fever [55]. An important drawback to diagnosis based on serology is that antibody production usually does not occur until a few weeks after onset of clinical symptoms. PCR on serum has been shown to have a high sensitivity (98%) for acute Q fever in seronegative patients and is therefore a useful diagnostic tool for early diagnosis [56].

Diagnosis of chronic Q fever can be difficult. As mentioned above, *C. burnetii* displays a unique antigenic variance in surface polysaccharides (phase 1 and phase 2 antigens). This can be used to distinguish between acute and chronic infection. In acute infection, mainly phase 2 antibodies develop and convalescent sera show low titers of phase 1 antibodies, whereas chronic infection is characterized by high titers of phase 1 antibodies. Most literature on diagnosis of chronic Q fever originates from the French National Reference Center for Rickettsial Diseases (NRC) and this group has proposed a cut off value for IgG to phase 1 proteins of 1:800 for the diagnosis of chronic Q fever [57]. However, multiple commercial immunofluorescence assays are available and cut off values for phase 1 antibodies indicating chronic infection are different for each assay as was recently shown in a case-report comparing different assays in serologic follow-up after acute Q fever [58]. Therefore, local cut off values need to be defined depending on the assay that is used by each individual microbiology department.

PCR can be performed on tissues obtained from valve replacement, vascular surgery or biopsies. Unfortunately PCR on peripheral blood is not always positive despite intravascular localisation of infection. Culture of *C. burnetii* is difficult, requires specific techniques and, because of its high infectivity, bio-safety level 3 laboratory facilities.

8 Therapy

Comparative trials of antibiotic treatments for acute Q fever are sparse, often retrospective and sometimes show conflicting results. Doxycycline is the preferred treatment for acute Q fever (200 mg per day for 2–3 weeks), but this agent is generally not recommended in pregnant women and children under the age of 8 years. However, although staining of permanent dentition sometimes does occur when treating young children with doxycycline, this is usually described after multiple courses of antibiotic therapy and limited use of this drug has a negligible effect on the color of permanent dentition [59, 60]. Doxycycline has been recommended as first line treatment in children with tick borne rickettsioses [61] and can therefore be considered in the treatment of acute Q fever (4 mg/kg/day with a maximum of 200 mg). An alternative, which can also be used during pregnancy, is cotrimoxazole (trimethoprim 18 mg/kg/day and sufamethoxazole 90 mg/kg/day divided in two doses).

Chronic Q fever requires prolonged antibiotic therapy and sometimes surgical intervention (e.g. valve replacement or vascular surgery). The number of chronic Q fever cases is very small, so prospective comparative trials are very difficult. A retrospective trial comparing a regimen of doxycycline plus ofloxacin with doxycycline and hydroxychloroquine, showed superiority of the latter combination [62]. Hydroxychloroquine alkalizes the intralysosomal pH, resulting in a bactericidal effect *in vitro* when combined with doxycycline, whereas doxycycline monotherapy is only bacteriostatic [63]. A minimum duration of 18 months of treatment and target levels of doxycycline of 5 mg per liter, showed the best results [62]. When doxycycline is contraindicated long term treatment with cotrimoxazole can be considered.

9 Prevention

The most important preventive measures consist of reducing transmission of *C. burnetii* from infected animals to humans. After a large scale culling of infected goats in The Netherlands in 2009, the number of human Q fever cases was markedly reduced in 2010.

An effective whole-cell vaccine is available in Australia and has been extensively used in persons with high occupational risks such as abattoir employees. In this population, it has been proven to be highly effective [64]. A recent study evaluating Australia's national Q fever vaccination program starting from 2002, showed a marked decline in Q fever notification rates in the subsequent years [65]. Almost 49,000 people were vaccinated. Adverse events were rare, occurring in 86 subjects (0.18%) and were mainly injection site reactions. One case of anaphylaxis was reported. No deaths occurred.

A major drawback of this vaccine is that administration to patients with pre-existing immunity can lead to serious local and systemic inflammatory reactions. Therefore prior infection needs to be excluded by skin testing and serology. This is very laborious in highly endemic areas and in The Netherlands it is reserved for high risk populations (such as patients with valvulopathy or vascular prosthesis). The vaccine is not registered for use under the age of 15, because of the lack of safety and efficacy data, but has nevertheless been used in children who have high risk of exposure [66].

10 Summary

Q fever is a worldwide zoonosis, caused by *C. burnetii*. Infection usually occurs through inhalation of infected aerosols. The reservoir mainly consists of dairy cattle. Clinical symptoms of acute Q fever are non-specific and resemble a mild flu-like illness. Children often present with gastrointestinal symptoms and rash. Rarely, chronic infection develops. This is usually manifested as endocarditis, vascular infection and, in children, osteomyelitis. Diagnosis is based on serology and nucleic acid amplification (PCR). Doxycycline is the treatment of choice for acute infection. An alternative for young children and pregnant women is cotrimoxazole. Chronic infection requires long term treatment usually with doxycycline combined with hydroxychloroquine.

References

1. Derrick EH. "Q" fever, a new fever entity: clinical features, diagnosis and laboratory investigation. Med J Aust 1937;2:281–99.
2. Marrie TJ. Q fever: The Disease. CRC Press; 1990.

3. Marrie TJ, Durant H, Williams JC, Mintz E, Waag DM. Exposure to parturient cats: a risk factor for acquisition of Q fever in Maritime Canada. J Infect Dis 1988;158(1):101–8.

4. Pinsky RL, Fishbein DB, Greene CR, Gensheimer KF. An outbreak of cat-associated Q fever in the United States. J Infect Dis 1991;164(1):202–4.

5. Hoover TA, Culp DW, Vodkin MH, Williams JC, Thompson HA. Chromosomal DNA deletions explain phenotypic characteristics of two antigenic variants, phase II and RSA 514 (crazy), of the Coxiella burnetii nine mile strain. Infect Immun 2002;70(12):6726–33.

6. Wiebe ME, Burton PR, Shankel DM. Isolation and characterization of two cell types of Coxiella burneti phase I. J Bacteriol 1972;110(1):368–77.

7. Tiggert WD, Benenson AS, Gochenour WS. Airborne Q fever. Bacteriol Rev 1961;25:285–93.

8. Cerf O, Condron R. Coxiella burnetii and milk pasteurization: an early application of the precautionary principle? Epidemiol Infect 2006;134(5):946–51.

9. Ossewaarde JM, Hekker AC. [Q fever infection probably caused by a human placenta]. Ned Tijdschr Geneeskd 1984;128(48):2258–60.

10. Q Fever transmitted by blood transfusion-United States [editorial]. Can Dis Wkly Rep 1977;3:210.

11. Milazzo A, Hall R, Storm PA, Harris RJ, Winslow W, Marmion BP. Sexually transmitted Q fever. Clin Infect Dis 2001;33(3):399–402.

12. Somma-Moreira RE, Caffarena RM, Somma S, Perez G, Monteiro M. Analysis of Q fever in Uruguay. Rev Infect Dis 1987;9(2):386–7.

13. Dupuis G, Petite J, Peter O, Vouilloz M. An important outbreak of human Q fever in a Swiss Alpine valley. Int J Epidemiol 1987;16(2):282–7.

14. Porten K, Rissland J, Tigges A, Broll S, Hopp W, Lunemann M, et al. A super-spreading ewe infects hundreds with Q fever at a farmers' market in Germany. BMC Infect Dis 2006;6:147.

15. van Duynhoven Y, Schimmer B, Van Steenbergen JE, van der Hoek W. The story of human Q fever in the Netherlands. 2010 Feb 25; 2010.

16. Van Steenbergen JE, Morroy G, Groot CA, Ruikes FG, Marcelis JH, Speelman P. [An outbreak of Q fever in The Netherlands – possible link to goats]. Ned Tijdschr Geneeskd 2007;151(36):1998–2003.

17. Schimmer B, ter Schegget R, Wegdam M, Zuchner L, de Bruin A, Schneeberger PM, et al. The use of a geographic information system to identify a dairy goat farm as the most likely source of an urban Q-fever outbreak. BMC Infect Dis 2010;10:69.

18. Orr HJ, Christensen H, Smyth B, Dance DA, Carrington D, Paul I, et al. Case-control study for risk factors for Q fever in southwest England and Northern Ireland. Euro Surveill 2006;11(10):260–2.

19. Raoult D, Marrie T, Mege J. Natural history and pathophysiology of Q fever. Lancet Infect Dis 2005;5(4):219–26.

20. Maltezou HC, Constantopoulou I, Kallergi C, Vlahou V, Georgakopoulos D, Kafetzis DA, et al. Q fever in children in Greece. Am J Trop Med Hyg 2004;70(5):540–4.

21. McQuiston JH, Holman RC, McCall CL, Childs JE, Swerdlow DL, Thompson HA. National surveillance and the epidemiology of human Q fever in the United States, 1978–2004. Am J Trop Med Hyg 2006;75(1):36–40.

22. Maltezou HC, Raoult D. Q fever in children. Lancet Infect Dis 2002;2(11):686–91.

23. Ruiz-Contreras J, Gonzalez MR, Ramos Amador JT, Giancaspro CE, Scarpellini VA. Q fever in children. Am J Dis Child 1993;147(3):300–2.

24. Terheggen U, Leggat PA. Clinical manifestations of Q fever in adults and children. Travel Med Infect Dis 2007;5(3):159–64.

25. Raoult D, Tissot-Dupont H, Foucault C, Gouvernet J, Fournier PE, Bernit E, et al. Q fever 1985–1998. Clinical and epidemiologic features of 1,383 infections. Medicine (Baltimore) 2000;79(2):109–23.

26. Tissot DH, Raoult D, Brouqui P, Janbon F, Peyramond D, Weiller PJ, et al. Epidemiologic features and clinical presentation of acute Q fever in hospitalized patients: 323 French cases. Am J Med 1992;93(4):427–34.

27. Chevalier P, Vandenesch F, Brouqui P, Kirkorian G, Tabib A, Etienne J, et al. Fulminant myocardial failure in a previously healthy young man. Circulation 1997;95(6):1654–7.

28. Hervas JA, de la Fuente MA, Garcia F, Reynes J, de Carlos JC, Salva F. Coxiella burnetii myopericarditis and rhabdomyolysis in a child. Pediatr Infect Dis J 2000;19(11):1104–6.

29. Fournier PE, Etienne J, Harle JR, Habib G, Raoult D. Myocarditis, a rare but severe manifestation of Q fever: report of 8 cases and review of the literature. Clin Infect Dis 2001;32(10):1440–7.

30. Baquero-Artigao F, del CF, Tellez A. Acute Q fever pericarditis followed by chronic hepatitis in a two-year-old girl. Pediatr Infect Dis J 2002;21(7):705–7.

31. Gomez-Aranda F, Pachon DJ, Romero AM, Lopez CL, Navarro RA, Maestre MJ. Computed tomographic brain scan findings in Q fever encephalitis. Neuroradiology 1984;26(4):329–32.

32. Drancourt M, Raoult D, Xeridat B, Milandre L, Nesri M, Dano P. Q fever meningoencephalitis in five patients. Eur J Epidemiol 1991;7(2):134–8.

33. Sawaishi Y, Takahashi I, Hirayama Y, Abe T, Mizutani M, Hirai K, et al. Acute cerebellitis caused by Coxiella burnetii. Ann Neurol 1999;45(1):124–7.

34. Ravid S, Shahar E, Genizi J, Schahor Y, Kassis I. Acute Q fever in children presenting with encephalitis. Pediatr Neurol 2008;38(1):44–6.

35. Kubota H, Tanabe Y, Komiya T, Hirai K, Takanashi J, Kohno Y. Q fever encephalitis with cytokine profiles in serum and cerebrospinal fluid. Pediatr Infect Dis J 2001;20(3):318–9.

36. Maltezou HC, Kallergi C, Kavazarakis E, Stabouli S, Kafetzis DA. Hemolytic-uremic syndrome associated with Coxiella burnetii infection. Pediatr Infect Dis J 2001;20(8):811–3.

37. Foucault C, Lepidi H, Poujet-Abadie JF, Granel B, Roblot F, Ariga T, et al. Q fever and lymphadenopathy: report of four new cases and review. Eur J Clin Microbiol Infect Dis 2004;23(10):759–64.

38. Chen TC, Chang K, Lu PL, Liu YC, Chen YH, Hsieh HC, et al. Acute Q fever with hemophagocytic syndrome: case report and literature review. Scand J Infect Dis 2006;38(11–12):1119–22.

39. Loudon MM, Thompson EN. Severe combined immunodeficiency syndrome, tissue transplant, leukaemia, and Q fever. Arch Dis Child 1988;63(2):207–9.

40. Heard SR, Ronalds CJ, Heath RB. Coxiella burnetii infection in immunocompromised patients. J Infect 1985;11(1):15–8.

41. Meis JF, Weemaes CR, Horrevorts AM, Aerdts SJ, Westenend PJ, Galama JM. Rapidly fatal Q-fever pneumonia in a patient with chronic granulomatous disease. Infection 1992;20(5):287–9.

42. Berkovitch M, Aladjem M, Beer S, Cohar K. A fatal case of Q fever hepatitis in a child. Helv Paediatr Acta 1985;40(1):87–91.

43. Fenollar F, Fournier PE, Carrieri MP, Habib G, Messana T, Raoult D. Risks factors and prevention of Q fever endocarditis. Clin Infect Dis 2001;33(3):312–6.

44. Wever PC, Arts CH, Groot CA, Lestrade PJ, Koning OH, Renders NH. [Screening for chronic Q fever in symptomatic patients with an aortic aneurysm or prosthesis.]. Ned Tijdschr Geneeskd 2010;154(28):A2122.

45. Fournier PE, Casalta JP, Habib G, Messana T, Raoult D. Modification of the diagnostic criteria proposed by the Duke Endocarditis Service to permit improved diagnosis of Q fever endocarditis. Am J Med 1996;100(6): 629–33.

46. Laufer D, Lew PD, Oberhansli I, Cox JN, Longson M. Chronic Q fever endocarditis with massive splenomegaly in childhood. J Pediatr 1986;108(4):535–9.

47. Beaufort-Krol GC, Van Leeuwen GH, Storm CJ, Richardus JH, Dumas AM, Schaap GJ. [Coxiella burnetii as a possible cause of endocarditis in a 7-year old boy]. Tijdschr Kindergeneeskd 1985;53(4):153–7.

48. Lupoglazoff JM, Brouqui P, Magnier S, Hvass U, Casasoprana A. Q fever tricuspid valve endocarditis. Arch Dis Child 1997;77(5):448–9.

49. al-Hajjar S, Hussain Qadri SM, al-Sabban E, Jager C. Coxiella burnetii endocarditis in a child. Pediatr Infect Dis J 1997;16(9):911–3.

50. Cottalorda J, Jouve JL, Bollini G, Touzet P, Poujol A, Kelberine F, et al. Osteoarticular infection due to Coxiella burnetii in children. J Pediatr Orthop B 1995;4(2):219–21.

51. Nourse C, Allworth A, Jones A, Horvath R, McCormack J, Bartlett J, et al. Three cases of Q fever osteomyelitis in children and a review of the literature. Clin Infect Dis 2004;39(7):e61–e66.

52. Carcopino X, Raoult D, Bretelle F, Boubli L, Stein A. Q Fever during pregnancy: a cause of poor fetal and maternal outcome. Ann N Y Acad Sci 2009;1166:79–89.

53. Rey D, Obadia Y, Tissot-Dupont H, Raoult D. Seroprevalence of antibodies to Coxiella burnetti among pregnant women in South Eastern France. Eur J Obstet Gynecol Reprod Biol 2000;93(2):151–6.

54. Baud D, Peter O, Langel C, Regan L, Greub G. Seroprevalence of Coxiella burnetii and Brucella abortus among pregnant women. Clin Microbiol Infect 2009;15(5):499–501.

55. Fournier PE, Marrie TJ, Raoult D. Diagnosis of Q fever. J Clin Microbiol 1998;36(7):1823–34.

56. Schneeberger PM, Hermans MH, van Hannen EJ, Schellekens JJ, Leenders AC, Wever PC. Real-time PCR with serum samples is indispensable for early diagnosis of acute Q fever. Clin Vaccine Immunol 2010;17(2): 286–90.

57. Dupont HT, Thirion X, Raoult D. Q fever serology: cutoff determination for microimmunofluorescence. Clin Diagn Lab Immunol 1994;1(2):189–96.

58. Ake JA, Massung RF, Whitman TJ, Gleeson TD. Difficulties in the diagnosis and management of a US service-member presenting with possible chronic Q fever. J Infect 2010;60(2):175–7.

59. Lochary ME, Lockhart PB, Williams WT, Jr. Doxycycline and staining of permanent teeth. Pediatr Infect Dis J 1998;17(5):429–31.

60. Volovitz B, Shkap R, Amir J, Calderon S, Varsano I, Nussinovitch M. Absence of tooth staining with doxycycline treatment in young children. Clin Pediatr (Phila) 2007;46(2):121–6.

61. Chapman AS, Bakken JS, Folk SM, Paddock CD, Bloch KC, Krusell A, et al. Diagnosis and management of tickborne rickettsial diseases: Rocky Mountain spotted fever, ehrlichioses, and anaplasmosis – United States: a practical guide for physicians and other health-care and public health professionals. MMWR Recomm Rep 2006;55(4):1–27.

62. Raoult D, Houpikian P, Tissot DH, Riss JM, Rditi-Djiane J, Brouqui P. Treatment of Q fever endocarditis: comparison of 2 regimens containing doxycycline and ofloxacin or hydroxychloroquine. Arch Intern Med 1999;159(2):167–73.
63. Maurin M, Benoliel AM, Bongrand P, Raoult D. Phagolysosomal alkalinization and the bactericidal effect of antibiotics: the Coxiella burnetii paradigm. J Infect Dis 1992;166(5):1097–102.
64. Ackland JR, Worswick DA, Marmion BP. Vaccine prophylaxis of Q fever. A follow-up study of the efficacy of Q-Vax (CSL) 1985–1990. Med J Aust 1994;160(11):704–8.
65. Gidding HF, Wallace C, Lawrence GL, McIntyre PB. Australia's national Q fever vaccination program. Vaccine 2009;27(14):2037–41.
66. Barralet JH, Parker NR. Q fever in children: an emerging public health issue in Queensland. Med J Aust 2004;180(11):596–7.

Rickettsioses in Children: A Clinical Approach

Emmanouil Galanakis and Maria Bitsori

1 Introduction

The term "rickettsiosis" has traditionally included not only diseases caused by pathogenic species of the genus *Rickettsia* but also diseases caused by *Orientia*, *Ehrlichia*, *Anaplasma* and even *Coxiella* species, which often present with similar clinical manifestations but are now known to belong to diverse genera, families, orders or even classes [1–3] (Fig. 1). Rickettsiae and related pathogens are obligate intracellular Gram-negative coccobacilli and their life-cycle involves small vertebrate hosts and arthropod vectors, with the exception of *C. burnetii*, which is a zoonosis infecting humans through contaminated soil. Humans are only incidental hosts and do not contribute to the persistence of the organisms in nature, with the exception of *R. prowazekii*. Identification and classification of these organisms occurred during the twentieth century and is still ongoing, however the diseases have plagued humans since antiquity and often shaped the history of mankind.

Rickettsioses have a global distribution with endemic foci worldwide and, despite the availability of effective and low-cost antibiotics, they continue to cause considerable morbidity and mortality [4]. Although children are among the most affected age groups [1], rickettsioses in childhood have not been thoroughly investigated. The main reason is their non-specific clinical presentation which makes them difficult to distinguish from viral infections. Thus, clinicians face the challenge of suspecting rickettsioses early in their clinical course, when antibiotics are most effective. In this review we summarise clinical aspects of childhood rickettsioses with focus on early diagnosis, which is crucial for successful treatment.

2 Epidemiology

The epidemiology of Rickettsiae and related pathogens is determined by the ecology of vectors, i.e., of ticks, mites, fleas and lice. Epidemiology in childhood is still unclear and based mostly on sporadic cases and small case series.

E. Galanakis (✉) • M. Bitsori
Department of Paediatrics, University of Crete, Heraklion, Crete, Greece
e-mail: emmgalan@med.uoc.gr; bitmar@hol.gr

N. Curtis et al. (eds.), *Hot Topics in Infection and Immunity in Children VIII*,
Advances in Experimental Medicine and Biology 719, DOI 10.1007/978-1-4614-0204-6_13,
© Springer Science+Business Media, LLC 2011

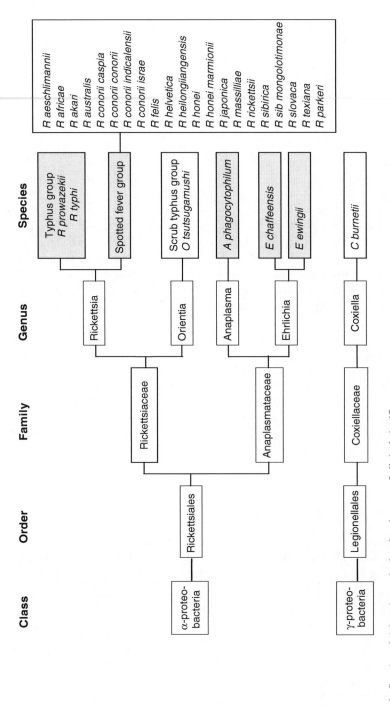

Fig. 1 Species of rickettsiae and related pathogens of clinical significance

2.1 Spotted Fever Rickettsial Infection

These tick-borne infections have a similar clinical presentation, and most of them affect children. *R. rickettsii* infection is prevalent in the American continent [1, 4]. Children 5–9 years of age are among the most commonly affected [5–9]. *R. conorii conorii* infection is usually encountered around the Mediterranean basin, commonly affecting children [10–13]. Increasing morbidity rates have been reported recently [14]. Fatal cases are mostly reported among individuals with G-6PD deficiency [15, 16]. Other subspecies of *R. conorii,* namely *R. conorii israelensis, R. conorii indica* and *R.conorii caspia* are responsible for similar clinical syndromes in Israel, India and the Asian part of Russia, respectively [14, 17–20].

R. africa has been recently recognised in sub-Saharan Africa as the cause of a tick-borne febrile infection [21]. Most of the described cases were in European travelers to sub-Saharan Africa, including children [22, 23]. *R. japonica* was recognised in the mid-eighties as the causative agent of a febrile spotted disease encountered in Japan and the Far East [24, 25]. Affected children have been reported [24, 26]. Among the newly recognised spotted fever rickettsial species, *R. slovaca* affects primarily children. It was first described in central Europe as a febrile tick-borne lymphadenitis, which was given the acronym TIBOLA [27, 28]. Recent paediatric reports indicate that it may be prevalent throughout the European continent [29].

Rickettsialpox caused by *R. akari* is an urban rickettsiosis affecting all ages and characterised by a generalised papulovesicular rash. It has been reported in many countries worldwide [1] and affected children were included in series from New York City and Mexico [30, 31]. *R. felis,* recently recognised as a world-wide human pathogen, is phenotypically classified among the spotted fever rickettsial species, although clinical characteristics and transmission route of the disease resemble typhus [32]. Seroprevalence studies indicate that the disease can occur in childhood [16, 33].

There are no specific reports for childhood infection from spotted fever rickettsioses endemic in Australia (*R. australis* and the recently identified *R. honei*), Northern Asia (*R. sibirica* and the recently identified subspecies *R. sibirica mongolotimonae*), Far East (*R. heilongiiangensis*), Southern Europe (*R. aeschlimannii, R. massiliae*) and America (*R. parkeri*) [20, 34–38].

2.2 Typhus Group Rickettsial Infection

Epidemic typhus, caused by *R. prowazekii*, has humans as its main reservoir. Being responsible for millions of deaths in the past, this infection is still associated with high fatality and continues to cause outbreaks or epidemics worldwide, when conditions of human misery favours the spread of body lice [20, 39–41]. Although mainly a disease of adults, children may be affected as well [42]. The late recrudescence of epidemic typhus is a similar clinical syndrome called Brill-Zinsser disease and occurs in areas of the world, which suffered louse-borne typhus epidemics in the past [43]. The clinically mild murine typhus due to *R. typhi* is one of the most common rickettsioses worldwide. Children are probably less frequently affected but they occasionally present with severe clinical manifestations [44–48].

2.3 Orientia Infections

The only known pathogenic species is *O. tsutsugamushi,* the causative agent of scrub typhus. The disease affects rural populations in tropical Asia and Australia and is occasionally imported in Europe and America by travellers or military personnel [49, 50]. It is common in children with occasionally severe presentation [51–54]. The interest on the disease is re-emerging because of its interaction with human immunodeficiency virus and the description of strains with reduced susceptibility to usual antirickettsial agents [49].

2.4 Ehrlichia and Anaplasma Infections

Human monocytic ehrlichiosis, human ewingii ehrlichiosis and human granulocytic anaplasmosis are caused by *E. chaffeensis*, *E. ewingii* and *A. phagocytophilum*, respectively [55]. The former two agents are limited in the American continent where their vector, *Amblyoma americanum*, exists. *A. phagocytophilum* infection occurs world-wide [55]. Data for childhood infection are derived from small case series and case reports; however, ehrlichial infections may often be misdiagnosed as common viral infections [56–64].

2.5 Coxiella Infections

The only known pathogenic species is *C. burnettii* and the infection, Q fever, is encountered worldwide except New Zealand. Cattle, sheep and goats are the main reservoirs of the organism, which is found in high densities in the placenta of infected animals, infecting humans via inhalation of contaminated aerosols [65]. Information about Q fever in children is obtained through small case series and case reports from different parts of the world [66–71]. *C. burnettii* may also cause chronic infection, mainly endocarditis, which usually affects individuals with immunodeficiencies or valvulopathies [72]. Only a few cases with paediatric chronic Q fever have been reported [65].

3 Early Suspicion: Clues from the Clinical History

Rickettsioses are often regarded as curiosities of travel medicine; their names remind exotic destinations, their connection to historical epidemics reminds diseases of a past era, and their association with animals makes them sound like unusual occupational infections. However, seroprevalence studies have shown that rickettsioses are quite common among children [73] and are an important cause of unspecified febrile illness [48, 74], but they escape early diagnosis due to the lack of specific initial symptoms and of laboratory methods. A thorough medical history might provide early suggestive clues (Table 1).

3.1 Tick Bite or Exposure – Animal Contact

A typical history of tick bite within 14 days of illness onset is often absent [75, 76], but questions about outdoor activities may elicit information relating to tick exposure. Children and parents may not recognize an attached tick and the bite is typically painless [4]. Possession of pets, especially dogs, findings of tick attachment to animals and potential contact with a rodent can also be important information. Not all rickettsioses are arthropod-borne, and occupational involvement of the family with cattle or sheep and goats or visit to farms, particularly after lambing season could be suggestive of *C. burnettii* infection [74]. Transmission routes can occasionally be even hard to guess. For example, epidemic typhus has long been known to be transmitted through body lice under conditions of crowding, but a cluster of cases in the East Coast of the United States was associated with flying squirrels. The vector and mode of acquisition had never been identified [42].

Table 1 Suspecting rickettsioses: clues from clinical history

Tick bite or exposure
- Outdoor activities, areas with low vegetation, warm months
- Bizarre cutaneous lesions, insect bites

Animal contact
- Possession of pets, particularly cats and dogs
- Contact with rodents
- Occupational involvement with cattle, sheep and goats

International travel
- Endemic countries

Similar symptoms in contacts
- Family members
- Pets
- Class mates
- Community members

Underlying diseases
- G-6PD deficiency
- Diabetes mellitus
- Immunodeficiency
- Congenital heart disease
- Antibiotic prophylaxis

3.2 Recent Travel to Endemic Areas

Rickettsioses have traditionally been associated with travel [23, 77–80]. Unusual destinations and ecoturism are increasingly popular, and children often participate in such activities, meaning more cases of imported rickettsioses are likely to be seen in the years to come. The most common causes of infection among travellers are *R. typhi, R. conorii, O. tsutsugamushi* and *R. Africa*. The incidence of the latter reaches rates of 4–5.3% in short-term foreign visitors to sub-Saharan Africa [80]. Other rickettsioses have been sporadically reported in returning travellers, such as *R. rickettsii, R. prowazekii* and *R. sibirica* infections, and among them the case of a teenager visitor to Eastern Australia who acquired *R. australis* infection visiting a crocodile farm [80]. Geographic name of a rickettsial disease does not define the area of endemicity. *R. conorii* infection, known as Mediterranean spotted fever, is also endemic in India and Africa under the names of Indian and Kenya tick typhus, respectively. *R. rickettsii* infection is encountered throughout the American continent and although it is known as Rocky Mountain spotted fever, only a minority of cases are currently reported from the Rocky Mountain area [81]. International travel is highly suggestive of rickettsial infection, but practising paediatricians should be equally familiar with the epidemiology of rickettsioses in their own area, as children might well acquire a rickettsial infection playing in their backyard [44].

3.3 Similar Symptoms in Family Members, Pets or Pals

Clinicians are often inclined to diagnose a viral infection when multiple family member present with similar symptoms. However, rickettsial infections are well known to present with clusters of cases [4], among family members [82, 83], co-workers, members of a community who share same outdoor activities [84], soldiers [49] or class-mates [85, 86]. Concurrent infections have also been observed in humans and dogs [87, 88].

3.4 Biological Warfare

Rickettsial agents such as *R. prowazekii, C. burnettii* and *R.typhi* have the properties of potential biological weapons: massive production and development of resistant strains is feasible, they can be released in aerosol form or through infected arthropods, they are highly virulent and they present with nonspecific manifestations leading to delayed diagnosis and treatment [89]. The possibility that large scale clusters could imply an attack scenario may sound unlikely but does not seem to be just hypothetical [40, 90].

3.5 Underlying Disease

G-6-PD deficiency and diabetes have been described as risk factors for fulminant *R. rickettsii* and *R. conorii* infection [91–94].

Immunocompromised patients might also be prone to severe illness. Fulminant ehrlichiosis in pediatric oncology patients and transplant recipients has been reported [62, 95] and children with congenital heart disease might be prone to chronic *C. burnetii* infection [65]. Attention should also be given in children who are on prophylaxis with cotrimoxazole, usually for urinary tract infections [96], as there are concerns that sulfonamides might increase the severity of *R. rickettsii, R. conorii* and *R. typhi* infection [97, 98].

4 Clinical Characteristics

The "classic" triad of fever, rash and headache is observed only in a minority of patients. Particularly in children, the initial presentation of a rickettsial infection is notoriously non-specific and frequently mimics common viral syndromes. However, certain clinical symptoms and signs might offer diagnostic clues.

4.1 Fever

Rickettsioses usually present with sudden onset of high fever, which is accompanied by chills and myalgias and lasts from days to weeks. Patients commonly seek medical advice after 3–4 days of disease, when fever exceeds the usual duration of a viral infection [4]. Fever can last for 1–2 weeks in spotted fever rickettsioses [16, 81], ehrlichioses, anaplasmosis [55, 59] and acute *C. burnetii* infection [72] and it can last longer in *R. typhi* and *O. tsutsugamushi* infection [49, 99, 100]. Fever of unknown origin (FUO) has for long been defined as a febrile syndrome that lasts more than 3 weeks and diagnosis remains elusive after 1 week of hospital investigation [101]. The term fever of intermediate duration (FID) has recently been proposed for fevers higher than 38°C, lasting 1–4 weeks and lacking diagnosis after an initial approach [102]. Rickettsioses are among the most common causes of both FID and FUO in children [48, 100, 102, 103] (Fig. 2).

4.2 Rash

Rickettsioses should be considered in febrile patients with bizarre rashes. A maculopapular rash, which begins on the wrists and ankles and spreads to involve the entire body, including palms and soles, 2–4 days after disease onset is typical for most spotted fever rickettsioses [4] and may also

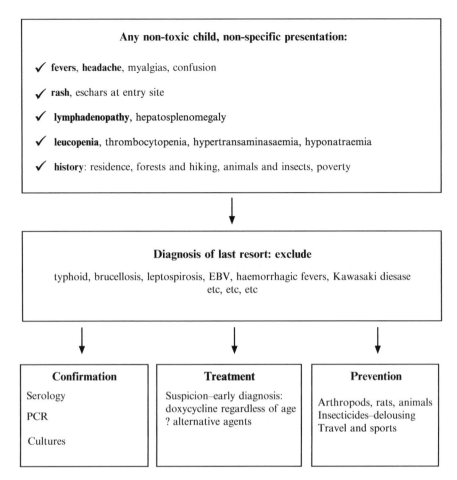

Any non-toxic child, non-specific presentation:

✓ **fevers**, **headache**, myalgias, confusion

✓ **rash**, eschars at entry site

✓ **lymphadenopathy**, hepatosplenomegaly

✓ **leucopenia**, thrombocytopenia, hypertransaminasaemia, hyponatraemia

✓ **history**: residence, forests and hiking, animals and insects, poverty

Diagnosis of last resort: exclude

typhoid, brucellosis, leptospirosis, EBV, haemorrhagic fevers, Kawasaki diesase
etc, etc, etc

Confirmation	**Treatment**	**Prevention**
Serology	Suspicion–early diagnosis: doxycycline regardless of age ? alternative agents	Arthropods, rats, animals Insecticides–delousing Travel and sports
PCR		
Cultures		

Fig. 2 Clinical approach to rickettsioses in childhood

be present in patients with endemic or epidemic typhus [41, 45], ehrlichioses and anaplasmoses [57, 59] and acute Q fever [72]. The rash can be pox-like or papulonodular in *R. akari* and *R. africa* infections [22, 31]. The classic spotted petechial rash seen in *R. rickettsii, R. conorii* and other spotted fever rickettsioses is usually not apparent until the completion of the first week from onset [14, 81]. Progression to necrosis or gangrene has been reported in severe cases of *R. rickettsii* and *R. prowazekii* infection [41, 104]. Rashes can be completely absent [4] and, some rickettsial infections included in the spotted fever group are spotless, such as *R. slovaca* infection [27]. A single painless eschar at the site of the initial tick inoculation, the "tache noire", is seen in 70% of children with *R. conorii* infection, and less commonly in other tick-borne rickettsioses. These lesions are reported to be multiple in *R. africa* infection [22]. They can however be either completely absent or go unrecognized, hidden in scalp, as is the case for *R. slovaca* infection [29].

4.3 Neurological Manifestations

Severe headache is commonly reported by adults and older children with a rickettsial infection [4, 40]. Altered mental status, lethargy, photophobia and menigism are not unusual during the course of *R. rickettsii, R. japonica*, and *O. tsutsugamushi* infection and have been reported less commonly for other spotted fever group rickettsioses and acute Q fever [24, 40, 105, 106]. Mild meningoencephalitis with subacute course and cerebrospinal fluid (CSF) pleocytosis is the most usual neurological involvement in children with *R. typhi*, *O. tsutsugamushi* and *E. chaffensis* infections [59, 107–109]. More severe presentation with seizures, focal neurological manifestations and findings of acute disseminating encephalomyelitis has occurred in children with *R. rickettsii* and *R. conorii* infections [10, 13, 110, 111]. Paediatricians in endemic areas should keep a high index of suspicion for rickettsioses in cases of subacute meningoencephalitis or unexplained neurologic manifestations.

4.4 Lymphadenopathy

Tender regional lymphadenopathy, close to the tick inoculation, has been reported either as typical in *R. conorii*, *R. africa* and *O. tsutsugamushi* infections [10, 22, 40] or as typically absent in *R. rickettsii* and *R. japonica* infection [40]. Lymphadenopathy accompanying a scalp tick-bite is the hallmark of *R. slovaca* infection, which is particularly prevalent among children, especially girls [27, 29]. This epidemiologic feature is attributed to the vector (*Dermacentor spp*) tropism for hairy mammals of accessible heights [29]. Generalised lymphadenopathy seems to be common among children with murine typhus [44].

4.5 Common Non-Specific Clinical Characteristics

Gastrointestinal and respiratory symptoms are common non-specific symptoms of rickettsioses in children and hepatosplenomegaly is among the most consistent findings. Nausea and vomiting, abdominal pain and diarrhoea often characterise the initial phase of infection in children [5, 11, 59]. Respiratory symptoms and atypical pneumonia are associated with *C. burnnetii* infection [65], but pneumonitis has also been reported in children with *O. tsutsugamushi* [112] and *E. chaffeensis* infections [59]. For the remaining rickettsioses respiratory symptoms probably signify clinical deterioration and progression to pulmonary oedema and respiratory distress syndrome [5, 10, 105, 113]. Other non-specific symptoms include conjunctivitis, arthritis and extremities oedema [5, 10, 11, 59, 105].

4.6 Uncommon Clinical Characteristics

The list of unusual clinical manifestations of rickettsioses is long enough to include almost every clinical symptom encountered in paediatrics. Among them it is worth mentioning myocarditis in spotted fever group rickettsioses [10, 26, 114], potentially fatal endocarditis in chronic *C. burnettii* infection [115, 116], and hemophagocytosis due to ehrlichial infection in paediatric oncology patients [63]. Paediatricians in endemic areas should be alert to consider rickettsioses as a potential diagnosis in every unusual clinical presentation for which there is no alternative explanation.

5 Laboratory Findings

5.1 First-Line and Non-Specific Laboratory Tests

Full blood count and biochemistry tests often are normal, but might offer early diagnostic clues when rickettsiosis is suspected [4, 105]. Leucopenia is observed in 25–50% of patients with predominance of neutrophils, left shift and thrombocytopenia, which is the most constant finding

[5, 10, 11, 59, 105]. Anaemia might also be present. Acute phase reactants are mildly elevated [4, 44, 65]. Other findings may include hypertransaminasaemia, hyponatraemia, hypoalbuminaemia and mildly elevated urea and creatinine [5, 10, 11, 59, 100, 105]. Coagulation studies may be impaired, with moderately increased PTT as the most common finding. Microscopic haematuria with or without mild proteinuria is commonly observed in spotted fever group rickettsioses, indicating glomerular involvement [10, 11, 16]. Mild CSF pleocytosis, usually <100 cells/mm^3, is observed, usually with lymphocytic predominance [105]. Neutrophillic pleocytosis is common in ehrlichial infections [59]. CSF protein is moderately elevated (up to 100–200 mg/dL) and glucose is normal [10, 59, 65, 105].

5.2 Specific Laboratory Tests

Serology is the commonest diagnostic approach. The reference method is immunofluorescence antibody (IFA) assay which can detect separately IgM and IgG antibodies and is commercially available for the most common rickettsial and associated pathogens [16]. Different cut-offs have been proposed for different diseases, single titres however are only indicative and confirmation of a case requires a fourfold increase in antibody titre in a sample obtained 2 weeks later [16, 105]. Although serological tests are sensitive and specific, both IgM and IgG antibodies can only be detected between 7 and 15 days, or even more, after the onset of the disease, and are therefore of no practical use for therapeutic decisions. The interpretation of serological data is often tricky and confounded by cross-reactivity and by seropositivity in endemic areas [5].

Direct visualisation of intracytoplasmic forms of the organism ("morulae") in Wright-stained peripheral blood smears may help in ehrlichia and anaplasma infections. The assay is insensitive but can offer a rapid diagnosis soon after disease onset [55]. The isolation of the organisms from blood and CSF specimens, skin biopsy or ticks, is possible but rather demanding. Because rickettsial and associated agents cannot be cultured on lifeless media, their isolation requires the application of antibiotic-free cell culture methods, direct inoculation in the culture medium and biosafety level 3, conditions that are not available in most clinical laboratories [16]. Molecular methods based on PCR can detect and identify rickettsial agents in EDTA blood specimens, skin biopsies and ticks and hold the promise for accurate and timely diagnosis. These methods are continuously developing, but are not widely available yet [38].

6 Differential Diagnosis

The list of differential diagnosis of rickettsioses in children can be very long as febrile syndromes are very common in paediatrics.

6.1 Non-Specific and "Viral" Presentation

Most rickettsioses are initially clinically indistinguishable from viral infections. They should therefore be considered in children with a viral syndrome that exceeds the usual duration of viral infections. Infectious mononucleosis due to Epstein-Barr virus (EBV) can last for several days and is associated with various and unusual clinical and laboratory findings [117, 118]. However, rashes are unusual in EBV infection and a history suggestive for zoonosis is missing.

6.2 Kawasaki-Like Presentation

The clinical manifestations of rickecttsioses and Kawasaki disease are so similar [119] that rickettsial agents have been investigated as possible causative agents for Kawasaki disease [120, 121]. As is the case for rickettsioses, therapeutic decisions for treatment of Kawasaki disease should be made on clinical grounds. First-line laboratory tests might offer help, as in Kawasaki disease leucocytosis, increased platelet counts (in convalescence) and acute phase reactants are usually observed, whereas rickettsioses are typically characterised by leucopenia and thrombocytopenia. In doubtful cases, treatment with IVIG is advisable.

6.3 Meningoencephalitis

Bacterial meningitis can usually be reliably distinguished from rickettsial CNS involvement, both on clinical and laboratory grounds [4]. However, for severely ill children with fever and purpuric rash, in whom neither disease can be ruled out, empiric treatment for both conditions is advisable [4].

6.4 Other Tick-Borne or Animal-Associated Infections

Various tick-borne and animal associated infections, such as Lyme disease, visceral leishmaniasis, leptospirosis, brucellosis and bartonellosis, share similar clinical and first-line laboratory findings to rickettsioses [40]. In such cases, knowledge of endemicity, detailed history and careful evaluation may help. Doxycycline is the antibiotic of choice for rickettsial and associated pathogens, but also for Lyme disease, leptospirosis and brucellosis and this facilitates therapeutical desicions.

6.5 Typhoid Fever

In past centuries, there was much confusion between typhoid and typhus fever [39]. For physicians of our time it is clear that typhoid or enteric fever is caused by *Salmonella enterica* seropypes *typhi* and *paratyphi*, and is prevalent in south-central and south-east Asia [122, 123]. Whereas rickettsioses should be considered in international travellers, typhoid fever should be considered particularly in immigrants from endemic countries [124, 125]. Cultures can offer diagnostic help, as *Salmonella* species can be easily isolated in ordinary blood culture media.

6.6 Treatment

Rickettsioses should be treated upon suspicion, as treatment is more effective when given early in the course of the disease. Milder cases may recover without treatment [38, 99], nevertheless complications and unfavourable outcome cannot be excluded. Case fatality rates have been reported to be up to 2–5% for *R. rickettsii*, 2% for *R. conorii*, 3% for *E. chaffeensis* and 1% for *R. typhi* infection and can reach 25% for *R. rickettsii*, 30% for *O. tsutsugamushi* and 70% for *R. prowazekii* infection if left untreated [4, 6, 41, 49].

6.7 Tetracyclines

Given their effective intracellular concentrations, tetracyclines have been the standard treatment for rickettsioses since their release in 1948. Rickettsiae and related pathogens remain so susceptible to tetracyclines, that rickettsioses have been termed "doxycycline-deficiency diseases" [126]. Doxycycline is the drug of choice for the treatment of most rickettsioses in adults and children. Minocycline is preferred for Japanese spotted fever as it was reported to be more effective *in vitro* against *R japonica* [26]. Tetracyclines were previously not used in paediatric practice due to the fear of bone toxicity and tooth discoloration [127]. However, the American Academy of Pediatrics Committee on Infectious Diseases recently revised its guidelines, suggesting doxycycline as the treatment of choice for *R. rickettsii* infection and ehrlichioses, regardless of age [1, 4]. The rationale is that morbidity outweighs the risk from a short course of doxycycline. The recommended dose for children is 2.2 mg/kg twice daily, orally or intravenously. Under current recommendations, antibiotics should be given for at least 3 days after clinical improvement, which means a minimum course of 5–7 days for most spotted fever group rickettsioses and *R. typhi* infection. Longer regimens of 10–14 days are recommended for ehrlichioses, epidemic typhus, acute Q fever or central nervous system involvement [1, 4].

6.8 Alternative Antibiotics

Chloramphenicol has been extensively used for childhood infections [128]. In the largest series with *R. conorii conorii* infection from Sicily, all the 645 children were successfully treated with oral or intravenous chloramphenicol [10]. Several issues of safety and efficacy have been raised, though: the use of chloramphenicol instead of tetracyclines for *R. rickettsii* infection was associated with higher mortality rates [76]; a fatal case of ehrlichial infection in a child treated with chloramphenicol was reported [58]; relapses of patients treated with chloramphenicol for *R. conorii* and *R. typhi* infection were noticed in Israel [129]; and irreversible side-effects may occur [128].

Newer macrolides, including azithromycin, clarithromycin, josamycin and roxithromycin, are characterised by excellent safety profile and have been successfully tested for the treatment of children with *R. conorii conorii* and *O. tsutsugamushi* [130–132], and *R. typhi* infection [45, 99]. However, macrolides are ineffective for ehrlichiosis, anaplasmosis [133] and Brill-Zinsser disease [43] and are not considered safe for the treatment of *R. rickettsii* infection. Despite the variable susceptibility of *C. burnetii* to macrolides *in vitro* [134] newer macrolides have been successfully used in acute Q fever [135] and have been suggested for chronic *C. burnetii* infection, especially in childhood where the optimal treatment for endocarditis and osteomyelitis remains an open question [65, 136].

Ciprofloxacin has been used to eradicate *R. japonica* infection in patients who did not respond to minocycline [24, 137] but a failure has been reported in a teenager with *R typhi* infection [138]. Fluoroquinolones are active against *C burnetii* and ehrlichiae [139–141], but they are not yet approved for routine use in children.

Rickettsiae are not susceptible *in vitro* to co-trimoxazole [140] and, more importantly, concerns have been raised that sulfonamides may increase the severity of *R. rickettsii*, *R. conorii conorii* and *R. typhi* infection [97, 98]. Furthermore, an adolescent boy and a young woman developed fulminant human monocytic ehrlichiosis, while on co-trimoxazole for acne and pyelonephritis [96, 142]. Currently, co-trimoxazole may only be considered for chronic *C. burnetii* infection [1].

Rifampin and rifabutin are active *in vitro* against rickettsial and associated agents (Rolain et al., 1998), but clinical experience in children is limited. Rifampin had little success against *R. conorii conorii* infection in children [143] but it was efficient against childhood ehrlichial infection [144]. It has also been used with encouraging results in adult *O. tsutsugamushi* cases in areas with strains resistant to doxycycline [49]. Rifampin is effective against *C. burnettii in vitro* [145] and has been suggested for the treatment of Q fever pneumonia, [72], but there are no reports of treatment of childhood Q fever yet.

Other antibiotics which are commonly used as first-line agents for childhood infections, including β-lactams, aminoglycosides, and clindamycin are ineffective.

7 Prevention

Prevention is strongly recommended and includes avoidance of ticks and arthropod bites, protective clothing, preferably impregnated with permethrin or other pyrethroid, frequent application of repellents on exposed skin and daily inspection and proper removal of ticks or mites. Additionally, children should be discouraged from touching rodents, dogs or domestic livestock [16, 80].

7.1 Vaccines

Recovery from a rickettsial infection confers strong and long-lasting immunity thus development of a vaccine against rickettsioses seems quite possible. However, no appropriate vaccines are available yet. Killed vaccines were tried for *R. rickettsii* with poor results [146]. The so-called Cox vaccine *for R. prowazekii* appeared at the outset of Word War II but was halted in 1944 because of limited efficacy and side-effects, and trials with live attenuated vaccines for *R. prowazekii* during 1950s and 1960s had high rates of side-effects [146]. Current research focuses on subunit vaccines which would be harmless and cross-protective against *R. prowazekii, R. typhi, R. rickettsii and R.conorii* [146]. Research on vaccines against ehrlichia and anaplasma species is still preliminary [147]. Currently, only acellular and whole-cell Q fever vaccines are available in Australia, the USA and some European countries. These vaccines are recommended for non-immune high-risk adults after pre-vaccination screening and have not been used in children [72].

7.2 Prophylactic Antibiotics

The possibility of acquiring rickettsiosos after a tick bite is small, given the rarity of infected ticks and the low affinity of ticks for humans [16]. Doxycycline in a weekly single dose of 200 mg for prolonged outdoor activities, such as for military personnel, is of questionable efficacy [80]. Asymptomatic individuals with recent tick bites seem not to benefit from prophylaxis, although a single dose of azithromycin appears to protect adults with a history of tick-bite against *R. conorii conorii* [148].

8 Conclusions

Rickettsioses are unlikely to be ever eradicated as a human hazard, as they are primarily infections of arthropods. Their hosts and humans are only incidental intruders in the natural cycle. On the contrary, they are re-emerging in a changing world, fulfilling Hans Zinsser's conclusion "Typhus is not dead. It will live on for centuries, and it will continue to break into the open whenever human stupidity and brutality give it a chance, as most likely they occasionally will." [149]. They should be considered in the differential diagnosis of children with FID, rashes, lymphadenopathy, constellation of unusual clinical symptoms, normal or nearly normal first-line laboratory tests and a suggestive history (Fig. 2). A short course of doxycycline is remarkably effective for most of these infections and should be given on clinical suspicion, without laboratory confirmation, which might come too late or never.

References

1. American Academy of Pediatrics, Committee on Infectious Diseases. Rickettsial diseases. In: Pickering LK, Baker CJ, Kimberlin DW, Long SS, eds. 2009 Red Book: report of the Committee on Infectious Diseases, 28th ed. Elk Grove Village, Illinois; 2009. p. 570–5.
2. Hechemy K, Oteo JA, Raoult D, Silverman DJ, Blanco JR. A century of rickettsiology: emerging, reemerging rickettsioses, clinical, epidemiological and molecular diagnostic aspects and emerging veterinary rickettsioses. Ann NY Acad Sci 2006;1078:1–14.
3. Raoult D, Fournier PE, Eremeeva M, Graves S, Kelly PJ, Oteo JA, et al. Naming of rickettsiae and rickettsial diseases. Ann NY Acad Sci 2005;1063:1–12.
4. Chapman AS, Bakken JS, Folk SM, Paddock CD, Bloch KC, Krusell A, et al. Tickborne Rickettsial Diseases Working Group; CDC. Diagnosis and management of tickborne rickettsial diseases: Rocky Mountain spotted fever, ehrlichioses, and anaplasmosis- United States: a practical guide for physicians and other health-care. MMWR Recomm Rep 2006;55:1–27.
5. Buckingham SC, Marshall GS, Schutze GE, Woods CR, Jackson MA, Patterson LE, et al. Tick-borne Infections in Children Study Group. Clinical and laboratory features, hospital course, and outcome of Rocky Mountain spotted fever in children. J Pediatr 2007;150:180–4.
6. Treadwell TA, Holman RC, Clarke MJ, Krebbs JW, Paddock CD, Childs JE. Rocky Mountain spotted fever in the United States, 1993–1996. Am J Trop Med Hyg 2000;157:21–6.
7. Chapman AS, Murphy SM, Demma LJ, Holman RC, Curns AT, McQuiston JH, et al. Rocky Mountain spotted fever in the United States, 1997–2002. Vector Borne Zoonotic Dis 2006;6:170–8.
8. Bradford WD, Hawkins HK. (1977). Rocky Mountain spotted fever in childhood. Am J Dis Child 1997;131;1228–32.
9. Haynes RE, Sanders DY, Cramblett HG. (1970). Rocky Mountain spotted fever in children. J Pediatr 1970;76:685–93.
10. Cascio A, Dones P, Romano A, Titone L. Clinical and laboratory findings of boutonneuse fever in Sicilian children. Eur J Pediatr 1998;157:482–6.
11. Colomba C, Saporito L, Frasca Polara V, Rubino R, Titone L. Mediterranean Spotted Fever: clinical and laboratory characteristics of 415 Sicilian children. BMC Infect Dis 2006;6: 60.
12. Moraga FA, Martinez-Roig A, Alonso JL, Boronat M, Domingo F. Boutonneuse fever. Arch Dis Child 1982;57:149–51.
13. Bitsori M, Galanakis E, Papadakis CE, Sbyrakis S. Facial nerve palsy associated with Rickettsia conorii infection. Arch Dis Child 2001;85:54–5.
14. Rovery C, Brouqui P, Raoult, D. Questions on Mediterranean Spotted Fever a century after its discovery. Emerg Infect Dis 2008;14:1360–7.
15. Ciceroni L, Pinto A, Ciarrocchi S, Ciervo A. Current knowledge of rickettsial diseases in Italy. Ann N Y Acad Sci 2006;1078:143–9.
16. Brouqui P, Parola P, Fournier PE, Raoult D. Spotted fever rickettsioses in Southern Europe. FEMS Immunol Med Microbiol 2007;49: 2–12.
17. Gross EM, Yagupsky P. Israeli rickettsial spotted fever in children. A review of 54 cases. Acta Trop 1987;44: 91–6.
18. Wolach B, Franco S, Bogger-Goren S, Drucker M, Goldwasser RA, Sadan N, et al. Clinical and laboratory findings of spotted fever in Israeli children. Pediatr Infect Dis J 1989;8:152–5.
19. Yagupsky P, Wolach B. (1993). Fatal Israeli spotted fever in children. Clin Infect Dis1993;17:850–3.
20. Tarasevich IV, Mediannikov OY. (2006). Rickettsial Diseases in Russia. Ann N Y Acad Sci 2006;1078:48–59.
21. Brouqui P, Harle JR, Delmont J, Frances C, Weiller PJ, Raoult D. (1997). African tick-bite fever. An imported spotless rickettsiosis. Arch Intern Med 1997;157:119–24.
22. Jensenius M, Fournier PE, Kelly P, Myrvang B, Raoult D. (2003). African tick-bite fever. Lancet Infect Dis 2003;3:557–64.
23. Jackson Y, Chappuis F, Loutan L. (2004). African tick-bite fever: four cases among Swiss travelers returning from South Africa. J Travel Med 2004;11:225–30.

24. Mahara F. Rickettsioses in Japan and the Far East. Ann N Y Acad Sci 2006;1078: 60–73.
25. Jang WJ, Hyun Kim JH, Choi YJ, Jung KD, Kim YG, Lee SH, et al. First serologic evidence of human spotted fever group rickettsiosis in Korea. J Clin Microbiol 2005;42:2310–3.
26. Fukuta Y, Mahara F, Nakatsu T, Yoshida T, Nishimura M. (2007). A case of japanese Spotted Fever complicated with myocarditis. Jpn J Infect Dis 2007;60:59–61.
27. Raoult D, Lakos A, Fenollar F, Beytout J, Brouqui P, Fournier PE. (2002). Spotless rickettsiosis caused by *Rickettsia slovaca* and associated with *Dermacentor* ticks. Clin Infect Dis 2002;34:1331–6.
28. Ibarra V, Oteo JA, Portillo A, Santibáñez S, Blanco JR, Metola L, et al. Rickettsia slovaca infection: DEBONEL/ TIBOLA. Ann N Y Acad Sci 2006;1078:206–14.
29. Porta FS, Nieto EA, Creus BF, Espín TM, Casanova FJ, Sala IS, et al. (2008). Tick-borne lymphadenopathy: a new infectious disease in children. Pediatr Infect Dis J 2008;27:618–22.
30. Kass EM, Szaniawski WK, Levy H, Leach J, Srinivasan K, Rives C. Rickettsialpox in a New York City hospital, 1980 to 1989. N Engl J Med 1994;331: 1612–7.
31. Zavala-Castro JE, Zavala-Velázquez JE, Peniche-Lara GF, Sulú Uicab JE. Human rickettsialpox, southeastern Mexico. Emerg Infect Dis 2009;15:1665–7.
32. Perez-Osorio CE, Zavala-Velasquez JE, Leon JJA, Zavala-Castro JE. (2008). Rickettsia felis as emergent global threat for humans. Emerg Infect Dis 2008;14:1019–23.
33. Bernabeu-Wittel M, del Toro MD, Nogueras MM, Muniain MA, Cardeñosa N, Márquez FJ, et al. (2006). Seroepidemiological study of Rickettsia felis, Rickettsia typhi, and Rickettsia conorii infection among the population of southern Spain. Eur J Clin Microbiol Infect Dis 2006;25:375–81.
34. Sexton DJ, Dwyer B, Kemp R, Graves S. Spotted fever group rickettsial infections in Australia. Rev Infect Dis 1991;13:876–86.
35. Graves S, Unsworth N, Stenos J. Rickettsioses in Australia. Ann N Y Acad 2006;1078: 74–9.
36. Graves S, Stenos J. Rickettsia honei: a spotted fever group rickettsia on three continents. Ann N Y Acad Sci 2003;990:62–6.
37. Fournier PE, Gouriet F, Brouqui P, Lucht F, Raoult D. Lymphangitis-associated rickettsiosis, a new rickettsiosis caused by *Rickettsia sibirica mongolotimonae*: seven new cases and review of the literature. Clin Infect Dis 2005;40:1435–44.
38. Parola P, Paddock CD, Raoult D. Tick-borne rickettsioses around the world: emerging diseases challenging old concepts. Clin Microbiol Rev 2005;18:719–56.
39. Quintal D. Historical aspects of the rickettsioses. Clin Dermatol 1996;14:237–42.
40. Wu JJ, Huang DB, Pang KR, Tyring SK. Rickettsial infections around the world, part 2: rickettsialpox, the typhus group, and bioterrorism. J Cutan Med Surg 2005;9:105–15.
41. Bechah Y, Capo C, Mege JL, Raoult D. Epidemic typhus. Lancet Infect Dis 2008;7:417–26.
42. Duma RJ, Sonenshine DE, Bozeman FM, Veazey JMJr, Elisberg BL, Chadwick DP, et al. Epidemic typhus in the United States associated with flying squirrels. JAMA 1981;245:2318–23.
43. Turcinov, D, Kuzman I, Herendi B. Failure of azithromycin in treatment of Brill-Zinsser disease. Antimicrob Agents Chemother 2000;44:1737–8.
44. Bitsori M, Galanakis E, Gikas A, Scoulica E, Sbyrakis S. Rickettsia typhi infection in childhood. Acta Paediatr 2002;91:59–61.
45. Whiteford SF, Taylor JP, Dumler JS. Clinical, laboratory, and epidemiologic features of murine typhus in 97 Texas children. Arch Pediatr Adolesc Med 2001;155:396–400.
46. Koliou M, Psaroulaki A, Georgiou C, Ioannou I, Tselentis Y, Gikas, A. Murine typhus in Cyprus: 21 paediatric cases. Eur J Clin Microbiol Infect Dis 2007;26:491–3.
47. Silpapojakul K, Chayakul P, Krisanapan S, Silpapojakul K. Murine typhus in Thailand: clinical features, diagnosis and treatment. *Q J Med* 1993;86:43–7.
48. Shalev H, Raissa R, Evgenia Z, Yagupsky P. Murine typhus is a common cause of febrile illness in Bedouin children in Israel. Scand J Infect Dis 2006;38: 451–5.
49. Watt G, Parola P. Scrub typhus and tropical rickettsioses. Curr Opin Infect Dis 2003;16: 429–36.
50. Faa AG, McBride WJ, Garstone G, Thompson RE, Holt P. Scrub typhus in the Torres Strait islands of north Queensland, Australia. Emerg Infect Dis 2003;9:480–2.
51. Chanta C, Triratanapa K, Ratanasirichup P, Mahaprom W. Hepatic dysfunction in pediatric scrub typhus: role of liver function test in diagnosis and marker of disease severity. J Med Assoc Thai 2007;90:2366–9.
52. Chanta C, Chanta S. Clinical study of 20 children with scrub typhus at Chiang Rai Regional Hospital. J Med Assoc Thai 2005;88:1867–72.
53. Ogawa M, Hagiwara T, Kishimoto T, Shiga S, Yoshida Y, Furuya Y, et al. Scrub typhus in Japan: epidemiology and clinical features of cases reported in 1998. Am J Trop Med Hyg 2002;67:162–5.
54. Silpapojakul K, Varachit B, Silpapojakul K. Paediatric scrub typhus in Thailand: a study of 73 confirmed cases. Trans R Soc Trop Med Hyg 2004;98:354–9.

55. Dumler JS, Madigan JE, Pusterla N, Bakken JS. Ehrlichioses in humans: epidemiology, clinical presentation, diagnosis, and treatment. Clin Infect Dis 2007;45 Suppl 1:S45–S51.
56. Arnez M, Petrovec M, Lotric-Furlan S, Zupanc TA, Strle F. First European pediatric case of human granulocytic ehrlichiosis. J Clin Microbiol 2001;39:4591–2.
57. Lantos P, Krause PJ. Ehrlichiosis in children. Semin Pediatr Infect Dis 2002;13:249–56.
58. Jacobs RF, Schutze GE. Ehrlichiosis in children. J Pediatr 1997;131:184–92.
59. Schutze GE, Buckingham SC, Marshall GS, Woods CR, Jackson MA, Patterson LE, et al. Tick-Borne Infections in Children Study Group. Human monocytic ehrlichiosis in children. Pediatr Infect Dis J 2007;26:475–9.
60. Martínez MC, Gutiérrez CN, Monger F, Ruiz J, Watts A, Mijares VM, et al. Ehrlichia chaffeensis in child, Venezuela. Emerg Infect Dis 2008;14:519–20.
61. Psaroulaki A, Koliou M, Chochlakis D, Ioannou I, Mazeri S, Tselentis, Y. Anaplasma phagocytophilum infection in a child. Pediatr Infect Dis J 2008;27:664–6.
62. Esbenshade A, Esbenshade J, Domm J, Williams J, Frangoul H. Severe ehrlichia infection in pediatric oncology and stem cell transplant patients. Pediatr Blood Cancer 2010;54: 776–8.
63. Burns S, Saylors R, Mian A. Hemophagocytic lymphohistiocytosis secondary to Ehrlichia chaffeensis infection: a case report. *J Pediatr Hematol Oncol* 2010;32:e142–3.
64. Hongo I, Bloch KC. Ehrlichia infection of the central nervous system. Curr Treat Options Neurol 2006;8:179–84.
65. Maltezou HC, Raoult D.Q fever in children. Lancet Infect Dis 2002;2:686–91.
66. Richardus JH, Dumas AM, Huisman J, Schaap GJ. Q fever in infancy: a review of 18 cases. Pediatr Infect Dis J 1985;4:369–73.
67. Ruiz-Contreras J, González Montero R, Ramos Amador JT, Giancaspro Corradi E, Scarpellini Vera A. Q fever in children. Am J Dis Child 1993;147:300–2.
68. To H, Kako N, Zhang GQ, Otsuka H, Ogawa M, Ochiai O, et al. Q fever pneumonia in children in Japan. J Clin Microbiol 1996;34:647–51.
69. Hervás JA, de la Fuente MA, García F, Reynés J, de Carlos JC, Salvá F. Coxiella burnetii myopericarditis and rhabdomyolysis in a child. Pediatr Infect Dis J 2000;19:1104–6.
70. Hufnagel M, Niemeyer C, Zimmerhackl LB, Tüchelmann T, Sauter S, Brandis M. Hemophagocytosis: a complication of acute Q fever in a child. Clin Infect Dis 1995;21: 1029–31.
71. Maltezou HC, Kallergi C, Kavazarakis E, Stabouli S, Kafetzis DA. Hemolytic-uremic syndrome associated with Coxiella burnetii infection. Pediatr Infect Dis J 2001;20:811–3.
72. Parker NR, Barralet JH, Bell AM. Q fever. Lancet 2006;367:679–88.
73. Marshall GS, Jacobs RF, Schutze GE, Paxton H, Buckingham SC, DeVincenzo JP, et al. Tick-Borne Infections in Children Study Group. Ehrlichia chaffeensis seroprevalence among children in the southeast and south-central regions of the United States. Arch Pediatr Adolesc Med 2002;156:166–70.
74. Hellenbrand W, Breuer T, Petersen L. Changing epidemiology of Q fever in Germany, 1947–1999. Emerg Infect Dis 2001;7:789–96.
75. Fishbein DB, Dawson JE, Robinson LE. Human ehrlichiosis in the United States, 1985–1990. Ann Intern Med 1994;120:736–43.
76. Dalton MJ, Clarke MJ, Holman RC, Krebs JW, Fishbein DB, Olson JG, et al. National surveillance for Rocky Mountain spotted fever, 1981–1992: epidemiologic summary and evaluation of risk factors for fatal outcome. Am J Trop Med Hyg 1995;52:405–13.
77. Raeber PA, Winteler S, Paget J. Fever in the returned traveler: remember rickettsial diseases. Lancet 1994;344:331.
78. Marschang A, Nothdurft H D, Kumlien S, von Sonnenburg F. Imported rickettsioses in German travelers. Infection 1995;23:94–7.
79. Rahman A, Tegnell A, Vene S, Giesecke J. Rickettsioses in Swedish travelers, 1997–2001. Scand j Infect Dis 2003;35:247–50.
80. Jensenius M, Fournier PE, Raoult D. Rickettsioses and the international traveler. Clin Infect Dis 2004;39:1493–9.
81. Dantas-Torres F. Rocky Mountain spotted fever. Lancet Infect Dis 2007;7:724–32.
82. Shazberg G, Moise J, Terespolsky N, Hurvitz H. Family outbreak of Rickettsia conorii infection. Emerg Infect Dis 1999;5:723–4.
83. Jones TF, Craig AS, Paddock CD, McKechnie DB, Childs JE, Zaki SR, et al. Family cluster of Rocky Mountain spotted fever. Clin Infect Dis 1999;28:853–859.
84. Standaert SM, Dawson JE, Schaffner W, Childs JE, Biggie KL, Singleton JJr, et al. Ehrlichiosis in a golf-oriented retirement community. N Engl J Med 1995;333:420–5.
85. Amitai Z, Bromberg M, Bernstein M, Raveh D, Keysary A, David D, et al. A large Q fever outbreak in an urban school in central Israel. Clin Infect Dis 2010;50:1433–8.

86. Jorm LR, Lightfoot NF, Morgan KL. An epidemiological study of an outbreak of Q fever in a secondary school. Epidemiol Infect 1990;104:467–77.

87. Elchos BN, Goddard J. Implications of presumptive fatal Rocky Mountain spotted fever in two dogs and their owner. J Am Vet Med Assoc 2003;223:1450–2.

88. Paddock CD, Brenner O, Vaid C, Boyd DB, Berg JM, Joseph RJ, et al. Short report: concurrent Rocky Mountain spotted fever in a dog and its owner. Am J Trop Med Hyg 2002;66:197–9.

89. Kelly DJ, Richards AL, Temenak J, Strickman, D, Dasch GA. The past and present threat of rickettsial diseases to military medicine and international public health. Clin Infect Dis 2002;34, Suppl 4:S145–9.

90. Madariaga M, Rezai K, Trenholme GM, Weinstein RA. Q fever: a biological weapon in your backyard. Lancet Infect Dis 2003;3:709–21.

91. Cappellini MD, Fiorelli G. Glucose-6-phosphate dehydrogenase deficiency. Lancet 2008; 371:64–74.

92. Cascio A, Iaria C. Epidemiology and clinical features of Mediterranean spotted fever in Italy. Parassitologia 2006;48:131–3.

93. Walker DH, Radisch DL, Kirkman HN. Haemolysis with rickettsiosis and glucose-6-phosphate dehydrogenase deficiency. Lancet 1983;2:217.

94. Regev-Yochay G, Segal E, Rubinstein E. Glucose-6-phosphate dehydrogenase deficiency: possible determinant for a fulminant course of Israeli spotted fever. Isr Med Assoc J 2000;10:781–2.

95. Safdar N, Love RB, Maki DG. (2002). Severe Ehrlichia chaffeensis infection in a lung transplant recipient: a review of ehrlichiosis in the immunocompromised patient. *Emerg Infect Dis* 2002;8:320–3.

96. Peters TR, Edwards KM, Standaert SM. Severe ehrlichiosis in an adolescent taking trimethoprim-sulfamethoxazole. Pediatr Infect Dis J 2000;19:170–2.

97. Steigman AJ. Rocky Mountain spotted fever and the avoidance of sulfonamides. J Pediatr 1977;91:163–4.

98. Ruiz Beltrán R, Herrero Herrero JI. Deleterious effect of trimethoprim-sulfamethoxazole in Mediterranean spotted fever. Antimicrob Agents Chemother 1992; 36: 1342–3.

99. Fergie JE, Purcell K, Wanat P, Wanat D. (2000). Murine typhus in South Texas Children. Pediatr Infect Dis J 2000;35:535–8.

100. Silpapojakul K, Chupuppakarn S, Yuthasompob S, Varachit B, Chaipak D, Borkerd T, Silpapojakul K. Scrub and murine typhus in children with obscure fever in the tropics. Pediatr Infect Dis J 1991;10:200–3.

101. Petersdorf RG, Beeson P. Fever of unexplained origin: report on 100 cases. Medicine Baltimore 1961;40:1–30.

102. Parra Ruiz J, Peña Monje A, Tomás Jiménez C, Parejo Sánchez MI, Vinuesa García D, Muñoz Medina L, et al. Clinical spectrum of fever of intermediate duration in the south of Spain. Eur J Clin Microbiol Infect Dis 2008;10:993–5.

103. Bernabeu-Wittel M, Pachón J, Alarcón A, López-Cortés LF, Viciana P, Jiménez-Mejías ME, et al. Murine typhus as a common cause of fever of intermediate duration: a 17-year study in the south of Spain. Arch Intern Med 1999;159:872–6.

104. Kirkland KB, Marcom PK, Sexton DJ, Dumler JS, Walker DH. (1993). Rocky Mountain spotted fever complicated by gangrene: report of six cases and review. Clin Infect Dis 1993;16:629–34.

105. Abramson JS, Ginver LB. Rocky Mountain spotted fever. (1999). Pediatr Infect Dis J 1999;18:539–40.

106. Kofteridis DP, Mazokopakis EE, Tselentis Y, Gikas A. Neurological complications of acute Q fever infection. Eur J Epidemiol 2004;19:1051–4.

107. Silpapojakul K, Ukkachoke C, Krisanapan S, Silpapojakul K. Rickettsial meningitis and encephalitis. Arch Intern Med 1991;151:1753–7.

108. Galanakis E, Gikas A, Bitsori M, Sbyrakis S. Rickettsia typhi infection presenting as subacute meningitis. J Child Neurol 2002;17:156–7.

109. Tikare NV, Shahapur PR, Bidari LH, Mantur BG. Rickettsial meningoencephalitis in a child-a case report. J Trop Pediatr 2010;56:198–200.

110. Horney LF, Walker DH. Meningoencephalitis as a major manifestation of Rocky Mountain spotted fever. South Med J 1988;81:915–8.

111. Wei TY, Baumann RJ. Acute disseminated encephalomyelitis after Rocky Mountain spotted fever. Pediatr Neurol 1999;21:503–5.

112. Sirisanthana V, Puthanakit T, Sirisanthana T.Epidemiologic, clinical and laboratory features of scrub typhus in thirty Thai children. Pediatr Infect Dis J 2003;22:341–5.

113. Donohue JF. Lower respiratory tract involvement in Rocky Mountain spotted fever. Arch Intern Med. 1980 Feb;140(2):223–7.

114. Marin-Garcia J, Gooch WM 3 rd, Coury DL. Cardiac manifestations of Rocky Mountain spotted fever. Pediatrics. 1981 Mar;67(3):358–61.

115. Lupoglazoff JM, Brouqui P, Magnier S, Hvass U Casasoprana A. Q fever tricuspid endocarditis. Arch Dis Child 1997;77:448–9.

116. Al-Hajjar S, Hussain Qadri SM, Al-Sabban E, Jager C. *Coxiella burnetii* endocarditis in a child. Pediatr Infect Dis J 1997;9:911–913.

117. Peter J, Ray CG. Infectious mononucleosis. Pediatr Rev 1998;19:276–9.

118. Macsween KF, Crawford DH. Epstein-Barr virus-recent advances. Lancet Infect Dis 2003;3:131–40.

119. Jenkins DR, Rees JC, Pollitt C, Cant A, Craft AW. Mediterranean spotted fever mimicking Kawasaki disease. BMJ 1997;314:655–6.

120. Rathore MH, Barton LL, Dawson JE, Regnery RL, Ayoub EM. Ehrlichia chaffeensis and Rochalimaea antibodies in Kawasaki disease. Clin Microbiol 1993;31:3058–9.

121. Kafetzis DA, Maltezou HC, Constantopoulou I, Antonaki G, Liapi G, Mathioudakis I. Lack of association between Kawasaki syndrome and infection with Rickettsia conorii, Rickettsia typhi, Coxiella burnetii or Ehrlichia phagocytophila group. Pediatr Infect Dis J 2001;20:703–6.

122. Crump JA, Mintz ED. Global trends in thyphoid and paratyphoid fever. Clin Infect Dis 2010; 50:241–6.

123. Lynch MF, Blanton EM, Bulens S, Polyak C, Vojdani J, Stevenson J, Medalla F, Barzilay E, Joyce K, Barrett T, Mintz ED. Typhoid fever in the United States, 1999–2006. JAMA 2009;302:859–65.

124. O' Brien DP, Leder K, Matchett E, Brown GV, Torresi J. Illness in returned travelers and immigrants/refugees: the 6-year experience of two Australian Infectious Diseases Units. J Travel Med 2006;13:145–52.

125. Elias AF, Peterson SN, Huntington MK. Typhoid fever in a young immigrant child: a case report and review of the literature. S D Med 2008;61:255–8.

126. Goodman JL. Ehrlichiosis-Ticks, Dogs and Doxycycline. New Engl J Med 1999;341:195–7.

127. (Anonymus). American Academy of Pediatrics. Committee on drugs. Requiem for tetracyclines. Pediatrics 1975;55:142–3.

128. Balbi HJ. Chloramphenicol: a review. Pediatr Rev 2004;25:284–8.

129. Shaked Y, Samra Y, Maeir MK, Rubinstein E. Murine typhus and spotted fever in Israel in the eighties: retrospective analysis. Infection 1988;16:283–7.

130. Meloni G, Meloni T. Azithromycin vs. doxycycline for Mediterranean spotted fever. Pediatr Infect Dis J 1996;15:1042–4.

131. Lee KY, Lee HS, Hong JH, Hur JK, Whang KT. Roxithromycin treatment of scrub typhus (tsutsugamushi disease) in children. Pediatr Infect Dis J 2003;22:130–3.

132. Cascio A, Colomba C, Di Rosa D, Salsa L, Di Martino L, Titone L. Efficacy and safety of clarithromycin as treatment for Mediterranean spotted fever in children: a randomized controlled trial. Clin Infect Dis 2001;33:409–11.

133. Branger S, Rolain JM, Raoult D. Evaluation of antibiotic susceptibilities of Ehrlichia canis, Ehrlichia chaffeensis, and Anaplasma phagocytophilum by real-time PCR. Antimicrob Agents Chemother 2004;48:4822–8.

134. Maurin M, Raoult D. In vitro susceptibilities of spotted fever group rickettsiae and Coxiella burnetti to clarithromycin. Antimicrob Agents Chemother 1993;37:2633–7.

135. Gikas A, Kofteridis DP, Manios A, Pediaditis J, Tselentis Y. Newer macrolides as empiric treatment for acute Q fever infection. Antimicrob Agents Chemother 2001;45:3644–6.

136. Nourse C, Allworth A, Jones A, Horvath R, McCormack J, Bartlett J, et al. Three cases of Q fever osteomyelitis in children and a review of the literature. Clin Infect Dis 2004;39:e61–6.

137. Seki M, Ikari N, Yamamoto S, Yamagata Y, Kosai K, Yanagihara K, et al. Severe Japanese spotted fever successfully treated with fluoroquinolone. Intern Med 2006;45:1323–6.

138. Laferl H, Fournier PE, Seiberl G, Pichler H, Raoult D. Murine typhus poorly responsive to ciprofloxacin: a case report. J Travel Med 2002;9:103–4.

139. Klein MB, Nelson CM, Goodman JL. Antibiotic susceptibility of the newly cultivated agent of human granulocytic ehrlichiosis: promising activity of quinolones and rifamycins. Antimicrob Agents Chemother 1997;41:76–9.

140. Rolain JM, Maurin M, Vestris G, Raoult D. In vitro susceptibilities of 27 rickettsiae to 13 antimicrobials. Antimicrob Agents Chemother 1998;42:1537–41.

141. Morovic M. Q Fever pneumonia: are clarithromycin and moxifloxacin alternative treatments only? Am J Trop Med Hyg 2005;73:947–8.

142. Brantley RK. Trimethoprim-sulfamethoxazole and fulminant ehrlichiosis. Pediatr Infect Dis J 2001;20:231.

143. Bella F, Espejo E, Uriz S, Serrano JA, Alegre MD, Tort J. Randomized trial of 5-day rifampin versus 1-day doxycycline therapy for Mediterranean spotted fever. J Infect Dis. 1991;164:433–4.

144. Krause PJ, Corrow CL, Bakken JS. Successful treatment of human granulocytic ehrlichiosis in children using rifampin. Pediatrics 2003;112:e252–3.

145. Brennan RE, Samuel JE. Evaluation of Coxiella burnetii antibiotic susceptibilities by real-time PCR assay. J Clin Microbiol 2003;41:1869–74.

146. Walker DH. (2009). The realities of biodefense vaccines against Rickettsia. Vaccine 2009;(27) Suppl 4: D52–5.

147. McBride JW, Walker DH. Progress and obstacles in vaccine development for the ehrlichioses. Expert Rev Vaccines 2010;9:1071–82.
148. Dzelalija B, Petrovec M, Avsic-Zupanc T, Strugar J, Milić TA. Randomized trial of azithromycin in the prophylaxis of Mediterranean spotted fever. Acta Med Croatica 2002;56:45–7.
149. Zinsser H. Rats, lice and history. New York: Little Brown & Co; 1935 Quoted in: Quintal D. Historical aspects of the rickettsioses. Clin Dermatol 1996;14:237–42.

What is the Evidence Behind Recommendations for Infection Control?

Christina Gagliardo and Lisa Saiman

1 Introduction

Effective infection control in healthcare settings can only be accomplished with a program that adopts evidence-based guidelines implemented by an interdisciplinary team responsible for monitoring implementation of these guidelines and for education of the practitioners involved in patient care. An effective infection control program can prevent acquisition and transmission of infectious pathogens among patients, staff and visitors. However, infection control practices can be time consuming, costly or tedious. Is there evidence to support infection control recommendations or are recommendations simply based on anecdotal experience and expert opinion? We present the rationale for grading infection control recommendations and provide relevant examples of graded recommendations for infection control practices relevant for pediatric healthcare settings.

2 Grading Evidence for Infection Control Strategies

The Centers for Disease Control and Prevention (CDC), the World Health Organization (WHO), as well as professional societies use a panel of experts to review the literature and develop evidence-based guidelines derived from existing scientific data, theoretical rationale, applicability, and economic impact. Table 1 shows the system used by the CDC for categorizing infection control recommendations [1].

The literature describing strategies to prevent acquisition and transmission of pathogens in healthcare settings largely consists of (a) case control or cohort studies assessing risk factors for infections or colonization with specific pathogens, (b) reports of outbreaks and the strategies used

C. Gagliardo
Department of Pediatrics, Division of Infectious Diseases, Morgan Stanley Children's Hospital
of New York-Presbyterian, Columbia University, New York, USA
e-mail: cg2406@columbia.edu

L. Saiman
Department of Pediatrics, Columbia University, Department of Infection Prevention
& Control, New York-Presbyterian Hospital, New York, USA
e-mail: ls5@columbia.edu

N. Curtis et al. (eds.), *Hot Topics in Infection and Immunity in Children VIII*,
Advances in Experimental Medicine and Biology 719, DOI 10.1007/978-1-4614-0204-6_14,
© Springer Science+Business Media, LLC 2011

Table 1 System for categorizing infection control recommendations

Category	Strength of evidence
IA	Strongly recommended for implementation and strongly supported by well-designed experimental, clinical or epidemiologic studies
IB	Strongly recommended for implementation and supported by some experimental, clinical or epidemiologic studies and strong theoretical rationale
IC	Required for implementation as mandated by federal and/or state regulation or standard
II	Suggested for implementation and supported by suggestive clinical or epidemiologic studies, theoretical rationale or consensus by panel of experts

to control them or (c) nonrandomized, pre- and post-intervention study designs, also known as "quasi-experimental" designs [2]. These studies can provide valuable information regarding the effectiveness of specific interventions. However, several factors can decrease the certainty of attributing improved outcomes to a specific intervention. These factors include: (a) difficulties in controlling for important confounding variables, (b) the use of multiple interventions during an outbreak, (c) regression to the mean, i.e., improvement over time without any interventions [3], and/or (d) over-estimation of treatment effect since each patient's 'treatment' is deliberately chosen rather than randomly assigned. While randomized treatment assignment would provide the most reliable, unbiased estimate of treatment effects, it is rarely feasible for infection control intervention studies to use this methodology due to cost (e.g. large sample size) and complexities of implementation (e.g. contamination between study groups) [4]. Thus, when crafting guidelines, it is critical to assess the quality of studies, potential biases, and consistency of results prior to assigning the previously mentioned evidence-based categories.

3 Hand Hygiene

Many patients may be colonized or less commonly, infected, with potentially transmissible pathogens. The hands of healthcare workers are often the conduit for the spread of such pathogens to other patients in their care. Fortunately, hand hygiene is a simple, effective, low-cost intervention to prevent the spread of pathogens that can cause healthcare-associated infections (HAI). Compliance with hand hygiene can dramatically enhance patient safety. The WHO and CDC have developed hand hygiene guidelines which emphasize the "moments" before, during and after patient encounters in which to perform hand hygiene [1, 5]. Table 2 specifies the "moments" during patient care when hand hygiene should occur, and rationale for the recommendation. These guidelines also aim to facilitate healthcare workers' understanding of the risks of pathogen transmission and to integrate hand hygiene practices into every interaction with a patient, their surroundings and equipment.

The preferred method of hand hygiene is an alcohol-based hand rub if hands are not visibly soiled (WHO and CDC Category IB). In contrast, hands should be washed with soap and water if visibly soiled or contaminated with proteinaceous material, blood or other body fluids (WHO Category 1B and CDC Category IA) or if there is known or suspected exposure to spore-forming organism such as *Clostridium difficile* (WHO Category 1B and CDC Category II).

Another important aspect of hand hygiene is the recommendation to ban the use of artificial fingernails or fingernail extenders (WHO and CDC Category IA). Artificial nails worn by HCWs are associated with healthcare-acquired infections particularly in the intensive care unit (ICU) population. Compared with natural nails, artificial nails have higher rates of colonization with Gram-negative flora and yeast [6] which are not removed with hand hygiene.

Table 2 "Moments" for hand hygiene during patient encounters

"Moment"	Rationale	Category[a] WHO	CDC
Before patient contact	Protect patient from potential pathogens on HCW hands	IB	IB
After removing gloves	Protect HCW from potential pathogens	IB	IB
Before aseptic task/procedure[b]	Protect patient from pathogen on HCW hands and/or own colonizing pathogen	IB	IB
After body fluid exposure risk	Protect HCW and environment from potential pathogens	IA	IA
When moving from contaminated to clean body site	Prevent contamination of non-infected/colonized body site	IB	II
After patient contact	Protect HCW and environment from potential pathogens	IB	IB
After contact with inanimate object/patient surroundings	Protect HCW and environment from potential pathogens	IB	II

HCW healthcare worker
[a]The WHO and CDC use the same classification system, although the grading of the recommendations varies slightly [1, 5]
[b]Aseptic tasks/procedures: involve contact with the patient's mucous membranes, non-intact skin, or invasive medical device

4 Transmission Routes and Precautions for Epidemiologically Significant Pathogens

Pathogens are transmitted by three major mechanisms: contact, droplet, and/or the airborne route. Recommendations for the type of isolation precaution, the environmental controls (e.g. environmental cleaning and room type) and the use of personal protective equipment by staff (e.g. gowns, gloves, and/masks) are based on the pathogen's mode of transmission. Table 3 summarizes the modes of transmission and provides relevant examples.

As new data become available, recommendations may change and become more (or less) stringent depending on specific clinical scenarios or situations. Gaps in knowledge leading to controversies do occur for recommendations. An example of such a controversy involved the use of personal protective equipment during the 2009 Influenza A (H1N1) pandemic. Under routine environmental conditions, seasonal influenza is spread by droplets and thus a surgical mask is indicated [7]. However, during the 2009 H1N1 pandemic, concerns about increased virulence and/or different transmission routes, led the Occupational Safety and Health Administration (OSHA) and the CDC to recommend the use of fit-tested disposable N95 respirators by healthcare workers who provided direct patient care or collected respiratory tract specimens from patients with suspected or confirmed 2009 H1N1. As the pandemic progressed and more knowledge was gained, it was determined that 2009 Influenza A (H1N1) was spread via droplet transmission [8] and that the virulence was comparable to that of seasonal influenza. Thus, the recommendation for the routine use of N95 respirators was modified both to account for this new knowledge and to address the rapidly diminishing stock of N95 respirators. As a result, many centers recommended the use of a surgical mask during routine patient care and the use of the N95 respirator during aerosol-generating procedures, e.g. bronchoscopy, open suctioning of airway secretions, endotracheal intubation. Updated recommendations for personal protective equipment for this year's seasonal influenza (Winter 2010–2011) were unavailable at the time of this publication.

Table 3 Routes of transmission and infection control precautions

	Transmission mechanism	Example of pathogens/illness	Environmental controls (category)[a]	Personal protective equipment (category)
Contact	Person to another susceptible person:	MRSA, VRE, C. *difficile*, Rotavirus, RSV, Lice, HSV (neonatal or disseminated)	Single-patient room (IB) Cohort patients with same infection (IB)	Gown (IB/IC) Gloves (IB/IC)
– Direct	– Without contaminated intermediate object or person		Clean/disinfect contaminated surfaces & patient care equipment (IB)	
– Indirect	– Through contaminated intermediate object (e.g. thermometer) or staff hands			
Droplet	Droplets ≥5 μ diameter Travel short distance (~3–6 ft) Cough/sneeze/talking or procedure generating droplets (e.g. suctioning, intubation)	Influenza, Pertussis, Meningococcemia	Single-patient room (II) Cohort patients with same infection (IB)	Mask (IB) Face-shield (unresolved)
Airborne	Small (< 5 μ) aerosolized particles Remain suspended in air, able to circulate/travel longer distance	Tuberculosis, Measles, Disseminated Herpes Zoster	AIIR Negative pressure (IA/IC) Six air changes/per hour (existing facility) or 12 (new construction/renovation) (IA/IC) Recirculation of air through a HEPA filter (IA/IC)	Fit-tested N95 mask or higher level respirator (IB)

MRSA methicillin-resistant *Staphylococcus aureus*, *VRE* vancomycin-resistant *enterococcus*, *RSV* respiratory syncytial virus, *HSV* herpes simplex virus, *AIIR* airborne infection isolation room, *HEPA* high efficiency particulate air
[a] CDC Grading categories [1]

5 Prevention of Central Line-Associated Bloodstream Infections and Use of Bundle Strategies

The cost of central line-associated bloodstream infections (CLABSIs) is substantial, in terms of morbidity and mortality and financial resources expended. To improve patient outcome and reduce health-care costs, strategies have been implemented to reduce CLABSIs.

There are multiple potential sources of pathogens that can cause infection of a percutaneous intravascular device [9]. These include, but are not limited to, organisms on the patient's skin at the device insertion site or on a contaminated catheter hub from the patient's own skin flora or those from healthcare workers' hands. Other potential (although less likely) sources of infection are from intrinsic contamination of the infusate (i.e., originating at the manufacturer), or from extrinsic contamination (i.e., originating during preparation within the healthcare setting).

A multidisciplinary team approach can reduce the incidence of CLABSIs. The team should include the healthcare providers who insert and maintain intravascular catheters, healthcare managers who allocate resources, and healthcare providers who educate patients and patients' families to properly care for central venous catheters. Several *individual* strategies have been studied and shown to be effective in reducing CLABSIs. More recently, however, care "bundles" have been introduced into patient care. Bundles are defined as a *group* of evidence-based interventions related to a disease process, that when implemented together, result in better outcomes than single interventions [10]. Table 4 delineates 10 core characteristics of care bundles [11]. In addition to CLABSIs, care bundles are being applied to reduce catheter-related urinary tract infections and ventilator associated pneumonias.

Table 5 provides examples of bundle elements for insertion and for maintenance of central venous catheters [12]. Initiation of bundles to reduce HAIs requires not only delineating the bundle elements, but extensive planning for implementation of the bundles (e.g. a central line cart) and monitoring adherence (e.g. checklists). Most recently, a 29 PICU collaborative demonstrated that adherence to *both* insertion and maintenance bundles was significantly associated with reducing CLABSIs [13, 14].

Table 4 Ten characteristics of care bundles[a]

1. Consistently improve reliability of processes required for well tolerated and effective care
2. Group of interventions implemented together that result in better clinical outcomes when compared with individual interventions
3. Each intervention is supported by quality scientific data, preferably a randomized-controlled trial
4. Each intervention is included because experts believe intervention is essential to improving clinical outcomes
5. Clinical value of interventions may be locally defined or change over time based on evolving research and experience
6. Care bundles do not include all possible therapies nor are intended to be a comprehensive list of all possible care
7. If clinically inappropriate or contraindicated, care bundle elements should not be "forced"
8. Care bundles reflect all-or-nothing-principle; reliable care requires completion of all bundle elements
9. Use quality improvement to document completion of each element and improve compliance
10. Care bundles improve teamwork and communication within the unit

[a]Modified from [11]

Table 5 Bundle elements for central venous catheter insertion and maintenance[a]

Insertion checklist elements	Maintenance checklist elements
• Hand hygiene before and after palpating site and before and after insertion of central line (IA)	• Hand hygiene before and after accessing catheter or changing dressing (IA)
• Maximal barrier precautions (IA)	• Evaluate insertion site daily (IB)
• Disinfect skin, e.g. chlorhexidine, alcohol product (IA)	• Change dressing if damp, soiled, loose (IA)
• Sterile transparent dressing or sterile gauze (IA)	• Standardize tubing set up and changes (IB)
	• Maintain aseptic technique, including scrub the hub (IA)

[a][1]

6 Antimicrobial Stewardship

An increasing proportion of hospital acquired infections can be attributed to antimicrobial-resistant organisms. Such infections are associated with increased morbidity, mortality, length of stay and cost. Many healthcare facilities are actively engaged in antimicrobial stewardship defined as "the optimal selection, dosage, and duration of antimicrobial treatment that results in the best outcome for the treatment or prevention of infection with minimal toxicity to the patient and minimal impact on subsequent resistance" [15]. It is estimated that as much as half of all antimicrobial use is inappropriate. Thus, one of the most critical strategies to reduce antimicrobial resistance in the healthcare setting is to improve appropriate antimicrobial use.

Antimicrobial resistance arises due to selective pressure from antibiotic use. However, studies utilize different methods to categorize the antimicrobial agents to assess the association between antibiotic use and the emergence of resistance, thereby making comparisons between studies difficult. Antimicrobial categories can include: specific agents, antibiotic class (e.g. β-lactam agents), spectrum of activity (e.g. anaerobic activity) or a combination of these. For example, there has not been a consistent approach to categorize agents to assess the emergence of extended-spectrum β-lactamase (ESBL) producing Gram-negative bacilli [16]. When analyzing the same data set, different conclusions could be drawn when using different antimicrobial categories, In the model using antibiotic class, prior use of vancomycin and duration of hospitalization were associated with ESBL infection, while third generation cephalosporin use was not independently associated with ESBL infection. In contrast, the model using antibiotic spectrum identified use of an agent with activity against Gram-negative bacilli as the only independent risk factor for infections caused by ESBL producing Gram-negative bacilli. Zaoutis, et al., found that receipt of extended-spectrum cephalosporins within 30 days was significantly associated with infection caused by ESBL among hospitalized children [17]. Thus, standardizing categorization strategies and elucidating these issues are critical to implement specific, effective recommendations to curb resistance. Further studies are needed not only to elucidate risk factors for antimicrobial resistance, but to better define previous antibiotic exposure.

To combat antimicrobial resistance, a comprehensive evidence-based stewardship program ideally includes "care bundles". Strategies that have been shown to be effective include: formulary restrictions for certain antibiotics and pre-authorization by infectious diseases experts, guidelines and clinical pathways, dose optimization such as converting from intravenous to oral agents, and prospective audit of antimicrobial usage with intervention and feedback to prescribers [18].

7 Summary

Effective infection control is an integral part of patient safety and quality. Recommendations for infection control practices are based on evidence from studies as well as the clinical experience of experts. The emergence of care bundles that incorporate several evidence-based practices together have led to further reduction of HAI and promise to improve antimicrobial stewardship.

References

1. Siegel J, Rhinehart E, Jackson M, Chiarello L, and the Healthcare Infection Control Practices Advisory Committee. Guideline for Isolation Precautions: Preventing Transmission of Infectious Agents in Healthcare Settings. 2007.
2. Harris AD, et al. The use and interpretation of quasi-experimental studies in infectious diseases. Clin Infect Dis 2004;38(11):1586–91.
3. Morton V and Torgerson DJ. Effect of regression to the mean on decision making in health care. BMJ 2003;326(7398):1083–4.
4. Pocock SJ and Elbourne DR. Randomized trials or observational tribulations? N Engl J Med 2000;342(25): 1907–9.
5. World Health Organization (WHO), WHO Guidelines on Hand Hygiene in Health Care (Advanced Draft). Global Patient Safety Challenge 2005–2006: "Clean Care is Safer Care." 2006.
6. Gupta A., et al. Outbreak of extended-spectrum beta-lactamase-producing Klebsiella pneumoniae in a neonatal intensive care unit linked to artificial nails. Infect Control Hosp Epidemiol 2004;25(3):210–5.
7. Brankston G, et al.. Transmission of influenza A in human beings. Lancet Infect Dis 2007;7(4):257–65.
8. Society for Healthcare Epidemiology of America (SHEA). Position Statement: Interim Guidance on Infection Control Precautions for Novel Swine-Origin Influenza A H1N1 in Healthcare Facilities. 2009.
9. Crnich CJ and Maki DG. The promise of novel technology for the prevention of intravascular device-related bloodstream infection. I. Pathogenesis and short-term devices. Clin Infect Dis 2002;34(9):232–42.
10. Bizzarro M.J et al. A quality improvement initiative to reduce central line-associated bloodstream infections in a neonatal intensive care unit. Infect Control Hosp Epidemiol 2010; 31(3):241–8.
11. Lachman P and Yuen S. Using care bundles to prevent infection in neonatal and paediatric ICUs. Curr Opin Infect Dis 2009;22(3):224–8.
12. O'Grady NP, et al. Guidelines for the Prevention of Intravascular Catheter–Related Infections. Infection Control and Hospital Epidemiology 2002;23(12):759–769.
13. Miller MR, et al. Decreasing PICU Catheter-Associated Bloodstream Infections: NACHRI's Quality Transformation Efforts. Pediatr 2010;125(2):206–213.
14. Shuman EK, et al. Analysis of central line-associated bloodstream infections in the intensive care unit after implementation of central line bundles. Infect Control Hosp Epidemiol 2010;31(5):551–3.
15. Owens RC, Jr, Fraser GL and Stogsdill P. Antimicrobial stewardship programs as a means to optimize antimicrobial use. Insights from the Society of Infectious Diseases Pharmacists. Pharmacotherapy 2004;24(7): 896–908.
16. MacAdam H. et al. Investigating the association between antibiotic use and antibiotic resistance: impact of different methods of categorising prior antibiotic use. Int J Antimicrob Agents 2006;28(4):325–32.
17. Zaoutis TE. et al. Risk factors for and outcomes of bloodstream infection caused by extended-spectrum beta-lactamase-producing Escherichia coli and Klebsiella species in children. Pediatrics 2005;115(4):942–9.
18. Dellit TH. et al. Infectious Diseases Society of America and the Society for Healthcare Epidemiology of America guidelines for developing an institutional program to enhance antimicrobial stewardship. Clin Infect Dis 2007;44(2):159–77.

What Do We Know About How to Treat Tuberculosis?

Ben J. Marais

1 Introduction

In the year 2000 it was estimated that 884 019 (11%) of the 8.3 million new tuberculosis (TB) cases diagnosed globally were children less than 15 years of age [1]. However, this figure almost certainly represent a gross underestimation of the current situation. The most recent World Health Organization (WHO) report estimated that 9.4 million new TB cases occurred during 2008, without specifying the number of children affected [2]. Poor case ascertainment and limited surveillance data hamper efforts to accurately quantify the global disease burden caused by childhood TB [3]. In a community-based survey performed in Cape Town, South Africa, children less than 13 years of age contributed 13.7% of the total disease burden. The calculated TB incidence was 407/100 000/ year; nearly 50% of the total adult TB incidence of 840/100 000/year during the same time period [4]. It is estimated that in areas where the TB epidemic is poorly controlled, children less than 15 years of age are likely to contribute 15–20% of the TB disease burden, with TB incidence rates roughly half of that reported for adults from the same community. In settings where transmission is limited and preventive therapy provided, the comparative burden of childhood TB would be greatly reduced.

A common misperception is that children usually develop "mild forms" of TB, as is seen in settings with diligent contact tracing and active case finding. However, most children diagnosed in TB endemic areas present with advanced disease, [5] and TB is a major (although frequently unrecognized) cause of death in young children. A necropsy study conducted in Zambia demonstrated that TB rivals bacterial pneumonia as a cause of death from respiratory disease, both in human immunodeficiency virus (HIV)-infected and -uninfected children [6]. A subsequent study that documented identifiable causes of respiratory disease in children with community acquired pneumonia not responding to first-line antibiotics, confirmed *Mycobacterium tuberculosis* as a common pathogen in southern Africa, irrespective of the child's HIV status [7]. This review focus on what we know about preventing and treating TB in children; issues related to TB diagnosis and management of TB/HIV co-infection have been recently reviewed [8, 9].

B.J. Marais (✉)
Department of Paediatrics and Child Health, Stellenbosch University, Tygerberg, South Africa
e-mail: bjmarais@sun.ac.za

N. Curtis et al. (eds.), *Hot Topics in Infection and Immunity in Children VIII*,
Advances in Experimental Medicine and Biology 719, DOI 10.1007/978-1-4614-0204-6_15,
© Springer Science+Business Media, LLC 2011

2 Preventing TB

In TB endemic areas the majority of the population develops primary *M. tuberculosis* infection during childhood, but ongoing transmission implies an ever present risk of re-infection. Therefore, preventive therapy should focus on those at greatest risk of progression to active TB following primary or re-infection. Age and immune status are the most important determinants of risk, with very young immune-immature children (less than 3 years of age) and those with significant immune compromise experiencing the greatest risk [10]. Due to the frequency and rapidity with which disease progression may occur in this high-risk group, documented TB exposure and/or infection warrants preventive chemotherapy. Isoniazid (INH) monotherapy for 6–9 months has proven efficacy, is widely available, cheap and associated with minimal safety concerns and/or drug-drug interactions, making it the most widely used preventive therapy regimen, also promoted by WHO [11–13].

Most national and international guidelines extend the preventive therapy age window to all children less than 5–6 years of age with documented TB exposure. However, in TB endemic countries implementation is extremely poor partly because provision of preventive therapy is not a measured outcome of TB control programme performance. It is also regarded as unfeasible in areas with limited resources where health care services are already overburdened. Prerequisite screening tests such as a tuberculin skin test (TST) and chest radiograph provides another barrier, but simple symptom-based screening offers a safe and feasible alternative [11, 14]. The perceived risk of generating drug-resistant TB when someone with undiagnosed active TB is inadvertently treated with INH monotherapy, is often raised as a concern. This risk does require careful consideration when treating HIV-infected adults, [12, 15] but children tend to develop paucibacillary disease with greatly reduced risk of acquiring drug resistance. The US public health trials of the 1950s also indicated that INH monotherapy seems sufficient even in the presence of disease, when this is minimal [16].

Older immunocompetent contacts are at low risk of progression to disease and are therefore not regarded as a priority for the provision of preventive therapy. However, it is important that all close contacts should be screened for symptoms suggestive of TB, since this presents an important opportunity for active case finding. All close contacts should be informed of their increased risk to develop TB in the coming months/years and encouraged to return for formal evaluation should they develop any symptoms suggestive of active TB disease. Non-endemic areas with sufficient resources may elect to provide preventive therapy to all patients with proven *M. tuberculosis* infection in an attempt to eliminate the pool of latent infection from which future cases may arise, since the risk of future re-infection is minimal in such a setting.

2.1 HIV-Infected Children

Apart from a marked increase in overall TB incidence the HIV epidemic significantly lowered the average age of adult TB patients with more women being affected than ever before [17]. In HIV-affected communities the highest TB incidence rates now occur among men and women aged 20–45 years, who are likely to be parents of young and vulnerable children. This is illustrated by the exceptionally high TB exposure rates recorded among South African infants born to HIV-infected mothers [18]. The importance of regular TB exposure screening and provision of post-exposure prophylaxis following each documented exposure event is undisputed. INH does not cause drug-drug interactions with antiretroviral therapy (ART) and no treatment adjustment is indicated; overlapping drug toxicity also seems extremely uncommon [12].

Given the high likelihood of exposure, the effect of routine pre-exposure prophylaxis provided to all HIV-infected infants in TB endemic areas has been explored. The first randomized controlled trial was conducted in children with advanced HIV disease and minimal access to ART. It was stopped early due to significantly reduced mortality in the INH group compared to placebo [19]. Interestingly differential survival benefit accrued mainly during the first 2–3 months of the trial, raising the possibility of undiagnosed or subclinical TB disease at trial entry. A second trial recruited very young HIV-exposed infants (3–4 months of age), strictly excluding all with documented TB exposure and meticulously monitored for subsequent TB exposure at which point they were taken off study drug and provided with INH. This study was discontinued for futility, since mortality and/or TB-related morbidity occurred with equal frequency in both groups [20]. These discrepant findings suggest that infants managed in programs where there is early ART initiation and constant vigilance for TB exposure do not benefit from pre-exposure INH prophylaxis. Early commencement of ART, as soon as possible following a confirmatory tests in infancy and irrespective of CD4 lymphocyte count or clinical disease staging, has significant survival benefit and reduces the risk of TB [21].

2.2 Adherence

Although INH preventive therapy for 6–9 months is regarded to be the standard of care, poor adherence is a major concern that limits "real life" effectiveness [22]. "Real life" effectiveness is determined both by efficacy (as determined in clinical trials) and adherence (under field conditions). Poor adherence with prolonged therapy demonstrates the need to consider alternative strategies that may improve "real life" effectiveness. Rifampicin (RMP) has strong sterilizing activity and its addition shortens the duration of preventive therapy required. The use of a 3–4 month INH and RMP regimen is associated with improved adherence and demonstrated equivalence compared to 6–9 months of INH prophylaxis [23, 24]. Disadvantages include increased drug cost, reluctance to use RMP given its central role in short-course treatment and numerous drug interactions particularly in HIV-infected children on ART. Theoretically RMP and pyrazinamide (PZA) represent an excellent combination to eradicate intra-cellular and dormant bacilli. Hepatic toxicity has been a major concern in adult studies, [25] but is rarely observed in children receiving standard 3-drug combination treatment (INH, RMP and PZA) and has not been observed during a preliminary preventive therapy trial using this combination in children [26]. The inclusion of PZA in short course preventive therapy regimens requires further exploration in children to identify strategies with improved effectiveness.

3 Curing TB

The aims of TB treatment are to (1) cure the individual patient, (2) terminate transmission and (3) prevent the development of drug resistance, all without causing severe adverse events [27]. Rapid reduction in the organism load is important since it improves clinical symptoms, limits disease progression, terminates transmission and protects companion drugs against the evolution of drug resistance, since the occurrence of resistance mutations are random events and its likelihood is directly related to the number of organisms present (bacillary load). The combination of multiple drugs with different modes of action ensures that organisms with newly acquired resistance to a single drug are killed before acquiring additional resistance mutations. In order to ensure permanent cure eradicating bacterial subpopulations that are less likely to be killed by potent bactericidal drugs, such as those with intermittent spurts of metabolism and/or residing in an acid pH environment, are as important. The need to ensure effective eradication of all organisms, including hypometabolic bacilli, provides additional justification for the use of multiple drugs and also explains the prolonged duration current therapy. Table 1 presents the mode and mechanism of action, main toxicities and dosage recommendations for current first- and second-line drugs.

Table 1 First- and second-line TB drugs, mode of action, main toxicities and recommended dosages in children Adapted from ref.[65] – with permission

First line drugs (discovery date)	Mode & mechanism of action	Main toxicities[a]	Single daily dose mg/kg (range); [maximum dose]
Isoniazid (1952)	Bactericidal	Hepatitis	10 (5–15)[b]
	Requires initial activation – inhibition of cell wall (mycolic acid) synthesis	Peripheral neuropathy	[maximum 300 mg]
Rifampicin (1963)	Bactericidal and sterilizing	Hepatitis	15 (10–20)[b]
	Inhibition of RNA synthesis	Orange discolouration of secretions	[maximum 600 mg]
		Drug interactions	
Pyrazinamide (1954)	Sterilizing	Hepatitis	35 (30–40)[b]
	Disrupt membrane energy metabolism	Arthralgia	[maximum 2000 mg]
Ethambutol (1961)	Bacteriostatic	Visual disturbance (acuity, colour vision)	20 (15–25)[b]
	Requires initial activation – inhibition of cell wall (arabinogalactan) synthesis		[maximum 1,200 mg]
Streptomycin[c] (1943)	Bacteriostatic	Oto and nephrotoxic	15 (12–18)
	Inhibition of protein synthesis		[maximum 1,000 mg]

Second line Drugs

Ethionamide (1956)	Bactericidal	Vomiting	15–20 [1,000 mg]
	Requires initial activation – inhibition of cell wall (mycolic acid) synthesis	Hypothyroidism	
		Hepatitis	
Fluoroquinolones[d] (1963)	Bactericidal	Arthralgia (rare)	15–20 [800 mg]
Ofloxacin[e]	Inhibition of DNA gyrase, essential for DNA replication, transcription and repair	Insomnia, confusion	7.5–10 [750 mg]
Levofloxacin		Oto and nephrotoxic	7.5–10 [400 mg]
Moxifloxacin	Bactericidal		
Aminoglycosides (1957)	Inhibition of protein synthesis		15–30 [1,000 mg]
Kanamycin			
Amikacin			
Polipeptides (1963)	Bacteriostatic	Oto and nephrotoxic	15–30 [1,000 mg]
Capreomycin	Inhibition of protein synthesis		
Cycloserine (1955) derivative Terizidone	Bacteriostatic	Psychosis, depression, convulsions	10–20 [1,000 mg]
	Inhibition of cell wall (peptidoglycan) synthesis		
Para-aminosalisylic acid (PAS) (1948)	Bacteriostatic	Diarrhoea and vomiting	150–200 [12 g] Divided in 2–3 doses/day
	Inhibition of folic acid and iron metabolism	Hypothyroidism	

[a] Hypersensitivity reactions and drug rashes may occur with any drug
[b] WHO-endorsed new recommendations for dosing of first-line TB drugs in children
[c] Streptomycin is rarely used, since there is no indication for using the retreatment regimen in children
[d] Ciprofloxacin has the weakest activity and is no longer indicated for TB treatment
[e] Thiacetazone (1946) is rarely used due to severe hypersensitivity reactions reported in HIV-infected children (114)

Of the current first-line drugs, INH has the most potent early bactericidal activity (EBA), killing the vast majority of rapidly metabolizing extra-cellular bacilli within the first few days of treatment, [27, 28] provided the organisms are drug-susceptible and adequate drug levels are reached. Slower growing intracellular bacilli are killed by INH, but only over a period of 12–18 months; RMP is more effective in eradicating this sub-population and do so within 9 months. The inclusion of both RMP and PZA enable 6-month combination therapy [28]. PZA contributes by killing extracellular bacilli that persist within the acidic centers of caseating granulomas and completes its action within the first 2 months of treatment. Ethambutol (EMB) kills actively growing bacilli but has limited potency. Its theoretical role in RMP containing regimens is to reduce the risk of acquired drug resistance [28]. Any drug given in isolation will select spontaneously occurring drug-resistant mutants that may subsequently multiply. The risk of acquired drug resistance is reduced by using multiple drugs in combination and ensuring strict adherence to treatment [29]. Table 2 presents a summary of the main advances in TB treatment regimen design over the past 60 years.

In addition to these theoretical concepts, operational issues remain crucial in everyday clinical practice. These include access to early and accurate diagnosis, ensuring an uninterrupted and quality assured drug supply that is stored and dispensed appropriately, use of optimal treatment regimens

Table 2 Advances in TB treatment and preventive chemotherapy during the past 60 years. Adapted from ref.[65] – with permission

Date/decade	Advances in TB treatment	Advances in preventive chemotherapy
1946	First randomized controlled trial (RCT) – compared streptomycin to placebo; documented rapid acquisition of resistance	
1950s	First use of combination regimens	H – for 9–12 months
	Optimal regimen included 3 drugs (S/H/PAS) – total treatment duration 24 months	
1960s	PAS replaced by ethambutol (S/H/E)	H – for 6–9 months
	Concept of intensive phase (to eliminate the bulk of bacilli) and continuation phase (to ensure permanent cure) – total treatment duration 18 months	
1970s	Addition of rifampicin (S/H/R/E), – total treatment duration 9–12 months	H – for 6–9 months
1980s	Streptomycin replaced by pyrazinamide	H – for 6–9 months
	Common regimens used:	
	1. H/R/Z/E intensive phase (2 months), H/R continuation phase (4 months) – total treatment duration 6 months	
	2. H/R/Z/E intensive phase (2 months), H/E continuation phase (6–7 months) – total treatment duration 8–9 months	
1990s	Launch of the Stop TB strategy inclusive of DOTS	H – for 6–9 months or
	Emphasis on global treatment access, adequate and quality assured drug supplies, use of standardized treatment regimens	HR for 3–4 months
2000s	Regimens that contain rifampicin for the full duration of treatment (6 months) shown to be superior	H – for 6–9 months or
	Only one regimen advised:	HR for 3 months or
	1. H/R/Z/E intensive phase (2 months), H/R continuation phase (4 months) – total treatment duration 6 months	R for 4 months
2010s	?Novel 2–4 month treatment regimens including Moxi- or Gatifloxacin, TMC207 and/or Rifapentine	?HRZ daily – 2 months
		?Weekly Rifapentine + H – 3 months
		?TMC207 for MDR/XDR contacts – 3–4 months

H isoniazid, *R* rifampicin, *Z* pyrazinamide, *E* ethambutol, *S* streptomycin, *PAS* para-aminosalisylic acid and the establishment of systems to enhance adherence. With proper implementation, the WHO's directly observed therapy, short-course (DOTS) strategy adequately addresses most of these issues. However, exclusive emphasis on sputum smear-positive disease excludes the vast majority of children and many HIV-infected adults with TB. Dogmatic application of direct observation may also neglect more comprehensive and effective forms of patient education and support. It is important to understand why patients default in particular settings and to identify optimal strategies to improve adherence, particularly among the poor and those living in rural and remote areas.

The main variables that influence the success of chemotherapy, apart from drug resistance, are the bacillary load and its anatomical distribution. Sputum smear-negative disease is usually paucibacillary and therefore the risk of acquired (previously treated) drug resistance is low. Drug penetration into the anatomical sites involved is good and the success of three drugs (INH, RMP, PZA) during the 2-month intensive phase and two-drugs (INH, RMP) during the 4-month continuation phase, is well established [27]. In the presence of extensive disease (excluding TB meningitis), HIV co-infection and/or suspicion of INH resistance, the addition of EMB during the intensive phase is advised to improve outcome and reduce the risk of acquiring drug resistance [27]. At the other end of the disease spectrum, the efficacy of shorter treatment durations in immune competent children with minimal disease requires further evaluation, since this would reduce treatment related adverse events and greatly reduce the burden placed on health care services.

Sputum smear-positive disease implies a high bacillary load and an increased risk for random drug resistance. Selecting multidrug-resistant (MDR; resistant to both INH and RMP) mutants is a particular concern, which explains the use of four drugs (INH, RMP, PZA, EMB) during the 2-month intensive phase of treatment. Once the bacillary load is sufficiently reduced, daily or intermittent (3x/week) therapy with INH and RMP during the 4-month continuation phase is sufficient to ensure organism eradication [30, 31]. WHO does not recommend 2x/week intermittent therapy and it should not be used if adherence cannot be ensured, especially in HIV-infected children or those with cavitary disease [32]. In general, children with drug-susceptible TB respond well to treatment, but until recently few countries had access to child friendly drug formulations. The Global Drug Facility (GDF) has made child-friendly formulations available to deserving countries since 2008, [33] but huge programmatic barriers continue to limit the availability and use of these drugs in many areas.

It is essential to consider the cerebrospinal fluid (CSF) penetration of drugs used in the treatment of TB meningitis or disseminated (miliary) disease. INH and PZA penetrate the CSF well [34]. RMP and streptomycin (SM) penetrate the CSF poorly, but may achieve therapeutic levels in the presence of meningeal inflammation [34]. In addition, the value of SM is limited by intramuscular administration. EMB hardly penetrates the CSF, even in the presence of meningeal inflammation, [32] but ethionamide shows good CSF penetration and has been used successfully as a fourth drug in short-course treatment of TB meningitis [35, 36]. Although there are no randomized clinical trials, several reports have illustrated the efficacy of short-course high-dose regimens in the treatment of TB meningitis [36, 37].

3.1 Drug Dosing

Previous drug dosing recommendations for children were derived from adult studies without taking into consideration the major age-related differences that exist in drug distribution, metabolism and excretion. For example, the serum level achieved using a similar mg/kg dose of INH is significantly lower in children than in adults [38]. Several recent studies have documented sub-optimal serum concentrations in children receiving standard mg/kg body weight dosages of INH, RMP, PZA and EMB [39–41]. Revised recommendations for first-line agents, which take account of the physiological and anatomical differences between adults and children, have recently been adopted by WHO [42]. The new dosage recommendations are included in Table 1. Regular weight-based dose adjustment is important, particularly in young and/or malnourished children during the intensive phase of treatment, when weight gain may be pronounced and rapid.

3.2 Adverse Events

Adverse events are less common in children than in adults. Hepatotoxicity is the most common severe adverse event and can be caused by the first-line agents, INH, RMP and PZA, and less commonly by the second-line agents such as ETH and para-aminosalisylic acid (PAS). Some elevation of liver enzymes (less than 5 times upper limit of normal values) is not unusual, but liver tenderness or jaundice are regarded as contraindications for the continuation of hepatotoxic drugs. Hepatic toxicity usually manifests within the first 2–4 weeks of therapy, but may occur at any time during the treatment period [43]. In many cases the TB drugs are not the cause of liver function derangement and children should be screened for other causes of hepatitis, such as hepatitis A, while continuing with non-hepatotoxic drugs in the interim. Hepatotoxic drugs (INH, RMP, PZA) should only be reintroduced after liver functions have normalized (less than 2x upper limit of normal). Since PZA is regarded as the most hepatotoxic agent its re-introduction is controversial. If 2 months of PZA has not been completed and it is not re-introduced, then it would be prudent to continue treatment with INH, RMP and EMB for a total duration of 9 months. If rechallenge of INH and RMP is unsuccessful, which it rarely is, then 9–12 months of treatment should be completed using liver friendly drugs that include a fluoroquinolone such as ofloxacin or levofloxacin.

INH-associated peripheral neuropathy is uncommon in children, but pyridoxine supplementation should be considered especially in HIV-infected children who have persistently low pyridoxine levels on TB treatment [44]. A recent comprehensive review of optic neuritis related to EMB suggests that optic neuritis is very unusual at the newly recommended dosage of 15–25 mg/kg/day; [41] its use is warranted in children with hepatotoxicity, cavitary or extensive pulmonary disease, severe extra-thoracic disease or resistance to other first-line agents. Ethionamide frequently causes vomiting, but this can usually be overcome by incrementally increasing the dose during the first 1–2 weeks of therapy and dividing the daily dose if required.

3.3 HIV-Infected Children

TB treatment recommendations are similar for HIV-infected and -uninfected children [11]. However, since severely immune compromised children are at increased risk of TB relapse, [45] consideration may be given to prolonging the treatment duration from 6 months to 9 months. Additional issues to consider include: higher incidence of co-infections with other pathogens, poor drug absorption and/ or poor treatment adherence due to chronic illness, high pill burden and parental illness/death [46].

TB is frequently the presenting disease in children with previously undiagnosed HIV infection. The Centers for Diseases Control and Prevention (CDC) has established guidelines for initiation and monitoring of ART in adults presenting with TB or initiating TB treatment in those already on ART. These guidelines are regularly revised; the CDC website has the latest recommendations [47]. No paediatric guidelines are provided but issues related to timing of initiation, drug-drug interactions, overlapping adverse events and adherence concerns are similar. According to the most recent WHO guidance, lifelong highly active ART (HAART) should be initiated within the first 2–8 weeks of TB therapy in all HIV-infected individuals, including children [48]. Significant drug interactions occur between the rifamycins (especially RMP), and the protease inhibitors. RMP is considered to be compatible with all nucleoside reverse transcriptase inhibitors (NRTI). However, the use of a triple NRTI regimen is not advised due to reduced efficacy and current recommendations are to combine a double NRTI backbone with efavirenz. In the absence of dosage recommendations for efavirenz in children less than 10 kg or 3 years of age, nevirapine at the upper limit of suggested dosages for weight or surface area performs well and may be used [48]. An alternative is to use superboosted lopinavir/ritonavir (additional ritonavir added to equal the lopinavir dose); [49] this is complicated by the short shelf life and poor taste of ritonavir syrup.

Trimethoprim-sulphamethoxazole prophylaxis reduces the risk of *Pneumocystis jiroveci* (PJP) pneumonia, invasive bacterial infections and malaria and is associated with significant survival benefit in HIV-infected adults and children [50]. TB treatment is unaffected by concomitant trimethoprim-sulphamethoxazole prophylaxis, which should be provided to all HIV-infected infants and older children with severe immune suppression [51]. Transient worsening of symptoms signs due to immune reconstitution inflammatory syndrome (IRIS) have been reported after ART initiation, manifesting either as "unmasking IRIS", where previous subclinical disease suddenly becomes evident, or as "paradoxical IRIS" where symptomatic deterioration occurs despite adequate treatment [52]. This temporary exacerbation does not indicate treatment failure and subsides spontaneously, although severe cases may require treatment with corticosteroids. IRIS due to *M. bovis* BCG has been recognized as the most common IRIS manifestation in young infants initiated on ART [53], with TB and other mycobacteria being more prevalent in older children [54].

4 TB Retreatment

TB treatment rarely fails in children. The most likely cause for treatment failure in settings where the prevalence of drug resistance is low is non-adherence. Following treatment interruption most children may be restarted on the original treatment regimen under close supervision, since the risk of developing drug resistance is small if they have paucibacillary disease. If a child presents with a new episode of TB more than 6 months after treatment completion, then it most likely represents a re-infection event and standard first-line treatment is appropriate. There is no indication in children to use the adult re-treatment regimen, which effectively adds a single drug (intramuscular streptomycin) to a "failing regimen". With genuine treatment failure (absence of clinical response to adherent treatment) it is important to critically review the TB diagnosis, take a comprehensive history of potential exposure to a drug-resistant source case (this should be done whenever a TB diagnosis is considered), re-culture and perform drug susceptibility testing, and also consider poor bioavailability of drugs and/or potential additional diagnoses.

5 Drug Resistant TB

It is evident that current control efforts are not containing the spread of the drug-resistant epidemic and that treatment efforts are inadequate; of an estimated 500,000 cases of MDR TB that occurred during 2008, only 30,000 (11%) were diagnosed and only 6,000 (1.2%) had access to Green Light Committee approved treatment [2]. Without greatly increased resources and committed action that includes enhanced infection control, early diagnosis and optimal management, the emergence of drug-resistant TB will continue to threaten the very fabric of TB control efforts.

Children with paucibacillary TB are unlikely to acquire drug resistance and contribute little to the creation and/or transmission of drug-resistant strains. However, since paediatric cases represent ongoing transmission, children are frequently affected in areas where drug resistance is common and transmission poorly controlled. As such, they present a valuable epidemiological perspective.

Children with drug-resistant TB provide an accurate estimation of transmitted (primary) drug resistance within communities [55]. Prospective surveillance data from Cape Town, collected over three time periods, suggest an upward trend of drug-resistant TB among children, [56] with an overrepresentation of specific genotypes such as Beijing and Haarlem [57].

5.1 Mono-Resistance

Resistance to INH is usually the first step to the development of MDR-TB. Evidence suggests that Regimen one (INH, RMP, PZA, EMB) should be sufficient to effect cure in most patients with INH mono-resistant TB, [58] although the risk of acquiring MDR-TB is increased in patients with high bacillary loads. In children we advise the addition of EMB (for the full treatment duration) and to consider continuing treatment for 8–9 months depending on the severity of involvement. Since minimum inhibitory concentrations (MICs) are not routinely done or reported to the clinician, it is rarely appreciated that INH resistance may be of a low or intermediate level [59]. The presence of an *inhA* mutation usually confers low level INH resistance with concomitant ethionamide resistance, while a *katG* mutation usually confers high level INH resistance without ethionamide resistance. [60]. The frequency with which low or intermediate level INH resistance occurs, probably explains the observation that MDR patients had better outcomes with the addition of high-dose INH to their regimen [61]. With genetic testing the presence of an *inhA* mutation can be used as an indicator that continuation of high-dose INH should be considered, despite reported INH resistance, and that ethionamide will probably not be effective [60]. In contrast, the presence of a *katG* mutation indicates that high-dose INH has no added value, but ethionamide is likely to be active. In the absence of genetic testing and/or MIC testing a combination of high-dose INH and ethionamide is frequently advised. Preventive treatment for high-risk contacts of infectious INH mono-resistant cases would be RMP mono-therapy for 4 months, which represents another strategy with improved adherence and reduced hepatotoxicity compared to INH for 9 months [62].

5.2 Multidrug Resistance (MDR)

MDR-TB implies resistance to the most potent drugs in the first-line regimen, INH and RMP, with/without resistance to other TB drugs. The principles guiding disease management remain unchanged. High-risk children in contact with MDR-TB should be considered for preventive therapy, while accurate disease classification and drug susceptibility test (DST) results, either of the presumed source case or the child's own isolates, should guide therapy. Second-line drugs are generally more toxic, but with correct dosing, few serious adverse events have been reported in children. Hearing loss is a major concern with prolonged use of injectable agents such as kanamycin or amikacin and careful monitoring for adverse events such as depression and/or hypothyroidism is indicated.

A rational guide for preventive therapy in high-risk MDR contacts would be to use at least two drugs to which the organism is susceptible, for at least 6 months [63]. Once MDR-TB has been identified it is essential to perform second-line DST to optimize treatment and prevent multiplication of resistance. The evidence base for the treatment of MDR TB is limited, but basic principles of treatment would include the following: (1) collect specimens for culture and DST prior to treatment initiation (don't delay treatment initiation with TB meningitis, but requisite specimens should still be collected), (2) treat the child according to the DST result of the likely source case, if an isolate from the child is not available, (3) give three or more drugs to which the isolate is susceptible and/or naïve depending on the severity of disease (never add one drug to a failing regimen; use a minimum of four drugs if the child has extensive or cavitary disease), (4) use only daily directly observed therapy (DOT), (4) schedule regular follow-up visits to monitor progress (including culture conversion) and adverse events and (5) treatment should continue for at least 12 months after the first negative culture, but 18 months after first negative culture in cavitary, extensive pulmonary or disseminated disease [64]. Optimal treatment should be discussed with an expert in the field, while parents and children require regular counseling and support to complete treatment.

5.3 *Extensive Drug Resistance (XDR)*

Extensive drug resistant (XDR) can be defined as MDR-TB with additional resistance to the fluoroquinolones and a second-line injectable agent such as kanamycin, amikacin or capreomycin. Currently, preventive therapy for XDR-TB contacts is not advised, but in the light of possible low or intermediate-level INH resistance, high-dose INH (15–20 mg/kg) may provide some protection to high-risk contacts. Treatment options for these patients are limited. In our experience, the addition of newer and alternative drugs such as linezolid should be considered and good outcomes are possible [65]. Novel drugs and new treatment options are urgently required to help individual patients and turn the tide against drug-resistant TB.

6 New TB Treatment Options

Major problems associated with current TB regimens include: (1) the long duration of treatment that complicates administration, increases cost and reduces adherence, (2) increasing rates of resistance to standard first-line drugs with amplification of drug resistance in the absence of adequate therapy, (3) general low potency and increased toxicity of second-line drugs, (4) high frequency of resistance to second-line drugs, especially the fluoroquinolones that are frequently misused and represent the most potent second-line agents available, and (5) pronounced drug-drug interactions especially with RMP-based regimens and ART. Large-scale financial investment is required to expand the pre-clinical pipeline and advance new products to human clinical trials.

Multiple new drugs were discovered during the 1940s, 1950s and 1960s, together with the development of creative study designs to test drug efficacy and identify optimal treatment regimens (Tables 2 and 3). Following this fruitful period, the discovery of novel compounds and development new regimens came to a near standstill, apart from minor refinement of existing regimens, development of fixed dose combination tablets and increased availability of quality-assured drugs. Within the last decade there has been a renewed focus on drug discovery by identifying promising new drug targets and systematic screening of existing compound libraries. Given the high attrition rate during pre-clinical development and the need to test multiple new compounds in different regimens combinations, the need to grow the pre-clinical pipeline becomes self evident. Experimental TB drugs that are in the discovery, pre-clinical or clinical phases of the development pipeline have been extensively reviewed [64, 66]. Table 3 summarizes novel TB drugs in clinical trials.

Children are usually excluded from drug trials due to perceived risk and limited financial gain. In fact most drugs used in paediatrics are not licensed for use in children and/or only become available for paediatric use long after initial registration. Difficulties in diagnosis and the monitoring of treatment response complicate children's inclusion in TB efficacy trials, but is this truly necessary? It seems reasonable to extrapolate efficacy data from adult studies since the disease causing organism is identical; in addition children tend to develop paucibacillary forms of disease compared to adults and demonstrate excellent treatment response with current drugs, even at the suboptimal dosages used previously. However, the existence of unique age-related pharmacokinetic and toxicity

Table 3 Novel TB drugs in clinical trials (those in discovery and preclinical phases are not included). Adapted from ref.[65] – with permission

Novel drug	Mode & mechanism of action	Summary of progress to date
Phase 3		
Moxi & Gatifloxacin	Bactericidal	EBA comparable to rifampicin
Newer fluoroquinolones	Inhibition of DNA gyrase; essential for DNA replication, transcription and repair	Randomized controlled trials ongoing to assess safety and efficacy of 4-month regimens where either moxi- or gatifloxacin replaces ethambutol or isoniazid during 2-month intensive phase
Phase 2		
TMC 207	Bactericidal, potentially sterilizing	Poor early bactericidal activity (EBA), but potent late bactericidal activity
Diarrylquinoline		
Tibotec Pharma Ltd.	ATP depletion, imbalance in pH homeostasis	Oral doses up to 400 mg/day well tolerated
		Additional dose escalation considered
		Randomized controlled trials ongoing in treatment of MDR patients; addition of TMC207 to background MDR-TB treatment increased sputum culture conversion rates at 2 months (48% vs 9%)
Nitroimidazoles		
PA-824	Bactericidal, potentially sterilizing	EBA consistent over a wide dosing range
Nitroimidazo-oxazine	Inhibition of cell wall (mycolic acid) synthesis	Randomized controlled trials ongoing in treatment of MDR patients
Chirion Corp. & TB Alliance	Generate reactive nitrogen species during anaerobic metabolism (bioreduction)	Publication of EBA data awaited
OPC-67683		Randomized controlled trials ongoing in treatment of MDR patients
Dihydroimidazo-oxazole	Bactericidal, potentially sterilizing	
Otsuka Pharma Ltd.	Inhibition of cell wall (mycolic acid) synthesis	
Rifapentine	Bactericidal and sterilizing	EBA comparable to rifampicin
Newer rifamycin	Inhibition of RNA synthesis, extended half life	Several trials planned and ongoing to assess safety and efficacy; eg. using a 3-month regimen of rifapentine, moxifloxacin and PZA as daily or 3x/week therapy
Phase 1		
SQ109	Potentially bactericidal	Currently being evaluated in a phase 1 trial
Ethylenediamine	Precise mechanism of action unknown, but affects cell wall synthesis	EBA data awaited
Sequella	Partial ethambutol analogue, BUT effective against resistant strains	
Linezolid & PNU-100480	Potentially bactericidal	Currently being evaluated in a phase 1 trial
Oxazolidinones	Inhibition of protein synthesis	EBA data awaited
Pfizer		

EBA early bactericidal activity, *MDR* multidrug-resistant, *PZA* pyrazinamide profiles, justifies the development of child-friendly formulations and inclusion of children in safety and dose ranging studies as soon as initial safety and efficacy trials have been concluded [67].

In conclusion, improved treatment access is of crucial importance to reduce the morbidity and mortality associated with childhood TB. While eagerly awaiting the development of more feasible diagnostic approaches and improved treatment regimens (both preventive and curative), much can be achieved using pragmatic approaches and available drugs.

References

1. Nelson LJ, Wells CD. Global epidemiology of childhood tuberculosis. Int J Tuberc Lung Dis 2004;8:636–647
2. World Health Organization. Global tuberculosis control: a short update to the 2009 report. WHO, Geneva, Switzerland. WHO/HTM/TB/2009.426
3. Newton SM, Brent AJ, Anderson S, Whittaker E, Kampmann B. Paediatric tuberculosis. Lancet Infect Dis 2008;8:499–510
4. Marais BJ, Hesseling AC, Gie RP, Schaaf HS, Beyers N. The burden of childhood tuberculosis and the accuracy of routine surveillance data in a high-burden setting. Int J Tuberc Lung Dis 2006;10:259–263
5. Marais BJ, Gie RP, Schaaf HS, Hesseling AC, Enarson DA, Beyers N. The spectrum of childhood tuberculosis in a highly endemic area. Int J Tuberc Lung Dis 2006;10:732–738
6. Chintu C, Mudenda V, Lucas S et.al. Lung diseases at necropsy in African children dying from respiratory illnesses: a descriptive necropsy study. Lancet 2002;360:985–990
7. McNally LM, Jeena PM, Gajee K, Thula SA, Sturm AW, Cassol S, Tomkins AM, Coovadia HM, Goldblatt D. Effect of age, polymicrobial disease, and maternal HIV status on treatment response and cause of severe pneumonia in South African children: a prospective descriptive study. Lancet. 2007;369:1440–1451
8. Marais BJ, Pai M. New Approaches and emerging technologies in the diagnosis of childhood tuberculosis. Paediatr Respir Rev 2007;8:124–133
9. Marais BJ, Cotton M, Graham S, Beyers N. Diagnosis and management challenges of childhood TB in the era of HIV. J Infect Dis 2007;196 (Suppl 1):S76–S85
10. Marais BJ, Gie RP, Schaaf HS, Donald PR, Beyers N, Starke J. Childhood pulmonary tuberculosis – Old wisdom and new challenges. Am J Resp Crit Care Med. 2006;173:1078–1090
11. World Health Organization. Guidance for National Tuberculosis Programmes on the Management of Tuberculosis in Children. WHO, Geneva, Switzerland. HO/HTM/TB/2006.371
12. Marais BJ, Ayles H, Graham SM, Godfrey-Faussett P. Screening and preventive therapy for tuberculosis. Clin Chest Med 2009;30:827–846
13. Smieja MJ, Marchettie CA, Cook DJ, Smaill FM. Isoniazid for preventing tuberculosis in non-HIV infected persons. The Cochrane Library 1999; Issue 4:1–20
14. Kruk A, Gie RP, Schaaf HS, Beyers N, Marais BJ. Symptom-based screening of child tuberculosis contacts: improved feasibility in resource-limited settings. Pediatrics 2008;121:e1646–1652
15. Balcells ME, Thomas SL, Godfrey-Faussett P, Grant AD. Isoniazid preventive therapy and risk for resistant tuberculosis. Emerg Infect Dis 2006;12:744–751
16. The United States Public Health Service Tuberculosis Prophylaxis Trial collaborators. Prophylactic effects of isoniazid on primary tuberculosis in children. Am Rev Tuberc 1957;76:942–963
17. Lawn SD, Bekker LG, Middelkoop K, Myer L, Wood R. Impact of HIV infection on the epidemiology of tuberculosis in a peri-urban community in South Africa: the need for age-specific interventions. Clin Infect Dis 2006;42:1040–1047
18. Cotton MF, Schaaf HS, Lottering G, Wever HL, Coetzee J, Nachman S. Tuberculosis exposure in HIV-exposed infants in a high-prevalence setting. Int J Tuberc Lung Dis 2008;12:225–227
19. Zar HJ, Cotton MF, Strauss S et.al. Effect of isoniazid prophylaxis on mortality and incidence of tuberculosis in children with HIV: randomized controlled trial BMJ 2007;334:136. doi: 10.1136/bmj.3900.486400.55
20. Mitchell C, McSherry G, Violari A, et al. Primary Isoniazid prophylaxis did not protect against tuberculosis or latent TB in HIV-exposed uninfected infants in South Africa. In: Congress of Retrovirology and Opportunistic Infection. Montreal; 2009
21. Violari A, Cotton MF, Gibb DM, Babiker AG, Steyn J, Madhi SA, Jean-Philippe P, McIntyre JA; CHER Study Team. Early antiretroviral therapy and mortality among HIV-infected infants. N Engl J Med 2008;359:2233–2244
22. Marais BJ, van Zyl S, Schaaf HS, van Aardt M, Gie RP, Beyers N. Adherence to isoniazid preventive chemotherapy in children: a prospective community based study. Arch Dis Child 2006;91:762–765

23. Ena J. Valls V. Short-course therapy with rifampin plus isoniazid, compared with standard therapy with isoniazid, for latent tuberculosis infection: a meta-analysis. Clin Infect Dis 2005;40:670–676

24. Spyridis NP, Spyridis PG, Gelesme A, Sypsa V, Valianatou M, Metsou F, Gourgiotis D, Tsolia MN. The effectiveness of a 9-month regimen of isoniazid alone versus 3- and 4-month regimens of isoniazid plus rifampin for treatment of latent tuberculosis infection in children: results of an 11-year randomized study. Clin Infect Dis. 2007;45:715–722

25. Priest DH, Vossel LF, Sherfy EA, Hoy DP, Haley CA. Use of intermittent rifampin and pyrazinamide therapy for tuberculosis infection in a targeted tuberculin-testing program. Clin Infect Dis 2004;15:1764–1771

26. Magdorf K, Arizzi Rusche AF, Geiter RJ, O'Brien RJ, Wahn U. Short-course preventive therapy for tuberculosis: A pilot study of rifampin and rifampin-pyrazinamide regimens in children. Am Rev Respir Dis 1991;143; Suppl A120

27. Marais BJ, Schaaf HS, Donald PR. Pediatric tuberculosis: Issues related to current and future treatment options. Future Microbiol 2009;4:661–675

28. Mitchison DA. Role of individual drugs in the chemotherapy of tuberculosis. Int J Tuberc Lung Dis 2000;4:796–806

29. Mitchison DA. How drug resistance emerges as a result of poor compliance during short course chemotherapy for tuberculosis. Int J Tuberc Lung Dis 1998;2:10–15

30. Te-Water-Naude JM, Donald PR, Hussey GD, Kibel MA, Louw A, Perkins DR, Schaaf HS. Twice weekly vs. daily chemotherapy for childhood tuberculosis. Pediatr Infect Dis J 2000;19:405–410

31. Al-Dossary FS, Ong LT, Correa AG, Starke JR. Treatment of childhood tuberculosis with a six month directly observed regimen of only two weeks of daily therapy. Pediatr Infect Dis J 2002;21:91–97

32. Rieder HL, Arnadottir A, Trebucq A, Enarson DA. Tuberculosis treatment: dangerous regimens? Int J Tuberc Lung Dis 2001;5:1–3

33. Gie RP, Matiru RH. Supplying quality-assured child-friendly anti-tuberculosis drugs to children. Int J Tuberc Lung Dis 2009;13:277–278

34. Ellard GA, Humphries MJ, Allen BW. Cerebrospinal fluid drug concentrations and the treatment of tuberculous meningitis. Am Rev Respir Dis 1993;148:650–655

35. Donald PR, Seifert HI. Cerebrospinal fluid concentrations of ethionamide in children with tuberculous meningitis. J Pediatr 1989;115:483–486

36. Donald PR, Schoeman JF, van Zyl LE, De Villiers JN, Pretorius M, Springer P. Intensive short course chemotherapy in the management of tuberculous meningitis. Int J Tuberc Lung Dis 1998;2:704–711

37. Alarcon F, Escalante L, Perez Y, Banda H, Chacon G, Duenas G. Tuberculous meningitis: short course therapy. Arch Neurol 1990;47:1313–1317

38. McIlleron H, Willemse M, Werely CJ, Hussey GD, Schaaf HS, Smith PJ, Donald PR. Isoniazid Plasma Concentrations in a Cohort of South African Children with Tuberculosis: Implications for International Pediatric Dosing Guidelines. Clin Infect Dis 2009;48:1547–1553.

39. Schaaf HS, Willemse M, Cilliers K, Labadarios D, Maritz JS, Hussey GD, McIlleron H, Smith P, Donald PR. Rifampin pharmacokinetics in children, with and without human immunodeficiency virus infection, hospitalized for the management of severe forms of tuberculosis. BMC Med. 2009 Apr 22;7(1):19.

40. Graham SM, Bell DJ, Nyirongo S, Hartkorn R, Ward SA, Molyneux EM. Low levels of pyrazinamide and ethambutol in children with tuberculosis and impact of age, nutritional status and human immunodeficiency virus infection. Antimicrob Agents Chemother 2006;50:407–413

41. Donald PR, Maher D, Maritz JS, Qazi S. Ethambutol dosage for the treatment of children: literature review and recommendations. Int J Tuberc Lung Dis 2006;10:1318–1330

42. http://www.who.int/selection_medicines/committees/expert/17/WEB_TRS_DEC_2009.pdf: 27–29

43. Ormerod LP, Skinner C, Wales J. Hepatotoxicity of antituberculosis drugs. Thorax 1996;51:111–113

44. Donald P, Cilliers K, Willemse M, Labadarios D, Schaaf H, Maritz J, Werely C. Pyridoxine serum concentrations in children hospitalized with tuberculosis. Int J Tuberc Lung Dis 2007;11 Suppl 1:S225

45. Schaaf HS, Krook S, Hollemans DW, Warren RM, Donald PR, Hesseling AC. Recurrent culture-confirmed tuberculosis in human immunodeficiency virus-infected children. Pediatr Infect Dis J 2005;24:685–691

46. Marais BJ, Rabie H, Cotton MF. TB and HIV in children – advances in prevention and management. Paediatr Respir Rev 2010; In press

47. Centers for Disease Control and Prevention. http://www.cdc.gov/nchstp/tb/

48. Prasitsuebsai W, Cressey TR, Capparelli E, et al. Pharmacokinetics of Nevirapine when Co-administered with Rifampin in HIV-Infected Thai Children with Tuberculosis. In: Congress of Retrovirology and Opportunistic Infection. Montreal, Canada; 2009, abstract 908

49. Ren Y, Nuttall JJ, Egbers C et.al. Effect of rifampicin on lopinavir pharmacokinetics in HIV-infected children with tuberculosis. J Acquir Immune Defic Syndr 2008;47:566–569

50. Chintu C, Bhat GJ, Walker AS. Co-trimoxazole as prophylaxis against opportunistic infections in HIV-infected Zambian children (CHAP): a double-blind randomised placebo-controlled trial. Lancet 2004:364:1865–71

51. Guidelines on co-trimoxazole prophylaxis for HIV-related infections among children, adolescents and adults in resource-limited settings. World Health Organization, Geneva, 2006. http://www.who.int/3by5/mediacentre/news32/en/index.html

52. Meintjes G, Lawn SD, Scano F et.al. Tuberculosis-associated immune reconstitution inflammatory syndrome: case definitions for use in resource-limited settings. Lancet Infect Dis 2008;8:516–23

53. Rabie H, Violari A, Madhi S et.al. Complications of BCG Vaccination in HIV-infected and -uninfected Children:CHER Study. 15th Conference On Retroviruses and Opportunistic Infections (CROI). Boston, Mass February 3–6, 2008

54. Puthanakit T, Oberdorfer PM, Akarathum N, Wannarit P, Sirisanthana T, Sirisanthana V. Immune Reconstitution Syndrome After Highly Active Antiretroviral Therapy in Human Immunodeficiency Virus-Infected Thai Children. Pediatr Infect Dis J 2006;25:53–8

55. Schaaf HS, Marais BJ, Hesseling AC, Gie RP, Beyers N, Donald PR. Childhood drug-resistant tuberculosis in the Western Cape Province of South Africa. Acta Paediatr 2006;95:523–528

56. Schaaf HS, Marais BJ, Hesseling AC, Brittle W, Donald PR. Surveillance of Antituberculosis Drug Resistance Among Children From the Western Cape Province of South Africa–An Upward Trend. Am J Public Health 2009;99:1486–1490

57. Marais BJ, Victor TC, Hesseling AC, Barnard M, Jordaan A, Brittle W, Reuter H, Beyers N, van Helden PD, Warren RM, Schaaf HS. Beijing and Haarlem genotypes are overrepresented among children with drug-resistant tuberculosis in the Western Cape Province of South Africa. J Clin Microbiol 2006;44:3539–3543

58. Schaaf HS, Victor TC, Engelke E, Brittle W, Marais BJ, Hesseling AC, van Helden PD, Donald PR. Minimal inhibitory concentration of isoniazid in isoniazid-resistant *Mycobacterium tuberculosis* isolates from children. Eur J Clin Microbiol Infect Dis 2007;26:203–205

59. Warren RM, Streicher EM, Gey van Pittius NC, van der Spuy GD, Marais BJ, Victor TC, Sirgel F, Donald PR, van Helden PD. The clinical relevance of mycobacterial pharmacogentics. Tuberculosis 2009;89:199–202

60. Katiyar SK, Bihari S, Prakash S, Mamtani M, Kulkarni H. A randomised controlled trial of high-dose isoniazid adjuvant therapy for multidrug-resistant tuberculosis. Int J Tuberc Lung Dis 2008;12:139–145

61. Ziakas PD, Mylonakis E. 4 months of rifampin compared with 9 months of isoniazid for the management of latent tuberculosis infection: a meta-analysis and cost-effectiveness study that focuses on compliance and liver toxicity. Clin Infect Dis 2009;15:1883–1889

62. World Health Organization. Guidelines for the programmatic management of drug-resistant tuberculosis. Emergency update 2008. WHO, Geneva, Switzerland, 2008. WHO/HTM/TB/2008.402.

63. Schaaf HS, Gie RP, Kennedy M, Beyers N, Hesseling PB, Donald PR. Evaluation of young children in contact with adult multidrug-resistant pulmonary tuberculosis: a 30-month follow-up. Pediatrics 2002;109:765–771.

64. Schaaf HS, Willemse M, Donald PR. Long-term linezolid treatment in a young child with extensively drug-resistant tuberculosis. Pediatr Infect Dis J 2009;28:748–750

65. Marais BJ, Schaaf HS, Donald PR. Management of tuberculosis in children and new treatment options. Infect Disord Drug Targets 2011;11:144–156

66. Ma Z, Lienhardt C. Toward an optimized therapy for tuberculosis? Drugs in clinical trials and in preclinical development. Clin Chest Med 2009;30:755–768

67. Burman WJ, Cotton MF, Gibb DM, Walker AS, Vernon AA, Donald PR. Ensuring the involvement of children in the evaluation of new tuberculosis treatment regimens. PLoS Med. 2008;5(8):e176

Bacterial Meningitis in Childhood

Manish Sadarangani and Andrew J. Pollard

1 Introduction

Meningitis describes inflammation of the meninges, although the arachnoid and pia mater are also usually inflamed, i.e. leptomeningitis. A wide variety of pathogens cause childhood meningitis and clinical features and outcomes vary according to the organism [1, 2]. Bacterial meningitis caused by *Neisseria meningitidis*, *Streptococcus pneumoniae* and *Haemophilus influenzae* type b (Hib) is of particular importance as it has a high mortality and morbidity [1], has been relatively common [3–5] and is treatable with antibiotics. In the last two decades widespread introduction of highly effective conjugate vaccines against these pathogens in many industrialised countries has led to a significant reduction in the incidence of bacterial meningitis [3, 5, 6], although it remains a major public health problem globally [7–9].

2 Causative Organisms

The predominant bacteria responsible for meningitis vary depending on age. In children over 3 months old, *N. meningitidis*, *S. pneumoniae* and Hib are the most common pathogens isolated [1]. In neonates (under 1 month of age), Group B Streptococcus (GBS) predominates, being responsible for 45% of cases. Other common organisms are *Escherichia coli* (20%), other Gram negative organisms (10%), *S. pneumoniae* (6%) and *Listeria monocytogenes* (5%) [10]. In children aged between 1 and 3 months, any of these pathogens can cause meningitis.

3 Epidemiology and Vaccines

3.1 Neisseria meningitidis

Invasive meningococcal disease is endemic globally and most cases are caused by 5 of the 13 meningococcal serogroups: A, B, C, Y and W135. The majority of disease in Europe, South and Central America and Australasia is caused by serogroup B and C organisms [3, 11, 12]. In North

M. Sadarangani (✉) • A.J. Pollard
Department of Paediatrics, Oxford Vaccine Group, University of Oxford, Oxford, UK
e-mail: manish.sadarangani@paediatrics.ox.ac.uk; andrew.pollard@paediatrics.ox.ac.uk

N. Curtis et al. (eds.), *Hot Topics in Infection and Immunity in Children VIII*,
Advances in Experimental Medicine and Biology 719, DOI 10.1007/978-1-4614-0204-6_16,
© Springer Science+Business Media, LLC 2011

America, bacteria from serogroups B, C and Y cause an approximately equal amount of disease [13]. Epidemics in Africa, which can reach an incidence of 1,000 per 100,000, are usually associated with serogroup A [14], and more recently serogroups W-135 and X [15, 16]. The peak incidence of meningococcal disease occurs in children aged 6 months to 2 years, with a second smaller peak at 15–19 years. The incidence of meningococcal disease across Europe is 2–89 per 100,000 per year in children under 1 year and 1–27 per 100,000 per year in 1–4 year olds, with high rates in the UK and Ireland; 60–90% have meningitis, with or without septicaemia [3].

There are now polysaccharide conjugate vaccines available for four of the five major serogroups of *N. meningitidis*: A, C, Y and W135. The serogroup C conjugate vaccine was introduced in the UK from 1999, and subsequently across Europe. This resulted in a tenfold drop in the incidence of serogroup C disease [3]. In addition to direct protection, there was a significant effect of herd immunity observed with this vaccine [17]. Serogroup B organisms now cause the majority of disease in these countries, including 85–90% of UK cases. Since serogroup Y organisms account for a substantial proportion of cases in North America, an ACYW-135 conjugate vaccine was introduced in the USA for adolescents in 2005. Early data suggest that the overall vaccine effectiveness is 75% (95% CI 17–93%) in healthy adolescents [18]. The serogroup B polysaccharide capsule is not immunogenic in humans, and a number of protein-based vaccines against serogroup B disease are in development. Two of these vaccines are currently undergoing phase II and III clinical trials, and a serogroup B vaccine may be available in the near future [19].

3.2 Streptococcus pneumoniae

In the year 2000, there were over 100,000 cases of pneumococcal meningitis worldwide, with a peak incidence in children under 2 years of age [9]. Incidence rates were highest in Africa at 38 cases per 100,000 per year, and lowest in Europe at six cases per 100,000 per year in children under 5 years of age. These data were gathered prior to widespread use of the 7-valent pneumococcal conjugate vaccine (PCV7).

PCV7 contains polysaccharides from serotypes 4, 6B, 9V, 14, 18 C, 19F, and 23F and was introduced into the routine USA immunisation schedule in 2000, and became routine in the UK in September 2006. Five years after introduction of PCV7 into the USA, in children under 2 years of age, the incidence of pneumococcal meningitis decreased by 64% overall, from 10 per 100,000 per year to 3.7 per 100,000 per year [6]. Furthermore, there was a decrease in PCV7-serotype meningitis of 93%. In 2005, the incidence of PCV7-serotype meningitis was 0.6 per 100,000 per year and of non PCV7-serotype meningitis was 2.9 per 100,000 per year. A change of predominant serotypes was also observed – serotypes 14 and 23F were the most common prior to widespread use of PCV7, and 19A and 22F afterwards. Two years after introduction of PCV7 for vaccination of 'high risk' children in France there was a 39% reduction in the incidence of pneumococcal meningitis in all children under 2 years of age [20]. New conjugate vaccines containing 10 (PCV10) or 13 (PCV13) polysaccharides have now been introduced into routine immunisation schedules in many countries around the world. PCV13 replaced PCV7 in the UK schedule in April 2010, leading to the possibility of further decreases in the incidence of pneumococcal disease in the near future.

3.3 Haemophilus influenzae type b

Most Hib disease occurs in children below 5 years of age. In the year 2000, there were approximately 173,000 cases of Hib meningitis worldwide, with the highest incidence in Africa (46 per 100,000 per year) and the lowest in Europe (16 per 100,000 per year) [7]. Before use of Hib conjugate vaccines, however, the incidence of invasive Hib disease in Europe was 12–54 per 100,000 per year in children less than 5 years, and approximately 60% of these had meningitis [5].

Most European countries implemented routine Hib conjugate vaccination between 1992 and 1996, leading to greater than 90% reduction of disease in all countries [5]. From 1999 there was a resurgence in the number of cases in the UK, predominantly in children aged 1–4 years. This may have been due to a waning effect of the initial catch-up campaign, interference with the co-administered acellular pertussis vaccine or serogroup C meningococcal vaccine conjugated to CRM197, or a consequence of the fact that no booster dose was in routine use at the time. In 2003 a further catch-up campaign occurred and a routine booster dose was introduced into the immunisation schedule in 2006, resulting in a decrease in disease.

3.4 Neonatal Bacterial Meningitis

The incidence of bacterial meningitis has been stable at 0.2–1 per 1,000 live births in developed countries since the 1980s [10]. Up to 30% of neonates with sepsis have associated bacterial meningitis [21], highlighting the importance of assessment for possible meningitis during evaluation of a septic neonate.

4 Predisposing Factors

There are a number of factors which increase a child's risk for acquiring bacterial meningitis, and it is important for these to be considered during the clinical assessment of possible cases. Some of these are risk factors for all pathogens, whereas others only increase the likelihood of infection with certain organisms. Predisposing factors include:

- Young age;
- Male gender;
- Malnutrition or chronic illness;
- Exposure to tobacco smoke (*N. meningitidis*);
- Recent head trauma, neurosurgery or presence of a ventriculo-peritoneal shunt;
- Presence of a cochlear implant (*S. pneumoniae*)
- Local anatomical defects;
- Close contact with:

    - A colonised carrier (N. meningitidis, S. pneumoniae, Hib);
    - An individual with disease (N. meningitidis, Hib, rarely S. pneumoniae);
    - Certain animals (e.g. reptiles – Salmonella, domestic animals – Listeria);
    - Consumption of unpasteurised dairy products in pregnancy (Listeria);

- Lack of immunisation (Hib, *S. pneumoniae, N. meningitidis*);
- Congenital or acquired immunosuppression:

    - Deficiencies in terminal complement components (N. meningitidis);
    - Hyposplenism, e.g. post-splenectomy, congenital asplenia (S. pneumoniae, Hib);
    - HIV infection (S. pneumoniae);

- Sickle cell disease (*S. pneumoniae*, Hib, *Salmonella*);
- Malignant neoplasia.

5 Clinical Presentation

The clinical manifestations of bacterial meningitis vary with age (Table 1) [22]. The classical symptoms and signs present in older children are rarely present in infants and young children. A number of studies, which have been mostly retrospective, have described the clinical features of children with bacterial meningitis [23–36].

In children over 2 years, the illness usually begins with fever, nausea and vomiting, photophobia, and severe headache. Occasionally the first sign is a seizure, which can also occur later. Irritability, delirium, and altered level of consciousness develop as CNS inflammation progresses. The most specific signs are neck stiffness, Kernig's sign (the inability to fully extend the knee while the hip is flexed due to contraction of the hamstring muscles and pain) and Brudzinski's sign (automatic flexion of the hips and knees after passive neck flexion), but these are often absent in children and especially infants. In addition, focal neurological abnormalities may occur, and in the absence of seizures they indicate cortical necrosis, occlusive vasculitis or cortical venous thrombosis. In infants and young children under 2 years of age, fever and irritability are the most common characteristics. Other symptoms are non-specific, and can include hypothermia, poor feeding, vomiting, lethargy, jaundice, respiratory distress or apnoea, and seizures. A bulging fontanelle is often present.

Some of the organisms causing meningitis can also cause other clinical manifestations. *N. meningitidis* causes septicaemia, which may co-exist with meningitis. Meningococcal septicaemia can cause leg pain and cold extremities early in the illness [37]. A petechial or purpuric rash is usually present in septicaemia, although the rash may be blanching, and can also be present with Hib or pneumococcal disease. As the disease progresses, septicaemic shock occurs. Joint involvement is suggestive of infection with *N. meningitidis* or Hib, and a chronically draining ear or history of head trauma is more likely to be associated with pneumococcal meningitis.

Table 1 Clinical features of bacterial meningitis in all children, and in those under 2 years of age [22]

Feature	All children (%)	Children <2 years (%)
Non-specific features		
Fever	66–97	69–96
Nausea and/or vomiting	18–70	31–60
Lethargy	13–87	28–54
Irritability or unsettled	21–79	>50
Refusing food and drink	26–76	
Muscle ache and joint pain	23	
Respiratory difficulties	13–49	29–38
Specific features		
Headache	3–59	
Photophobia	5–16	7
Stiff neck	13–74	13–56
Kernig's sign	10–53	10–36
Brudzinski's sign	11–66	11–68
Back rigidity		
Bulging fontanelle	13–45	41–45
Seizures	14–38	22–55
Focal neurological deficit	6–47	
Toxic/moribund state	3–49	
Reduced conscious level	60–87	
Coma	4–18	3–6
Shock	8–16	
Rash	9–62	

It is not possible to reliably distinguish bacterial and viral meningitis on clinical features alone. Children with bacterial meningitis, however, are more likely to have non-specific features and many have neck stiffness or an altered conscious level, and children with viral meningitis are less likely to have shock, an altered conscious level or seizures.

6 Differential Diagnosis

The main differential diagnosis for bacterial meningitis is other infectious causes of meningitis. Viruses are the commonest cause of meningitis overall in developed countries, most frequently enteroviruses. Other viral causes include mumps, herpes simplex virus (HSV), cytomegalovirus (CMV), Epstein Barr virus (EBV), varicella zoster virus (VZV), adenoviruses, HIV, measles, rubella, influenza, parainfluenza, and rotavirus. Other infectious causes of aseptic meningitis are:

* Partially-treated bacterial meningitis;
* Non-pyogenic bacteria, e.g. Mycobacteria, *Leptospira*, *Treponema pallidum*, *Borrelia*, *Nocardia*, *Bartonella*, and *Brucella;*
* Atypical organisms, e.g. *Chlamydia*, *Rickettsia*, and *Mycoplasma;*
* Fungi, e.g. *Candida*, *Cryptococcus, Histoplasma*, and *Coccidioides;*
* Protozoa and helminths, e.g. roundworms, tapeworms, flukes, amoebae, and *Toxoplasma.*

The differential diagnosis also includes other CNS infections, such as encephalitis and intracranial abscesses (cerebral, subdural or epidural), as well as generalised sepsis from another focus. Non-infectious causes that should be considered are leukaemia and solid CNS tumours, connective tissue disorders (e.g. systemic lupus erythematosus, Behçet's disease), Kawasaki disease, sarcoidosis, and drugs and toxins, including intravenous immunoglobulin and heavy metal poisoning.

7 Investigations

A lumbar puncture (LP) is required to obtain a definitive diagnosis of bacterial meningitis, including identification of the organism. Additional investigations can be helpful to support the diagnosis of bacterial infection, and to indicate severity of disease.

7.1 Lumbar Puncture

Cerebrospinal fluid (CSF) should ideally be obtained prior to commencing treatment, but initiation of anti-microbial therapy should never be delayed if an immediate LP cannot be performed. CSF analysis by microscopy, Gram stain, culture and polymerase chain reaction (PCR) is the definitive method of diagnosis, and biochemistry analysis for protein and glucose (with a plasma glucose) should also be undertaken (Table 2).

An LP should be performed in any child in whom meningitis is suspected and any drowsy or ill infant, in the absence of any of the following contra-indications [22]:

* Signs of raised intra-cranial pressure:

  – Reduced conscious level (Glasgow Coma Score <9, or a drop of 3 or more);
  – Relative bradycardia and hypertension;

- – Abnormal posture;
- – Unequal, dilated or poorly responsive pupils;
- – Papilloedema (an uncommon finding in acute meningitis and its presence should prompt consideration of venous sinus occlusion, subdural empyema or brain abscess);
- – Abnormal 'doll's eye' movements;

- Abnormal focal neurological signs;
- Shock;
- Extensive or spreading purpura;
- After convulsions, until stabilised;
- Abnormal clotting studies (if available) or concurrent anticoagulant therapy;
- Severe thrombocytopenia (platelet count $<100 \times 10^9$/l);
- Localised infection at the site of LP.

If contra-indications are present, LP should be delayed and performed when contra-indications are no longer applicable. Once CSF has been obtained it should be examined as soon as possible because white blood cells (WBCs) start to degrade after approximately 90 min [38]. Initial Gram staining of CSF reveals an organism in 68–80% of bacterial meningitis cases [39]. It is rare for CSF microscopy to be normal and a pathogen identified later, although this occurs most often in meningococcal meningitis (up to 8%) and in neonates. If CSF variables are normal and the child is seriously unwell, it is important to consider alternative diagnoses. CSF cultures become negative rapidly following the administration of antibiotics to a child with suspected meningitis. CSF cultures are negative 2 h after parenteral antibiotics are given in meningococcal meningitis, after 6 h in pneumococcal meningitis and after 8 h in neonatal GBS meningitis [39]. A delayed LP can still provide useful information, however, since CSF cellular and biochemical changes remain for up to 44–68 h after the start of treatment [40].

Table 2 CSF white blood cell (WBC) count, protein and glucose values in normal children and changes that occur with meningitis (Adapted from [38, 56–58])

	Macroscopic appearance	CSF WBC count (per µl)[a]	CSF neutrophil count (per µl)[a]	CSF protein (g/l)	CSF glucose (% of plasma glucose)
Normal CSF					
Neonate	clear and colourless	0–20[b]	0–4[b]	20–130	>60
>1 month		0–5[c]	0	0–40	60–70
Children with meningitis					
Bacterial meningitis	turbid or purulent	↑↑↑[d]	↑↑↑[d]	↑↑↑	↓↓
Viral meningitis	usually clear	↑[e]	N/↑[e]	N/↑	↓/N
TB meningitis	yellow or cloudy	↑↑[f]	N/↑[f]	↑↑↑	↓
Fungal meningitis	usually clear	↑[f]	N/↑[f]	↑↑	↓

[a] In the case of a traumatic LP, one WBC per 500 red blood cells can be subtracted from the total CSF WBC count
[b] WBCs in neonatal CSF is predominantly lymphocytes, although neutrophils may be present
[c] Outside the neonatal period all CSF WBCs should be lymphocytes and the presence of any neutrophils is abnormal
[d] CSF WBCs in bacterial meningitis are usually mostly neutrophils, although lymphocytes can be predominant in early disease
[e] CSF WBCs in viral meningitis are usually mostly lymphocytes, although neutrophils can be predominant
[f] In TB or fungal meningitis, the majority of CSF WBCs are lymphocytes

7.2 Cranial Computed Tomography (CT) and Magnetic Resonance Imaging (MRI)

The main indication for cranial imaging is when the diagnosis is uncertain or to detect other possible intracranial pathology. The use of anti-microbial therapy should not be delayed while imaging is awaited. Importantly, a normal CT scan does not mean it is safe to do an LP – this decision should be based on clinical assessment. However, if a scan shows evidence of raised intra-cranial pressure an LP should not be performed. If a CT scan is required, it should be undertaken urgently after stabilisation of the child. While CT is widely available and very useful for rapid assessment of hydrocephalus, mass lesions, haemorrhage or cerebral oedema, MRI will detect more subtle findings. It should be noted that non-contrast CT or MRI can be normal in early cases of meningitis. CT in cerebral oedema may show slit-like lateral ventricles, areas of low attenuation, and absence of basilar and suprachiasmatic cisterns [41].

7.3 Other Investigations

All children with suspected meningitis should have a blood culture (which is positive in 80–90% of antibiotic-naïve children), blood taken for PCR (see below), and full blood count (FBC), C-reactive protein (CRP), clotting, urea and electrolytes, and glucose [41]. Bacterial meningitis is likely in those with abnormal CSF parameters who have a significantly raised WBC count and/or CRP. A normal CRP and WBC count, however, do not rule out bacterial meningitis. If bacterial meningitis is suspected clinically and an LP has not been performed, children should be managed as such regardless of blood results.

7.4 Molecular Techniques

It is increasingly common in developed countries that children are treated with parenteral antibiotics prior to their arrival at hospital. Molecular techniques are therefore becoming an important adjunct to blood and CSF culture to identify an aetiological agent in cases of meningitis. PCR for *N. meningitidis, S. pneumoniae* and Hib using CSF or for *N. meningitidis* and *S. pneumoniae* using blood can be obtained in the UK from the meningococcal reference laboratory in Manchester. For *N. meningitidis*, PCR from blood has a sensitivity of 87% and specificity of 100%, whereas for *S. pneumoniae*, PCR is sensitive and specific on CSF, but false positive results may be obtained from blood due to the high nasopharyngeal carriage rate in young children [41]. Rapid antigen latex agglutination tests on CSF or blood (which can be used to detect *N. meningitidis, S. pneumoniae, H. influenzae, E. coli* or GBS) can be done locally and rapidly, but the lack of sensitivity (and specificity) limits their clinical use.

8 Clinical Decision Rules

It has been estimated that over 90% of children presenting with meningitis in the highly immunised populations of developed countries will have aseptic meningitis [42–44], so clinical decision rules have been developed since the introduction of the Hib conjugate vaccine to distinguish bacterial from aseptic meningitis, to reduce antibiotic and corticosteroid use and hospitalisation rates. The 'Bacterial Meningitis Score' is the only rule which has been sufficiently validated in a large number of children, and classifies patients with CSF pleocytosis (WBC count >10 per mm^3) as very low risk of bacterial meningitis if they fulfil the following criteria: negative CSF Gram stain; CSF neutrophil count <1,000 per mm^3; CSF protein <80 mg/dl; blood neutrophil count <10×10^9/l; no seizure prior to presentation [45]. Importantly, this score cannot be used to identify those who do have bacterial meningitis. In a Hib immunised population, this score had a negative predictive value of 100% (95% CI 97–100%) prior to routine use of PCV7, and 99.9% (95% CI 99.6–100%) when validated after introduction of the vaccine [42, 45]. In this very large study only 1.3% of children with a CSF WBC count <300 per mm^3 had bacterial meningitis, increasing to 10% and 28% for those with a CSF WBC count >500 per mm^3 and >1,000 per mm^3, respectively.

It is likely that these studies underestimated the prevalence of bacterial meningitis in children with CSF pleocytosis because they excluded 23% of eligible children, including those with critical illness, purpura, immunosuppression, and previous antibiotic administration. Morever, no data are available from countries where the meningococcal serogroup C conjugate vaccine is routinely used. Clinical decision rules provide a potentially useful tool to identify the diminishing proportion of children with meningitis who have a bacterial cause, but these tools need further validation before they can be routinely implemented to determine treatment of children with suspected meningitis.

9 Management

Many children with meningitis, particularly those with meningococcal meningitis, will have co-existing septicaemia and shock. Resuscitation following standard guidelines is therefore the priority for initial management of these children, with the expectation that prompt and adequate fluid resuscitation may be required. Any child with suspected meningitis should, therefore, be transferred to a hospital immediately and be rapidly assessed for dehydration, shock and raised intracranial pressure. Once initial resuscitation has commenced, appropriate antibiotics should be administered as soon as possible, with corticosteroids if appropriate (see below).

9.1 Anti-microbial Therapy

In cases of suspected meningococcal disease (i.e. presence of a purpuric or petechial rash), antibiotic therapy with parenteral benzylpenicillin is often given before admission to hospital and is recommended in the UK (Table 3) [22]. There is no reliable evidence to support or refute this practice and the priority of transfer to hospital should remain [46]. In cases of suspected meningitis where there is no rash (to indicate meningococcal disease) use of pre-hospital antibiotics is not

Table 3 Current UK guidelines for empirical and specific therapy in bacterial meningitis [22]

Age group	Empirical therapy	Specific therapy
>3 months	Ceftriaxone or Cefotaxime	Ceftriaxone 7 days for *N. meningitidis* 10 days for Hib
	± Vancomycin	14 days for *S. pneumoniae* ≥10 days for unconfirmed organism
<3 months	Amoxicillin/Ampicillin	GBS: ≥14 days Cefotaxime/Penicillin
	+ Cefotaxime	Gram negative organisms: 21 days Cefotaxime
	(or Ceftriaxone or Meropenem)	*Listeria*: 21 days Amoxicillin/Ampicillin with Gentamicin for first 7 days
	± Vancomycin	Unconfirmed: ≥14 days Amoxicillin/Ampicillin plus Cefotaxime
	Consider acyclovir	

recommended unless there is a delay in transfer to hospital. In hospital, antibiotic therapy for suspected acute bacterial meningitis must be started immediately, before the results of CSF culture and antibiotic sensitivity are available. At all ages antibiotics should be initiated if the CSF WBC count is abnormal (Table 2). In neonates bacterial meningitis should still be considered if other clinical features are present, irrespective of CSF WBC count.

Intravenous antibiotics are required to achieve adequate serum and CSF levels. The choice of empirical agent(s) should take into account local bacterial antibiotic sensitivity patterns. Once a pathogen has been identified and antibiotic susceptibility results are available, specific therapy may need to be adjusted. Current recommendations in the UK are provided below and summarised in Table 3 [22].

9.1.1 Empirical Therapy for Children Aged >3 Months

Monotherapy with a third generation cephalosporin (e.g. ceftriaxone or cefotaxime) is recommended, with ceftriaxone sometimes preferred as it is once-daily dosing. These drugs have a broad spectrum of activity against Gram-positive and Gram-negative organisms, are highly resistant to β-lactamase activity, and penetrate the blood–brain barrier. In a meta-analysis of studies mostly from the 1980s, there was no difference in outcome (i.e. death or deafness) following treatment with third generation cephalosporins compared to conventional antibiotics (i.e. penicillin, ampicillin, chloramphenicol and gentamicin). However, cephalosporins were more likely to have sterilised the CSF at 10–48 h following the onset of treatment, and were more likely to cause diarrhoea [47].

Neonatal deaths have been reported due to an interaction between ceftriaxone and calcium-containing products [48], so ceftriaxone should not be administered simultaneously with calcium-containing infusions in any age group. In this situation cefotaxime should be used. In the USA, and a number of other countries, there is an increasing problem with penicillin-resistant pneumococci. Overall 20–45% of *S. pneumoniae* strains worldwide are resistance to penicillin [1]. In regions where there is a high prevalence of resistance, or in children with recent prolonged or multiple exposure to antibiotics, or those who have recently travelled to an area with a high rate of pneumococcal resistance, vancomycin should be added.

9.1.2 Specific Therapy for Children Aged >3 Months

Specific therapy with ceftriaxone is recommended for convenience and cost-effectiveness [22]. The overall cost of ceftriaxone, which includes costs of the drug, labour required and consumables for administration, are less than either cefotaxime or benzylpenicillin, mainly due to the lower labour costs associated with once-daily dosing. The difference in cost of these treatments is also dependent on the weight of the child. Ceftriaxone is the cheapest treatment in all children under approximately 50 kg, and is most cost-effective in young children. The duration of antibiotic therapy depends upon the infecting organism. There are no adequate studies of treatment duration, but 7 days is usually considered sufficient for *N. meningitidis*, 10 days for Hib and 14 days for *S. pneumoniae*. Unconfirmed, uncomplicated but clinically suspected bacterial meningitis should be treated with ceftriaxone for at least 10 days depending on clinical features and course.

9.1.3 Empirical Therapy for Children Aged <3 Months

Cefotaxime is usually the treatment of choice, plus amoxicillin or ampicillin to cover *Listeria*. Some centres use an aminoglycoside in addition to this regime. Ceftriaxone may be used instead of cefotaxime, but should be avoided in infants who are jaundiced, hypoalbuminaemic, acidotic or born prematurely as it may exacerbate hyperbilirubinaemia [1, 10]. Ceftriaxone should not be administered at the same time as calcium-containing infusions (see above). In addition, vancomycin should be used for indications as above. In settings with high rates of community acquired extended spectrum β-lactamase (ESBL) producing Gram-negative organisms, meropenem should be considered instead of cefotaxime.

9.1.4 Specific Therapy for Children <3 Months

There are no controlled clinical trials to guide duration of therapy in this age group. For GBS meningitis, cefotaxime or penicillin should be continued for at least 14 days after initiation, but treatment extended to at least 21 days in complicated cases. Data from *in vitro* studies suggest a synergy between gentamicin and penicillins, although there are no clinical trials to support this [49]. In infections with Gram-negative organisms, treatment for 21 days is recommended with cefotaxime or an alternative antibiotic based on local resistance patterns and sensitivities of the specific organism. If *L. monocytogenes* infection is confirmed, therapy is recommended for 21 days with amoxicillin, adding gentamicin for at least the first 7 days. In cases of unconfirmed but clinically suspected meningitis, cefotaxime plus amoxicillin/ampicillin should be administered for at least 14 days. If the course is complicated, extending the duration of treatment should be considered in consultation with a specialist in paediatric infectious diseases. Repeat LP should be performed in neonates after 48–72 h only if there is worsening or no improvement of clinical condition and/or laboratory parameters. An LP at the end of treatment is unnecessary if the clinical response is otherwise uneventful, since CSF findings at this stage are not useful to predict a relapse [50].

9.2 Corticosteroid Therapy

Corticosteroids reduce meningeal inflammation and modulate cytokine secretion to reduce pro-inflammatory responses. In clinical trials corticosteroids reduced the rate of severe hearing loss in bacterial meningitis from 11.0% to 6.6% [51]. The majority of children in these trials had meningitis due to Hib and a CSF WBC count >1,000 per mm^3, but there was also a trend for better outcome in non-*Haemophilus* meningitis. In adults, steroids appears to reduce mortality in addition to the improved hearing outcomes [51]. Data from studies conducted in low-income settings showed no significant impact of steroids on hearing loss. This may be due to delay in presentation to the hospital, differences in the causative organisms, or differences in the antibiotics used.

The use of corticosteroid therapy in bacterial meningitis remains controversial, however, principally because of the lack of data relevant to the post-conjugate vaccine era. Most guidelines emphasise the need to target steroid use to children who are most likely to have bacterial meningitis. The recently published UK guidelines recommend that children over 3 months of age should receive corticosteroids if they have frankly purulent CSF, bacteria on CSF Gram stain, a CSF WBC count >1,000 per mm^3, or CSF pleocytosis and a CSF protein >100 mg/dl [22]. Corticosteroids should ideally be administered before or with the first antibiotic dose, but they may be beneficial up to 12 h later [51].

9.3 Ongoing Fluid Management

Fluid therapy should be guided by clinical assessment of hydration status, signs of raised intra-cranial pressure and shock and regular electrolyte measurements, since both over- and under-hydration are associated with adverse outcomes. Over 50% of children have hyponatraemia at presentation attributed to increased concentrations of anti-diuretic hormone (ADH), and this is a marker of severe disease. There are differing opinions as to whether ADH production is due to dehydration and is therefore *appropriate*, or is part of the syndrome of *inappropriate* ADH secretion (SIADH). In general, enteral fluids or feeds should be used where appropriate, and isotonic fluid when intravenous therapy is required. After correction of dehydration, full maintenance fluid should be given to prevent hypoglycaemia and maintain electrolyte balance. In settings with high mortality and where children present late, full maintenance fluid therapy was associated with reduced spasticity, seizures, and chronic severe neurological sequelae [52]. Where children present early and mortality rates are lower there is insufficient evidence available, so fluid restriction should not be employed routinely. In cases where there is evidence of raised intracranial pressure or circulatory failure, emergency management should be initiated and ongoing fluid management discussed with a paediatric intensivist.

9.4 Other Supportive Treatment

It is likely that some children with meningitis will require supportive treatment in addition to antibiotics, corticosteroids and appropriate fluid therapy. It is important that any possible need for management in an intensive care setting is considered early in the illness. Children should be adequately oxygenated to prevent further brain injury, and hypoglycaemia should be actively sought and treated. Children with seizures will require anti-convulsant therapy. If there is clinically-evident raised intra-cranial pressure, or if there is a suggestion on a CT scan, then the appropriate management of this is required. This may include elevation of the head of the bed to an angle of 30°, maintenance of normal pCO_2 through mechanical ventilation, and treatment with mannitol and furosemide. Children with severe sepsis will require circulatory support with fluid replacement and inotropes.

10 Prevention of Secondary Cases

Both *N. meningitidis* and Hib are carried in the nasopharynx prior to causing invasive disease, so there is a high risk of transmission from the index case to close contacts. Chemoprophylaxis and/or vaccination are therefore recommended for close contacts of cases when these organisms are suspected. Chemoprophylaxis against meningococcal disease with rifampicin, ciprofloxacin or ceftriaxone should be given as soon as possible and ideally within 24 h of diagnosis to household members who have had prolonged close contact with the index case and those who have had transient close contact with the index case if they have been directly exposed to large particles or respiratory droplets/secretions (e.g. healthcare workers) [53]. Chemoprophylaxis against Hib disease is only indicated if the index case is less than 10 years old or there is a vulnerable individual (defined as immunosuppressed, asplenic or under 10 years of age) in the household [54]. In such cases rifampicin should be given to the index case, and all household contacts if there is a vulnerable individual in the household. In cases of Hib disease, Hib immunisation should additionally be given to the index case if under 10 years of age and incompletely immunised or convalescent antibody levels <1 μg/ml or hyposplenic, as well as all incompletely immunised children under 10 years of age in the same household.

11 Complications and Prognosis

Outcome depends on multiple factors, including age, time and clinical stability prior to treatment, organism and host inflammatory response. Early complications are seizures and SIADH. Subdural effusions occur in one-third of cases, and are often asymptomatic with spontaneous resolution.

Table 4 Risk of sequelae from bacterial meningitis in adults and children globally [55]

Sequelae	Risk after		
	S. pneumoniae (%)	Hib (%)	*N. meningitidis* (%)
Hearing loss	7.5	3.2	2.6
Motor deficit	5.8	2.2	1.0
Behavioural problems	4.6	2.1	0.6
Cognitive difficulties	4.2	1.1	1.6
Epilepsy	2.5	1.5	0.5
Visual disturbance	1.1	0.5	1.5
At least one sequelae	34.7	14.5	9.5
Multiple impairments	5.7	2.6	1.3

They may also manifest with enlargement of head circumference, vomiting, seizures, bulging fontanelle, focal neurological signs or persistent fever. Other complications that can occur later during acute disease are focal neurological abnormalities, hydrocephalus (more often in younger infants) and brain abscesses, especially in newborns infected with *Citrobacter diversus* or *Proteus*.

Long-term complications occur in 10–30% overall (with higher rates in resource-poor countries), and are summarised in Table 4 [1, 55]. The most common complication of bacterial meningitis is sensorineural hearing loss, so all children should have hearing screening performed soon after discharge from hospital. Other common complications are epilepsy, motor and cognitive impairment, blindness and optic atrophy, and learning and behavioural problems.

In the developed world, case fatality rates are less than 10% overall and under 5% for meningitis due to *N. meningitidis* or Hib [1]. However, case-fatality rates in developing countries are much higher, and are around 70% for pneumococcal and Hib meningitis in Africa [7, 9]. For neonatal bacterial meningitis, mortality is approximately 5–10% overall. Disability at 5 years is 50% for GBS and *E. coli*, and 78% following infection with other Gram-negative organisms [10].

12 Conclusions

The introduction of vaccines against the encapsulated bacteria which predominantly cause bacterial meningitis has significantly reduced the incidence in developed countries, making cases of bacterial meningitis sometimes difficult to distinguish clinically from other sick children with febrile illnesses. There is still a substantial global burden of disease, predominantly occurring in resource-poor countries, where case-fatality rates are higher and long-term sequelae are more common. There remains a limited evidence base for many aspects of what is considered to be the standard of clinical management, most notably regarding the duration of antibiotic therapy, indications for corticosteroid treatment and optimum fluid therapy. Future challenges include the development of vaccines against serogroup B *N. meningitidis*, prevention of neonatal infection through maternal vaccination or use of specific antibody preparations, enabling routine use of more sensitive tests, such as amplification of 16S rRNA gene by PCR, for diagnosis in antibiotic pretreated patients, and the assessment of new antimicrobial agents against resistant pneumococcal strains.

References

1. Saez-Llorens X, McCracken GH, Jr. Bacterial meningitis in children. Lancet. 2003;361:2139–48.
2. Rorabaugh ML, Berlin LE, Heldrich F, Roberts K, Rosenberg LA, Doran T, et al. Aseptic meningitis in infants younger than 2 years of age: acute illness and neurologic complications. Pediatrics. 1993;92:206–11.
3. Ramsay M, Fox A. EU-IBIS Network. Invasive Neisseria meningitidis in Europe 2006. London: Health Protection Agency; 2007.
4. McIntosh ED, Fritzell B, Fletcher MA. Burden of paediatric invasive pneumococcal disease in Europe, 2005. Epidemiology and Infection. 2007;135:644–56.
5. Watt JP, Levine OS, Santosham M. Global reduction of Hib disease: what are the next steps? Proceedings of the meeting Scottsdale, Arizona, September 22–25, 2002. The Journal of Pediatrics. 2003;143:S163-87.
6. Hsu HE, Shutt KA, Moore MR, Beall BW, Bennett NM, Craig AS, et al. Effect of pneumococcal conjugate vaccine on pneumococcal meningitis. The New England Journal of Medicine. 2009;360:244–56.
7. Watt JP, Wolfson LJ, O'Brien KL, Henkle E, Deloria-Knoll M, McCall N, et al. Burden of disease caused by Haemophilus influenzae type b in children younger than 5 years: global estimates. Lancet. 2009;374:903–11.
8. Harrison LH, Trotter CL, Ramsay ME. Global epidemiology of meningococcal disease. Vaccine. 2009;27 Suppl 2:B51-63.
9. O'Brien KL, Wolfson LJ, Watt JP, Henkle E, Deloria-Knoll M, McCall N, et al. Burden of disease caused by *Streptococcus pneumoniae* in children younger than 5 years: global estimates. Lancet. 2009;374:893–902.
10. Heath PT, Nik Yusoff NK, Baker CJ. Neonatal meningitis. Archives of Disease in Childhood. 2003;88:F173-8.
11. Baker RC, Kummer AW, Schultz JR, Ho M, Gonzalez del Rey J. Neurodevelopmental outcome of infants with viral meningitis in the first three months of life. Clinical Pediatrics. 1996;35:295–301.
12. Baethgen LF, Weidlich L, Moraes C, Klein C, Nunes LS, Cafrune PI, et al. Epidemiology of meningococcal disease in southern Brazil from 1995 to 2003, and molecular characterization of Neisseria meningitidis using multilocus sequence typing. Trop Med Int Health. 2008;13:31–40.
13. Cohn AC, MacNeil JR, Harrison LH, Hatcher C, Theodore J, Schmidt M, et al. Changes in Neisseria meningitidis disease epidemiology in the United States, 1998–2007: implications for prevention of meningococcal disease. Clin Infect Dis. 2010;50:184–91.
14. Lapeyssonnie L. La méningite cérébro-spinale en Afrique. Bull World Health Organ. 1963;28:449–74.
15. Mayer LW, Reeves MW, Al-Hamdan N, Sacchi CT, Taha MK, Ajello GW, et al. Outbreak of W135 meningococcal disease in 2000: not emergence of a new W135 strain but clonal expansion within the electophoretic type-37 complex. J Infect Dis. 2002;185:1596–605.
16. Djibo S, Nicolas P, Alonso JM, Djibo A, Couret D, Riou JY, et al. Outbreaks of serogroup X meningococcal meningitis in Niger 1995–2000. Trop Med Int Health. 2003;8:1118–23.
17. Ramsay ME, Andrews NJ, Trotter CL, Kaczmarski EB, Miller E. Herd immunity from meningococcal serogroup C conjugate vaccination in England: database analysis. BMJ (Clinical research ed.) 2003;326:365–6.
18. MacNeil J, Cohn AC, Mayer R, Zell ER, Clark TA, Messonnier NE, et al. Interim analysis of the effectiveness of quadrivalent meningococcal conjugate vaccine (MenACWY-D): a matched case-control study. 17th International Pathogenic Neisseria Conference. Banff, Canada 2010.
19. Sadarangani M, Pollard AJ. Serogroup B meningococcal vaccines-an unfinished story. Lancet Infect Dis. 2010;10:112–24.
20. Ruckinger S, van der Linden M, Reinert RR, von Kries R, Burckhardt F, Siedler A. Reduction in the incidence of invasive pneumococcal disease after general vaccination with 7-valent pneumococcal conjugate vaccine in Germany. Vaccine. 2009;27:4136–41.
21. Kumar P, Sarkar S, Narang A. Role of routine lumbar puncture in neonatal sepsis. Journal of Paediatrics and Child Health. 1995;31:8–10.
22. National Collaborating Centre for Women's and Children's Health. Bacterial meningitis and meningococcal septicaemia: management of bacterial meningitis and meningococcal septicaemia in children and young people younger than 16 years in primary and secondary care. London, UK: Royal College of Obstetricians and Gynaecologists. Available at http://www.nice.org.uk/CG102; 2010.
23. Kirkpatrick B, Reeves DS, MacGowan AP. A review of the clinical presentation, laboratory features, antimicrobial therapy and outcome of 77 episodes of pneumococcal meningitis occurring in children and adults. The Journal of Infection. 1994;29:171–82.
24. Riordan FA, Thomson AP, Sills JA, Hart CA. Bacterial meningitis in the first three months of life. Postgraduate Medical Journal. 1995;71:36–8.
25. Valmari P, Peltola H, Ruuskanen O, Korvenranta H. Childhood bacterial meningitis: initial symptoms and signs related to age, and reasons for consulting a physician. European Journal of Pediatrics. 1987;146:515–8.
26. Miner JR, Heegaard W, Mapes A, Biros M. Presentation, time to antibiotics, and mortality of patients with bacterial meningitis at an urban county medical center. The Journal of Emergency Medicine. 2001;21:387–92.

27. De Cauwer HG, Eykens L, Hellinckx J, Mortelmans LJ. Differential diagnosis between viral and bacterial meningitis in children. Eur J Emerg Med. 2007;14:343–7.

28. Casado-Flores J, Aristegui J, de Liria CR, Martinon JM, Fernandez C. Clinical data and factors associated with poor outcome in pneumococcal meningitis. European Journal of Pediatrics. 2006;165:285–9.

29. Kornelisse RF, Westerbeek CM, Spoor AB, van der Heijde B, Spanjaard L, Neijens HJ, et al. Pneumococcal meningitis in children: prognostic indicators and outcome. Clin Infect Dis. 1995;21:1390–7.

30. Rothrock SG, Green SM, Wren J, Letai D, Daniel-Underwood L, Pillar E. Pediatric bacterial meningitis: is prior antibiotic therapy associated with an altered clinical presentation? Annals of Emergency Medicine. 1992;21:146–52.

31. Molyneux E, Riordan FA, Walsh A. Acute bacterial meningitis in children presenting to the Royal Liverpool Children's Hospital, Liverpool, UK and the Queen Elizabeth Central Hospital in Blantyre, Malawi: a world of difference. Annals of Tropical Paediatrics. 2006;26:29–37.

32. Walsh-Kelly C, Nelson DB, Smith DS, Losek JD, Melzer-Lange M, Hennes HM, et al. Clinical predictors of bacterial versus aseptic meningitis in childhood. Annals of Emergency Medicine. 1992;21:910–4.

33. Kallio MJ, Kilpi T, Anttila M, Peltola H. The effect of a recent previous visit to a physician on outcome after childhood bacterial meningitis. JAMA. 1994;272:787–91.

34. Ostergaard C, Konradsen HB, Samuelsson S. Clinical presentation and prognostic factors of *Streptococcus pneumoniae* meningitis according to the focus of infection. BMC Infectious Diseases. 2005;5:93.

35. Lee BE, Chawla R, Langley JM, Forgie SE, Al-Hosni M, Baerg K, et al. Paediatric Investigators Collaborative Network on Infections in Canada (PICNIC) study of aseptic meningitis. BMC Infectious Diseases. 2006;6:68.

36. Michos AG, Syriopoulou VP, Hadjichristodoulou C, Daikos GL, Lagona E, Douridas P, et al. Aseptic meningitis in children: analysis of 506 cases. PloS One. 2007;2:e674.

37. Thompson MJ, Ninis N, Perera R, Mayon-White R, Phillips C, Bailey L, et al. Clinical recognition of meningococcal disease in children and adolescents. Lancet. 2006;367:397–403.

38. Feigin RD, McCracken GH, Jr., Klein JO. Diagnosis and management of meningitis. The Pediatric Infectious Disease Journal. 1992;11:785–814.

39. Riordan FA, Cant AJ. When to do a lumbar puncture. Arch Dis Child. 2002;87:235–7.

40. Blazer S, Berant M, Alon U. Bacterial meningitis. Effect of antibiotic treatment on cerebrospinal fluid. Am J Clin Pathol. 1983;80:386–7.

41. El Bashir H, Laundy M, Booy R. Diagnosis and treatment of bacterial meningitis. Arch Dis Child. 2003;88:615–20.

42. Nigrovic LE, Kuppermann N, Macias CG, Cannavino CR, Moro-Sutherland DM, Schremmer RD, et al. Clinical prediction rule for identifying children with cerebrospinal fluid pleocytosis at very low risk of bacterial meningitis. JAMA. 2007;297:52–60.

43. Pierart J, Lepage P. [Value of the "Bacterial Meningitis Score" (BMS) for the differential diagnosis of bacterial versus viral meningitis]. Revue Medicale De Liege. 2006;61:581–5.

44. Dubos F, Lamotte B, Bibi-Triki F, Moulin F, Raymond J, Gendrel D, et al. Clinical decision rules to distinguish between bacterial and aseptic meningitis. Arch Dis Child. 2006;91:647–50.

45. Nigrovic LE, Kuppermann N, Malley R. Development and validation of a multivariable predictive model to distinguish bacterial from aseptic meningitis in children in the post-Haemophilus influenzae era. Pediatrics. 2002;110:712–9.

46. Sudarsanam T, Rupali P, Tharyan P, Abraham OC, Thomas K. Pre-admission antibiotics for suspected cases of meningococcal disease. Cochrane Database of Systematic Reviews (Online). 2008:CD005437.

47. Prasad K, Singhal T, Jain N, Gupta PK. Third generation cephalosporins versus conventional antibiotics for treating acute bacterial meningitis. Cochrane Database of Systematic Reviews (Online). 2004:CD001832.

48. Bradley JS, Wassel RT, Lee L, Nambiar S. Intravenous ceftriaxone and calcium in the neonate: assessing the risk for cardiopulmonary adverse events. Pediatrics. 2009;123:e609-13.

49. Swingle HM, Bucciarelli RL, Ayoub EM. Synergy between penicillins and low concentrations of gentamicin in the killing of group B streptococci. J Infect Dis. 1985;152:515–20.

50. Schaad UB, Nelson JD, McCracken GH, Jr. Recrudescence and relapse in bacterial meningitis of childhood. Pediatrics. 1981;67:188–95.

51. van de Beek D, de Gans J, McIntyre P, Prasad K. Corticosteroids for acute bacterial meningitis. Cochrane Database of Systematic Reviews (Online). 2007:CD004405.

52. Maconochie I, Baumer H, Stewart ME. Fluid therapy for acute bacterial meningitis. Cochrane Database of Systematic Reviews (Online). 2008:CD004786.

53. Health Protection Agency Meningococcus Forum. Guidance for public health management of meningococcal disease in the UK. London, UK2006.

54. Ladhani S, Neely F, Heath PT, Nazareth B, Roberts R, Slack MP, et al. Recommendations for the prevention of secondary Haemophilus influenzae type b (Hib) disease. The Journal of Infection. 2009;58:3–14.

55. Edmond K, Clark A, Korczak VS, Sanderson C, Griffiths UK, Rudan I. Global and regional risk of disabling sequelae from bacterial meningitis: a systematic review and meta-analysis. Lancet Infect Dis. 2010;10:317–28.
56. Connell T, Curtis N. How to interpret a CSF--the art and the science. Adv Exp Med Biol. 2005;568:199–216.
57. Shah SS, Ebberson J, Kestenbaum LA, Hodinka RL, Zorc JJ. Age-specific reference values for cerebrospinal fluid protein concentration in neonates and young infants. J Hosp Med. 2011;6:22–7.
58. Ahmed A, Hickey SM, Ehrett S, Trujillo M, Brito F, Goto C, et al. Cerebrospinal fluid values in the term neonate. The Pediatric Infectious Disease Journal. 1996;15:298–303.

Mycobacterium marinum Infection

Marc Tebruegge and Nigel Curtis

1 Introduction

Mycobacterium marinum was first identified as a causative organism of tuberculosis in fish in 1926 [1], but its pathogenic role in causing skin disease in humans was only identified more than two decades later [2]. The organism is prevalent in natural aquatic environments world-wide and can be found in fresh, brackish and salt water [3].

M. marinum causes disease in fish, amphibians and aquatic mammals [4–9]. Human infection is commonly the result of trauma in an aquatic environment that results in a breach of the skin barrier, such as abrasions or lacerations. The injury is often minor, and may therefore not be recalled by the patient [10, 11]. Aquaria and swimming pools used to be commonly reported sources of infection – which led to the disease being referred to as 'fish tank granuloma' or 'swimming pool granuloma' [11–15]. However, due to the now widespread use of chlorination and other disinfection measures, public swimming pools have become an unusual source of infection [13, 16, 17]. During the last two decades contact with contaminated aquarium water has been the most common source of infection in practically all published case series. *M. marinum* infection can also be acquired by handling infected fish or shellfish or, less commonly, as a result of injuries actively inflicted by marine animals (i.e. fish spine or bite injuries) [17]. Consequently, at risk groups include fishermen and other individuals who professionally handle fish and other seafood (including chefs, fishmongers and oyster shuckers), fish fanciers and water sports enthusiasts [11, 12, 18–24]. No cases of human-to-human transmission have been reported.

2 Microbiology

M. marinum is a nontuberculous, acid-fast bacillus that belong to the Runyon I group of photochromogens, which are characterised by slow growth and production of a yellow pigment on exposure to light (Table 1) [25, 26]. In culture, *M. marinum* grows best at temperatures between 30°C and

M. Tebruegge (✉) • N. Curtis
Department of Paediatrics, The University of Melbourne, Parkville, Australia

Department of General Medicine, Infectious Diseases Unit, Parkville, Australia

Murdoch Childrens Research Institute, Parkville, Australia

Royal Children's Hospital Melbourne, Parkville, Australia
e-mail: marc.tebruegge@rch.org.au; nigel.curtis@rch.org.au

N. Curtis et al. (eds.), *Hot Topics in Infection and Immunity in Children VIII*,
Advances in Experimental Medicine and Biology 719, DOI 10.1007/978-1-4614-0204-6_17,
© Springer Science+Business Media, LLC 2011

Table 1 Runyon classification of nontuberculous mycobacteria

Group		Pigment production	Organisms
Group I	Photochromogens	Nonpigmented colonies when cultured in the dark. Yellow pigment production when cultured in light	*M. marinum* *M. kansasii* *M. simiae*
Group II	Scotochromogens	Yellow-orange pigmented colonies when cultured in light or the dark	*M. gordonae* *M. scrofulaceum* *M. szulgai*
Group III	Non-chromogens	Non-pigmented colonies	*M. avium* complex *M. haemophilum* *M. ulcerans*
Group IV	Rapid growers	–	*M. abscessus* *M. chelonei* *M. fortuitum*

32°C, and poorly or not at all at 37°C [27–29]. *M. marinum* is an aerobic bacterium and requires oxygen for optimal growth [27].

M. marinum is resistant to isoniazid and pyrazinamide and only intermediately susceptible to streptomycin [22, 30]. Most clinical isolates are susceptible to rifampicin, rifabutin, ethambutol, amikacin, tetracyclines (doxycycline, minocycline), clarithromycin, cotrimoxazole and some quinolones [12, 22, 31, 32]. However, some strains have been found to be resistant to ciprofloxacin, resulting in recent treatment guidelines discouraging the use of this agent [30]. Results of a relatively recent *in vitro* study involving 43 clinical isolates of *M. marinum* that investigated the activity of gatifloxacin, gemifloxacin, levifloxacin and moxifloxacin have cast further doubt on the potential role of quinolones in *M. marinum* infections [33].

The sequencing and assembly of the *M. marinum* genome has been completed in 2008 [34]. The genome of the M strain comprises a 6.6 Mb circular chromosome, and is therefore 1.2 Mb larger than that of *Mycobacterium tuberculosis* (H37Rv). A comparison with the genome of *M. tuberculosis* confirmed a close genetic relationship with the two species sharing approximately 3,000 orthologous genes with an amino acid identity of 85%. A phylogenetic analysis that included further mycobacterial species revealed that *M. marinum* is most closely related to *Mycobacterium ulcerans*, followed by *M. tuberculosis*. *M. ulcerans* strains are genetically less diverse than *M. marinum* strains, indicating that the former species is likely to have evolved from the latter [35].

3 Epidemiology

M. marinum infections in humans are rare, although solid epidemiological data remain relatively limited. A national survey from France estimated the incidence to be approximately 0.04 cases per 100,000 inhabitants per year [12]. A retrospective study conducted over a 10-year-period at the largest dermatological referral center in Hong Kong identified only 17 patients with *M. marinum* infection, which equated to less than 0.005% of the 345,394 new patients seen during this period [10]. Nevertheless, in most published case series of nontuberculous mycobacterial skin and soft tissue infections, *M. marinum* is the most commonly isolated organism, followed by *M. abcessus, M. fortuitum* and *M. chelonae* [10, 22, 36, 37].

The majority of infections occur in previously healthy adults [10, 21, 23, 37, 38]. Infections in children and teenagers have been described, but are less common [12, 19, 39–42]. In most larger studies there is a moderate male preponderance, likely reflecting occupational exposure [12, 22, 23, 43].

4 Clinical Features

The average duration between inoculation and the onset of symptoms is 2–4 weeks [12, 17, 44], although considerably longer incubation times – in a few instances exceeding 6 months – have been reported [12, 17].

In most cases *M. marinum* infection is confined to the skin and subcutaneous tissue. The skin lesions are typically papular, nodular or plaque-like in appearance. A sporotrichoid pattern (i.e. a linear distribution of multiple lesions along lymphatic vessels) is seen in about a quarter to a third of cases [12, 22, 43, 44]. Ulcerative lesions, pustules and abscesses are less common manifestations [10, 12, 22, 23]. In the majority of patients the lesions are painless, but some patients report mild to moderate pain and tenderness [45]. Co-existing regional lymphadenitis is present in 10–20% of patients [12]. Systemic symptoms, such as fever and malaise, are rare [14, 45].

The most commonly affected sites are the upper limbs, accounting for close to 90% or more of the cases in most larger case series, followed by the lower limbs [10, 12, 22, 38, 46, 47]. Truncal lesions have rarely been reported. There are two potential explanations for these observations: (1) the upper limbs are most prone to exposure or injury in an occupational context as well as during the cleaning of fish tanks, and (2) as *M. marinum* grows best at temperatures below 37°C the comparatively lower temperature of the extremities may favour bacterial replication.

Locally invasive disease, including tenosynovitis, arthritis and osteomyelitis, is relatively uncommon. Data from the Centers for Disease Control over a 7-year-period revealed that synovial or osteoarticular infection was present in 5 (8.5%) out of 59 cases [16]. A review of published cases with invasive *M. marinum* infection suggests that some form of immunocompromise can be identified in the vast majority of these patients [21]. Of the 35 cases summarised in this review, 14 (40%) had received steroid injections at the site of infection, while another nine (26%) had received systemic steroids (with or without other immunosuppressive medication). In addition, one patient had severe combined immunodeficiency, while another had acquired immunodeficiency syndrome; in both cases the outcome was fatal [48, 49]. Another review highlights solid organ transplant recipients as an at-risk group for severe disease [50]. Several recent case reports have illustrated that patients treated with anti-TNF-α agents (infliximab or etanercept) for rheumatological conditions or Crohn's disease are also at increased risk of severe *M. marinum* disease [51–56]. This is perhaps not surprising, given the critical role that TNF-α plays in the host immune response to mycobacterial infections, and in view of the fact that an increase in risk for tuberculosis, as well as other nontuberculous mycobacterial infections, is well-documented in patients receiving anti-TNF-α therapy [57–61].

True disseminated infection (i.e. systemic, rather than localised spread of the infection) is very rare, and almost exclusively occurs in immunodeficient or immunosupressed individuals [14, 49, 62].

5 Diagnosis

The diagnosis of *M. marinum* infection is frequently considerably delayed. The average interval between the onset of symptoms and the correct diagnosis in many published case series is close to or exceeds 12 months [10, 11, 16, 21, 23]. There are several potential explanations for this. The injury resulting in inoculation with *M. marinum* is often minor and may not be recalled by the patient. Also, there is often a significant delay between the inoculation and the onset of symptoms, which further increases the difficulty of making a connection between the skin lesion and the preceding aquatic injury. Notably, one literature review found that in 35% of the published cases the incubation period was 1 month or longer [17]. Furthermore, *M. marinum* skin lesions evolve slowly over several weeks, and are frequently painless, which may reduce the urgency to seek medical attention [44]. Notably, one study from Australia, which included 29 patients, reported that the average delay between onset of symptoms and consultation with a medical practitioner was 5 months [43]. Also, given the rarity of *M. marinum* infection many medical practitioners will be unfamiliar with this condition, and might fail to conduct specific tests that would help identify the causative infection. Importantly, *M. marinum* grows slowly in culture, and therefore escapes detection in routine, short-term bacterial cultures.

On histological examination with conventional Ziehl-Neelsen staining, acid-fast bacilli are identified in only the minority of cases [23, 38], with detection rates below 10% being reported in several case series [10, 11, 37, 46]. In contrast, granulomatous inflammatory changes, which can be caseating or non-caseating, are observed in the vast majority of cases [11, 38, 63].

Biopsy or abscess material should be cultured in parallel in solid (e.g. Lowenstein-Jensen or Middlebrook 7H10 and 7H11 media) [64] and liquid media (i.e. broth media), such as the commonly used mycobacteria growth indicator tube (MGIT) system (Becton Dickenson, Sparks, MD, USA) [30]. Broth cultures produce a more rapid result, but can be hampered by bacterial overgrowth. Solid cultures serve as a backup in instances where liquid cultures have been contaminated, and can also be used for susceptibility testing [30]. Dessication of biopsy material prior to processing must be avoided; if the tissue sample is very small it can be immersed in a small amount of sterile saline [30]. Communication with the laboratory is vital, so that cultures are incubated at a temperature that promotes optimal growth of *M. marinum* (i.e. 30–32°C, instead of 35–37°C used for most other nontuberculous mycobacteria) [27]. In some studies, between 70% and 80% of the cases were reported to be culture-positive [10, 44], although considerably lower figures have been reported by other authors [23]. On average, growth can be detected in culture after 3–4 weeks of incubation [22].

Routine susceptibility testing of clinical isolates is not recommended as the susceptibility pattern of *M. marinum* is well documented, and acquired mutational resistance is rare [31]. However, susceptibility testing should be considered in patients with treatment failure (i.e. those cases who fail to show improvement despite several weeks of therapy), and those with persistently positive culture results after more than 3 months of therapy [30].

Nucleic acid amplification techniques (NAATs) are further useful tools for the diagnosis of *M. marinum* infection. The clear advantage of NAATs is that results are available considerably faster compared with culture, while their greatest disadvantage is that they do not provide any information about the susceptibilities of the organism identified. A number of NAATs, such as polymerase chain reaction (PCR) coupled with restriction fragment length polymorphism analysis and 16S rRNA gene-based PCR techniques, which enable distinction between different nontuberculous mycobacteria at species level have been developed [65–69]. Several reports have described the use of conventional PCR for the diagnosis or confirmation of *M. marinum* infection in clinical cases [19, 70–72]. Although PCR-based methods are likely to be highly sensitive and specific, the number of published cases is currently too small to allow an accurate estimate of these test characteristics. However, a recent publication highlights that real-time PCR (based on amplification of 16S rRNA genes) performed on paraffin-embedded skin biopsies may increase the diagnostic yield in patients with *M. marinum* and other nontuberculous mycobacterial skin infections which are culture-negative [73].

The majority of cases have a positive tuberculin skin test (Mantoux), with induration of greater than or equal to 10 mm in diameter reported to occur in between 60% and 100% of patients [11, 14, 74, 75].

Interferon-gamma release assays (IGRA) are further adjunctive tools that can support the presumptive diagnosis of *M. marinum* infection. *M. marinum* has been shown to express early secretory antigenic target 6 (ESAT-6) and culture filtrate protein 10 (CFP-10) [76, 77], the two RD1 antigens that form the basis of both currently commercially available IGRA licensed for the diagnosis of tuberculosis – the QuantiFERON-TB Gold assay (Cellestis, Carnegie, Australia) and the T-SPOT.*TB* assay (Oxford Immunotec, Oxford, United Kingdom). Arend et al. reported significant interferon-gamma production in response to stimulation with ESAT-6 and CFP-10 in whole blood samples from patients with *M. marinum* infection using in-house ELISA and ELISPOT assays [75]. A study by Kobashi et al. that evaluated the performance of the second generation QuantiFERON-TB Gold assay (incorporating ESAT-6 and CFP-10) in patients with nontuberculous mycobacterial infections confirmed that commercial IGRA can be a useful adjunctive tool for the diagnosis of *M. marinum* infection [78]. In this report seven (58%) out of 12 patients diagnosed with *M. marinum* infection (and without a history of previous tuberculosis) were found to have a positive IGRA result. Another report described an adolescent with *M. marinum* infection confirmed by PCR testing, in whom a third generation QuantiFERON-TB Gold-in-tube assay (incorporating TB7.7, in addition to ESAT-6 and CFP-10) was also positive [19]. Importantly, this patient was very unlikely to have been exposed to *M. tuberculosis*. Further cases of *M. marinum* infection with a positive QuantiFERON-TB Gold assay result have been described [79, 80].

6 Differential Diagnosis

Table 2 provides a comparison of bacterial infections that occur commonly as a result of aquatic trauma or contact with fish and other aquatic animals [81–91]. Other differential diagnoses include skin infections caused by other nontuberculous mycobacteria, sporotrichosis, nocardiosis, pyoderma gangrenosum, cutaneous leishmaniasis, herpetic whitlow, psoriasis, sarcoidosis, skin cancer and vasculitis.

Table 2 Comparison of common bacterial skin and soft tissue infections following aquatic injury

	Vibrio species[a]	Aeromonas hydrophila	Erysipelothrix rhusiopathiae	Mycobacterium marinum
Typical source	Salt and brackish water; contact with fish/seafood	Fresh water; contact with fish/seafood	Salt water; contact with fish/seafood	Salt, brackish and fresh water; contact with fish/seafood
Incubation period	Hours to days	Hours to days	Days	Weeks
Typical manifestations	Painful cellulitis, vesicles/bullae, necrotising soft tissue infection	Painful cellulitis, abscess	Painful, pruritic ring-shaped lesions (erysipeloid)	Painless skin/soft tissue lesion
Regional lymphadenitis	Common	Common	Up to 30%	Rare
Systemic symptoms	Common	Common	Rare	Very rare
Progression	Rapid (hours to days)	Rapid (hours to days)	Slow (days to weeks)	Very slow (weeks to months)
Fatal outcome	5–20%	Rare	Rare	Rare
References	[81–85, 88]	[81, 86–88]	[81, 89–91]	

[a]Commonly implicated species include *V. vulnificus, V. alginolyticus* and *V. parahaemolyticus*

7 Treatment

Spontaneous resolution has been reported in a few cases, although this appears to be relatively rare [92]. Conversely, there are cases reported in the literature in whom skin lesions have persisted for more than two decades [93].

The optimal treatment of *M. marinum* infection remains uncertain. There are currently no published controlled trials that have investigated the treatment of *M. marinum* infections. Cases of success and failure have been reported with practically any combination of antibiotics with anti-mycobacterial activity. In addition, several publications suggest that antibiotic *in vitro* activity does not accurately predict clinical effectiveness [12, 16].

There are a considerable number of published cases of *M. marinum* infection in which treatment with a single antibiotic agent has been successful [10, 14, 23, 46, 94, 95]. However, there is some evidence suggesting that monotherapy is less likely to be adequate for the treatment of invasive *M. marinum* infections [12]. Based on the limited data available from case series and data extrapolated from studies on the treatment of other nontuberculous mycobacterial infections, most experts recommend treatment of *M. marinum* infections (including isolated skin and soft tissue infections) with two antibiotics. Treatment should be continued for 1–2 months after resolution of all lesions [14, 46, 96]. This strategy has been recently also recommended in the latest American Thoracic Society (ATS)/Infectious Diseases Society of America (IDSA) statement on the management of nontuberculous mycobacterial disease [30]. Some authors have proposed adding a third active agent in patients with disseminated or extensive invasive infection [14]. Rifampicin has been suggested as a suitable addition for patients with bone and joint involvement [14, 30, 96], primarily on the basis of its good bone penetration, rather than compelling clinical data.

An accurate assessment of the clinical efficacy of single antibiotic agents and combination regimens is practically impossible. Although treatment regimens have been described in many reports, it is often not possible to link these to the reported outcomes. Also, in most larger studies a considerable number of patients were switched from one regimen to another, the rationale for which is often unclear (i.e. side effects vs. poor treatment response). In addition, in many cases simultaneous surgical intervention complicates the interpretation of the response to antibiotic treatment.

The latest ATS/IDSA guidelines recommend using a combination regimen comprising clarithromycin together with either ethambutol or rifampicin [30]. Nevertheless, it is worth noting that in the largest published study, seven (18%) of 39 patients who received a regimen containing clarithromycin failed to show adequate response to treatment or relapsed, despite the fact that all corresponding isolates were found to be susceptible to this agent on *in vitro* testing [12]. Similar failure rates were described for ethambutol- and rifampicin-containing regimens in the same report (23% and 24%, respectively), while treatment only failed in 5% of the patients receiving tetracyclines. Doxycycline and minocycline have also been successfully used as monotherapy, with or without surgery, in several other reports [10, 11, 43, 46, 97, 98]. Nevertheless, monotherapy is discouraged by most experts, and tetracyclines are not suitable for the treatment of paediatric cases. Cotrimoxazole monotherapy has also been reported to be successful in a few reports; however, most of these patients had disease limited to skin and soft tissue only [11, 43]. Recent, encouraging reports have shown that linezolid has excellent *in vitro* activity against *M. marinum* [33, 99, 100], but there are no published cases that were treated with this agent.

The average treatment duration required to achieve cure in comparatively large studies including all forms of disease (i.e. skin/soft tissue and invasive disease) ranges between 3 and 6 months [10, 12]. Patients with extensive invasive disease generally require longer treatment courses [12, 21].

The role of surgery in the treatment of *M. marinum* infection remains poorly defined [45]. However, several authors have suggested that surgical debridement is indicated in cases with invasive infection [3, 14, 21, 30].

8 Prognosis

The prognosis of *M. marinum* infection largely depends on the extent of the disease at presentation and co-existing morbidity (i.e. immunosupression or immunodeficiency). In the majority of immunocompetent patients who present with isolated skin and soft tissue infection, cure can be achieved with antibiotic treatment alone. However, resolution frequently takes several months, and residual scaring or hyperpigmentation of the skin are not uncommon. One study that included 24 patients with tenosynovitis of the hand or wrist, some of which were treated with antibiotics alone while others simultaneously underwent surgical debridement, suggests that full functional recovery can be achieved in the majority of these cases [45]. Fatal outcome, as a result of disseminated infection, is very rare and has been reported almost exclusively in immunocompromised patients.

9 Future Prospects

There is a need for better data on the optimal antibiotic treatment of *M. marinum* infection. However, given the rarity of this condition a randomised controlled trial of adequate size would require a multi-center study. In such a trial, cases would need to be stratified into different severity groups (i.e. skin/soft tissue only vs. invasive disease) to provide meaningful data. Further studies are also required to better define the role of surgical intervention.

References

1. Aronson JD. Spontaneous tuberculosis in salt water fish. J Inf Dis 1926;39(4):315–20.
2. Norden A, Linell F. A new type of pathogenic *Mycobacterium*. Nature 1951;168(4280):826.
3. Petrini B. *Mycobacterium marinum*: ubiquitous agent of waterborne granulomatous skin infections. Eur J Clin Microbiol Infect Dis 2006;25(10):609–13.
4. Ramakrishnan L, Valdivia RH, McKerrow JH, Falkow S. *Mycobacterium marinum* causes both long-term subclinical infection and acute disease in the leopard frog (*Rana pipiens*). Infect Immun 1997;65(2):767–73.
5. Kaattari IM, Rhodes MW, Kaattari SL, Shotts EB. The evolving story of *Mycobacterium tuberculosis* clade members detected in fish. J Fish Dis 2006;29(9):509–20.
6. Sato T, Shibuya H, Ohba S, Nojiri T, Shirai W. Mycobacteriosis in two captive Florida manatees (*Trichechus manatus latirostris*). J Zoo Wildl Med 2003;34(2):184–8.
7. Cosma CL, Swaim LE, Volkman H, Ramakrishnan L, Davis JM. Zebrafish and frog models of *Mycobacterium marinum* infection. Curr Protoc Microbiol 2006;Chapter 10:Unit 10B 2.
8. Swaim LE, Connolly LE, Volkman HE, Humbert O, Born DE, Ramakrishnan L. *Mycobacterium marinum* infection of adult zebrafish causes caseating granulomatous tuberculosis and is moderated by adaptive immunity. Infect Immun 2006;74(11):6108–17.
9. Gauthier DT, Rhodes MW. Mycobacteriosis in fishes: a review. Vet J 2009;180(1):33–47.
10. Ho MH, Ho CK, Chong LY. Atypical mycobacterial cutaneous infections in Hong Kong: 10-year retrospective study. Hong Kong Med J 2006;12(1):21–6.
11. Kullavanijaya P, Sirimachan S, Bhuddhavudhikrai P. *Mycobacterium marinum* cutaneous infections acquired from occupations and hobbies. Int J Dermatol 1993;32(7):504–7.
12. Aubry A, Chosidow O, Caumes E, Robert J, Cambau E. Sixty-three cases of *Mycobacterium marinum* infection: clinical features, treatment, and antibiotic susceptibility of causative isolates. Arch Intern Med 2002;162(15): 1746–52.
13. Dailloux M, Hartemann P, Beurey J. Study on the relationship between isolation of mycobacteria and classical microbiological and chemical indicators of water quality in swimming pools. Zentralbl Bakteriol Mikrobiol Hyg B 1980;171(6):473–86.
14. Lewis FM, Marsh BJ, von Reyn CF. Fish tank exposure and cutaneous infections due to *Mycobacterium marinum*: tuberculin skin testing, treatment, and prevention. Clin Infect Dis 2003;37(3):390–7.

15. Mollohan CS, Romer MS. Public health significance of swimming pool granuloma. Am J Public Health Nations Health 1961;51:883–91.
16. Feldman M, Long MW, David HL. *Mycobacterium marinum*: a leisure-time pathogen. J Inf Dis 1974;129(5): 618–21.
17. Jernigan JA, Farr BM. Incubation period and sources of exposure for cutaneous *Mycobacterium marinum* infection: case report and review of the literature. Clin Infect Dis 2000;31(2):439–43.
18. Beecham HJ, 3rd, Oldfield EC, 3rd, Lewis DE, Buker JL. *Mycobacterium marinum* infection from shucking oysters. Lancet 1991;337(8755):1487.
19. Tebruegge M, Connell T, Ritz N, Orchard D, Curtis N. *Mycobacterium marinum* infection following kayaking injury. Int J Infect Dis 2010; 14 Suppl 3:e305-306.
20. Cennimo DJ, Agag R, Fleegler E, Lardizabal A, Klein KM, Wenokor C, et al. *Mycobacterium marinum* Hand Infection in a "Sushi Chef". Eplasty 2009;9:e43.
21. Lahey T. Invasive *Mycobacterium marinum* infections. Emerg Infect Dis 2003;9(11):1496–8.
22. Dodiuk-Gad R, Dyachenko P, Ziv M, Shani-Adir A, Oren Y, Mendelovici S, et al. Nontuberculous mycobacterial infections of the skin: A retrospective study of 25 cases. J Am Acad Dermatol 2007;57(3):413–20.
23. Ang P, Rattana-Apiromyakij N, Goh CL. Retrospective study of *Mycobacterium marinum* skin infections. Int J Dermatol 2000;39(5):343–7.
24. Walker HH, Shinn MF, Higaki M, Ogata J. Some characteristics of "swimming pool" disease in Hawaii. Hawaii Med J 1962;21:403–9.
25. Runyon EH. Pathogenic mycobacteria. Bibl Tuberc 1965;21:235–87.
26. Runyon EH. Anonymous mycobacteria in pulmonary disease. Med Clin North Am 1959;43(1):273–90.
27. Gao LY, Manoranjan J. Laboratory maintenance of *Mycobacterium marinum*. Curr Protoc Microbiol 2005; Chapter 10:Unit 10B 1.
28. Kent ML, Watral V, Wu M, Bermudez LE. In vivo and in vitro growth of *Mycobacterium marinum* at homoeothermic temperatures. FEMS Microbiol Lett 2006;257(1):69–75.
29. Clark HF, Shepard CC. Effect of Environmental Temperatures on Infection with *Mycobacterium Marinum* (Balnei) of Mice and a Number of Poikilothermic Species. J Bacteriol 1963;86:1057–69.
30. Griffith DE, Aksamit T, Brown-Elliott BA, Catanzaro A, Daley C, Gordin F, et al. An official ATS/IDSA statement: diagnosis, treatment, and prevention of nontuberculous mycobacterial diseases. Am J Respir Crit Care Med 2007;175(4):367–416.
31. Woods GL. Susceptibility testing for mycobacteria. Clin Infect Dis 2000;31(5):1209–15.
32. Brown BA, Wallace RJ, Jr., Onyi GO. Activities of clarithromycin against eight slowly growing species of nontuberculous mycobacteria, determined by using a broth microdilution MIC system. Antimicrob Agents Chemother 1992;36(9):1987–90.
33. Braback M, Riesbeck K, Forsgren A. Susceptibilities of *Mycobacterium marinum* to gatifloxacin, gemifloxacin, levofloxacin, linezolid, moxifloxacin, telithromycin, and quinupristin-dalfopristin (Synercid) compared to its susceptibilities to reference macrolides and quinolones. Antimicrob Agents Chemother 2002;46(4):1114–6.
34. Stinear TP, Seemann T, Harrison PF, Jenkin GA, Davies JK, Johnson PD, et al. Insights from the complete genome sequence of *Mycobacterium marinum* on the evolution of *Mycobacterium tuberculosis*. Genome Res 2008;18(5):729–41.
35. Stinear TP, Jenkin GA, Johnson PD, Davies JK. Comparative genetic analysis of *Mycobacterium ulcerans* and *Mycobacterium marinum* reveals evidence of recent divergence. J Bacteriol 2000;182(22):6322–30.
36. Street ML, Umbert-Millet IJ, Roberts GD, Su WP. Nontuberculous mycobacterial infections of the skin. Report of fourteen cases and review of the literature. J Am Acad Dermatol 1991;24(2 Pt 1):208–15.
37. Abbas O, Marrouch N, Kattar MM, Zeynoun S, Kibbi AG, Rached RA, et al. Cutaneous non-tuberculous mycobacterial infections: a clinical and histopathological study of 17 cases from Lebanon. J Eur Acad Dermatol Venereol. 2011;25(1):33–42.
38. Liao CH, Lai CC, Ding LW, Hou SM, Chiu HC, Chang SC, et al. Skin and soft tissue infection caused by nontuberculous mycobacteria. Int J Tuberc Lung Dis 2007;11(1):96–102.
39. King AJ, Fairley JA, Rasmussen JE. Disseminated cutaneous *Mycobacterium marinum* infection. Arch Dermatol 1983;119(3):268–70.
40. Doedens RA, van der Sar AM, Bitter W, Scholvinck EH. Transmission of *Mycobacterium marinum* from fish to a very young child. Pediatr Infect Dis J 2008;27(1):81–3.
41. Feddersen A, Kunkel J, Jonas D, Engel V, Bhakdi S, Husmann M. Infection of the upper extremity by *Mycobacterium marinum* in a 3-year-old boy--diagnosis by 16S-rDNA analysis. Infection 1996;24(1):47–8.
42. Lacaille F, Blanche S, Bodemer C, Durand C, De Prost Y, Gaillard JL. Persistent *Mycobacterium marinum* infection in a child with probable visceral involvement. Pediatr Infect Dis J 1990;9(1):58–60.
43. Iredell J, Whitby M, Blacklock Z. *Mycobacterium marinum* infection: epidemiology and presentation in Queensland 1971–1990. Med J Aust 1992;157(9):596–8.
44. Gluckman SJ. *Mycobacterium marinum*. Clin Dermatol 1995;13(3):273–6.

45. Chow SP, Ip FK, Lau JH, Collins RJ, Luk KD, So YC, et al. *Mycobacterium marinum* infection of the hand and wrist. Results of conservative treatment in twenty-four cases. J Bone Joint Surg Am 1987;69(8):1161–8.
46. Edelstein H. *Mycobacterium marinum* skin infections. Report of 31 cases and review of the literature. Arch Intern Med 1994;154(12):1359–64.
47. Casal M, Casal MM. Multicenter study of incidence of *Mycobacterium marinum* in humans in Spain. Int J Tuberc Lung Dis 2001;5(2):197–9.
48. Parent LJ, Salam MM, Appelbaum PC, Dossett JH. Disseminated *Mycobacterium marinum* infection and bacteremia in a child with severe combined immunodeficiency. Clin Infect Dis 1995;21(5):1325–7.
49. Tchornobay AM, Claudy AL, Perrot JL, Levigne V, Denis M. Fatal disseminated *Mycobacterium marinum* infection. Int J Dermatol 1992;31(4):286–7.
50. Pandian TK, Deziel PJ, Otley CC, Eid AJ, Razonable RR. *Mycobacterium marinum* infections in transplant recipients: case report and review of the literature. Transpl Infect Dis 2008;10(5):358–63.
51. Ramos JM, Garcia-Sepulcre MF, Rodriguez JC, Padilla S, Gutierrez F. *Mycobacterium marinum* infection complicated by anti-tumour necrosis factor therapy. J Med Microbiol;59(Pt 5):617–21.
52. Chopra N, Kirschenbaum AE, Widman D. *Mycobacterium marinum* tenosynovitis in a patient on etanercept therapy for rheumatoid arthritis. J Clin Rheumatol 2002;8(5):265–8.
53. Fallon JC, Patchett S, Gulmann C, Murphy GM. *Mycobacterium marinum* infection complicating Crohn's disease, treated with infliximab. Clin Exp Dermatol 2008;33(1):43–5.
54. Danko JR, Gilliland WR, Miller RS, Decker CF. Disseminated *Mycobacterium marinum* infection in a patient with rheumatoid arthritis receiving infliximab therapy. Scand J Infect Dis 2009;41(4):252–5.
55. Dare JA, Jahan S, Hiatt K, Torralba KD. Reintroduction of etanercept during treatment of cutaneous *Mycobacterium marinum* infection in a patient with ankylosing spondylitis. Arthritis Rheum 2009;61(5):583–6.
56. Rallis E, Koumantaki-Mathioudaki E, Frangoulis E, Chatziolou E, Katsambas A. Severe sporotrichoid fish tank granuloma following infliximab therapy. Am J Clin Dermatol 2007;8(6):385–8.
57. Winthrop KL, Chang E, Yamashita S, Iademarco MF, LoBue PA. Nontuberculous mycobacteria infections and anti-tumor necrosis factor-alpha therapy. Emerg Infect Dis 2009;15(10):1556–61.
58. Swart RM, van Ingen J, van Soolingen D, Slingerland R, Hendriks WD, den Hollander JG. Nontuberculous mycobacteria infection and tumor necrosis factor-alpha antagonists. Emerg Infect Dis 2009;15(10):1700–1.
59. Harris J, Keane J. How tumour necrosis factor blockers interfere with tuberculosis immunity. Clin Exp Immunol 2010;161(1):1–9.
60. Keane J, Gershon S, Wise RP, Mirabile-Levens E, Kasznica J, Schwieterman WD, et al. Tuberculosis associated with infliximab, a tumor necrosis factor alpha-neutralizing agent. N Engl J Med 2001;345(15):1098–104.
61. Mohan AK, Cote TR, Block JA, Manadan AM, Siegel JN, Braun MM. Tuberculosis following the use of etanercept, a tumor necrosis factor inhibitor. Clin Infect Dis 2004;39(3):295–9.
62. Streit M, Bohlen LM, Hunziker T, Zimmerli S, Tscharner GG, Nievergelt H, et al. Disseminated *Mycobacterium marinum* infection with extensive cutaneous eruption and bacteremia in an immunocompromised patient. Eur J Dermatol 2006;16(1):79–83.
63. Travis WD, Travis LB, Roberts GD, Su DW, Weiland LW. The histopathologic spectrum in *Mycobacterium marinum* infection. Arch Pathol Lab Med 1985;109(12):1109–13.
64. Middlebrook G, Cohn ML. Bacteriology of tuberculosis: laboratory methods. Am J Public Health Nations Health 1958;48(7):844–53.
65. Ringuet H, Akoua-Koffi C, Honore S, Varnerot A, Vincent V, Berche P, et al. hsp65 sequencing for identification of rapidly growing mycobacteria. J Clin Microbiol 1999;37(3):852–7.
66. Devallois A, Goh KS, Rastogi N. Rapid identification of mycobacteria to species level by PCR-restriction fragment length polymorphism analysis of the hsp65 gene and proposition of an algorithm to differentiate 34 mycobacterial species. J Clin Microbiol 1997;35(11):2969–73.
67. Telenti A, Marchesi F, Balz M, Bally F, Bottger EC, Bodmer T. Rapid identification of mycobacteria to the species level by polymerase chain reaction and restriction enzyme analysis. J Clin Microbiol 1993;31(2):175–8.
68. Talaat AM, Reimschuessel R, Trucksis M. Identification of mycobacteria infecting fish to the species level using polymerase chain reaction and restriction enzyme analysis. Vet Microbiol 1997;58(2–4):229–37.
69. Kirschner P, Springer B, Vogel U, Meier A, Wrede A, Kiekenbeck M, et al. Genotypic identification of mycobacteria by nucleic acid sequence determination: report of a 2-year experience in a clinical laboratory. J Clin Microbiol 1993;31(11):2882–9.
70. Cai L, Chen X, Zhao T, Ding BC, Zhang JZ. Identification of *Mycobacterium marinum* 65 kD heat shock protein gene by polymerase chain reaction restriction analysis from lesions of swimming pool granuloma. Chin Med J (Engl) 2006;119(1):43–8.
71. Witteck A, Ohlschlegel C, Boggian K. Delayed diagnosis of atypical mycobacterial skin and soft tissue infections in non-immunocompromised hosts. Scand J Infect Dis 2008;40(11–12):877–80.
72. Sanal HT, Zor F, Kocaoglu M, Bulakbasi N. Atypical mycobacterial tenosynovitis and bursitis of the wrist. Diagn Interv Radiol 2009;15(4):266–8.

73. van Coppenraet LS, Smit VT, Templeton KE, Claas EC, Kuijper EJ. Application of real-time PCR to recognize atypical mycobacteria in archival skin biopsies: high prevalence of *Mycobacterium haemophilum*. Diagn Mol Pathol 2007;16(2):81–6.
74. Jolly HW, Jr., Seabury JH. Infections with *Myocbacterium marinum*. Arch Dermatol 1972;106(1):32–6.
75. Arend SM, van Meijgaarden KE, de Boer K, de Palou EC, van Soolingen D, Ottenhoff TH, et al. Tuberculin skin testing and in vitro T cell responses to ESAT-6 and culture filtrate protein 10 after infection with *Mycobacterium marinum* or *M. kansasii*. J Infect Dis 2002;186(12):1797–807.
76. Tan T, Lee WL, Alexander DC, Grinstein S, Liu J. The ESAT-6/CFP-10 secretion system of *Mycobacterium marinum* modulates phagosome maturation. Cell Microbiol 2006;8(9):1417–29.
77. Gao LY, Guo S, McLaughlin B, Morisaki H, Engel JN, Brown EJ. A mycobacterial virulence gene cluster extending RD1 is required for cytolysis, bacterial spreading and ESAT-6 secretion. Mol Microbiol 2004;53(6):1677–93.
78. Kobashi Y, Mouri K, Yagi S, Obase Y, Miyashita N, Okimoto N, et al. Clinical evaluation of the QuantiFERON-TB Gold test in patients with non-tuberculous mycobacterial disease. Int J Tuberc Lung Dis 2009;13(11):1422–6.
79. Perez-Jorge EV, Burdette SD. Chronic ulceration from *Mycobacterium marinum* infection and the diagnostic value of T-cell interferon-gamma release assays. Mol Diagn Ther 2010;14(2):119–22.
80. Kobashi Y, Obase Y, Fukuda M, Yoshida K, Miyashita N, Oka M. Clinical reevaluation of the QuantiFERON TB-2G test as a diagnostic method for differentiating active tuberculosis from nontuberculous mycobacteriosis. Clin Infect Dis 2006;43(12):1540–6.
81. Noonburg GE. Management of extremity trauma and related infections occurring in the aquatic environment. J Am Acad Orthop Surg 2005;13(4):243–53.
82. Howard RJ, Bennett NT. Infections caused by halophilic marine *Vibrio* bacteria. Ann Surg 1993;217(5): 525–30
83. Dechet AM, Yu PA, Koram N, Painter J. Nonfoodborne *Vibrio* infections: an important cause of morbidity and mortality in the United States, 1997–2006. Clin Infect Dis 2008;46(7):970–6.
84. Kumamoto KS, Vukich DJ. Clinical infections of *Vibrio vulnificus*: a case report and review of the literature. J Emerg Med 1998;16(1):61–6.
85. Tacket CO, Brenner F, Blake PA. Clinical features and an epidemiological study of *Vibrio vulnificus* infections. J Infect Dis 1984;149(4):558–61.
86. Gold WL, Salit IE. *Aeromonas hydrophila* infections of skin and soft tissue: report of 11 cases and review. Clin Infect Dis 1993;16(1):69–74.
87. Kelly KA, Koehler JM, Ashdown LR. Spectrum of extraintestinal disease due to *Aeromonas* species in tropical Queensland, Australia. Clin Infect Dis 1993;16(4):574–9.
88. Tsai YH, Hsu RW, Huang TJ, Hsu WH, Huang KC, Li YY, et al. Necrotizing soft-tissue infections and sepsis caused by *Vibrio vulnificus* compared with those caused by *Aeromonas* species. J Bone Joint Surg Am 2007;89(3):631–6.
89. Brooke CJ, Riley TV. *Erysipelothrix rhusiopathiae*: bacteriology, epidemiology and clinical manifestations of an occupational pathogen. J Med Microbiol 1999;48(9):789–99.
90. King PF. Erysipeloid Survey of 115 cases. Lancet 1946;248(6416):196–8.
91. Reboli AC, Farrar WE. *Erysipelothrix rhusiopathiae*: an occupational pathogen. Clin Microbiol Rev 1989;2(4): 354–9.
92. Dorronsoro I, Sarasqueta R, Gonzalez AI, Gallego M. [Cutaneous infections by *Mycobacterium marinum*. Description of 3 cases and review of the literature]. Enferm Infecc Microbiol Clin 1997;15(2):82–4.
93. Lee MW, Brenan J. *Mycobacterium marinum*: chronic and extensive infections of the lower limbs in south Pacific islanders. Australas J Dermatol 1998;39(3):173–6.
94. Kern W, Vanek E, Jungbluth H. [Fish breeder granuloma: infection caused by *Mycobacterium marinum* and other atypical mycobacteria in the human. Analysis of 8 cases and review of the literature]. Med Klin (Munich) 1989;84(12):578–83.
95. Huminer D, Pitlik SD, Block C, Kaufman L, Amit S, Rosenfeld JB. Aquarium-borne *Mycobacterium marinum* skin infection. Report of a case and review of the literature. Arch Dermatol 1986;122(6):698–703.
96. Harris DM, Keating MR. *Mycobacterium marinum*: current recommended pharmacologic therapy. J Hand Surg Am 2009;34(9):1734–5.
97. Cummins DL, Delacerda D, Tausk FA. *Mycobacterium marinum* with different responses to second-generation tetracyclines. Int J Dermatol 2005;44(6):518–20.
98. Hurst LC, Amadio PC, Badalamente MA, Ellstein JL, Dattwyler RJ. *Mycobacterium marinum* infections of the hand. J Hand Surg Am 1987;12(3):428–35.
99. Molicotti P, Ortu S, Bua A, Cannas S, Sechi LA, Zanetti S. In vitro efficacy of Linezolid on clinical strains of *Mycobacterium tuberculosis* and other mycobacteria. New Microbiol 2006;29(4):275–80.
100. Brown-Elliott BA, Crist CJ, Mann LB, Wilson RW, Wallace RJ, Jr. In vitro activity of linezolid against slowly growing nontuberculous mycobacteria. Antimicrob Agents Chemother 2003;47(5):1736–8.

Index

N. Curtis et al. (eds.), *Hot Topics in Infection and Immunity in Children VIII*, 211
Advances in Experimental Medicine and Biology 719, DOI 10.1007/978-1-4614-0204-6,
© Springer Science+Business Media, LLC 2011